Rationis Defensor

STUDIES IN HISTORY AND PHILOSOPHY OF SCIENCE

VOLUME 28

For other titles published in this series, go to
http://www.springer.com/series/5671

James Maclaurin
Editor

Rationis Defensor

Essays in Honour of Colin Cheyne

Springer

Editor
James Maclaurin
Department of Philosophy
University of Otago
364 Leith Walk
Dunedin, New Zealand

ISSN 0929-6425
ISBN 978-94-007-3982-6 ISBN 978-94-007-3983-3 (eBook)
DOI 10.1007/978-94-007-3983-3
Springer Dordrecht Heidelberg New York London

Library of Congress Control Number: 2012935858

Printed on acid-free paper

Springer is part of Springer Science+Business Media (www.springer.com)

For Colin and Liz

Foreword

"Ask a mathematician whether there are numbers bigger than 100, and he will say 'Of course there are'. Ask a mathematician whether numbers exist, and he will wonder what you are talking about". So Colin Cheyne once said to me, shortly after he was first bitten by the philosophy bug.

That happened to him later in life than most who are so bitten. Colin's first career was in mathematics and as a High School teacher of mathematics, both in New Zealand and in the UK. Then, in the 1980s, he returned to the University of Otago as a part-time student of philosophy in general and of mathematical logic in particular. His talent for philosophy soon became apparent. He completed his BA and a Postgraduate Diploma in Philosophy, and embarked upon his PhD in 1990. By that time he had abandoned his career as a High School teacher to join the Philosophy Department, where he was appointed first as a Teaching Fellow and then a Lecturer.

Colin completed a stunning doctoral thesis in 1994. It formed the basis of his major publication, Knowledge, Cause, and Abstract Objects: Causal Objections to Platonism, which appeared in 2001. As its title suggests, it returns to the puzzle with which I began. Do numbers exist, and if they do, what kind of thing are they? Colin was skeptical about the Platonist view that numbers are abstract objects that do not exist in space or time and do not do anything. He was skeptical about this, above all, because it makes it wholly mysterious how mathematical knowledge is possible, how we can know anything about numbers so conceived. He claims that we can know of the existence of an object only if we can somehow causally interact with it, however remotely. And he argues that numbers and other abstract (a-causal and non-spatio-temporal) objects fail this test.

These anti-Platonist views were at odds with the views of Colin's main teacher in mathematical logic, the late Pavel Tichý, who was a resolute Platonist regarding mathematical and logical objects. Tichý's tragic death in 1994 left a gaping hole in the teaching resources of the Philosophy Department for a specialist logician. Who better to fill that gap than Colin Cheyne, who had excelled in that area? Accordingly Colin was appointed chiefly to take over Tichý's series of high-powered mathematical logic courses. This he did for nearly 20 years. His philosophical disagreements with Tichý did not prevent him from teaching logic the 'Tichý way'. Nor did they prevent him from co-editing the posthumous publication of Pavel Tichý's Collected Papers in Logic and Philosophy, which appeared in 2004. This work brought him into contact with Tichý's former colleagues and disciples in the Czech Republic, and he became a regular participant in logic conferences held there.

Colin's editorial labours on behalf of his colleagues did not end there. In 2006 he co-edited Rationality and Reality: Conversations with Alan Musgrave, a Festschrift to mark my own retirement from the Philosophy Department. The volume proved to be a little premature – or as Colin put it, "this retirement theory was refuted".

Meanwhile, Colin's own work blossomed. He was a demanding and rigorous teacher. Most of our most talented students passed through his hands and benefitted from his logic courses. One of the most talented of them all once complained to me that "grumpy Colin" had found mistakes in her latest logic exercise and had given her a poor mark. She later told me how grateful she was to "grumpy Colin" for teaching her the true meaning of logical and intellectual rigor. These same qualities were manifested in a steady stream of high-quality articles that Colin wrote on a variety of issues in epistemology and metaphysics. Especially important among these were his demolition of the appeal to intuition as a source of knowledge of abstract objects, and his discovery of a new paradox regarding reasonable belief.

Over the years Colin took an active interest in all the affairs of the Philosophy Department. He took a special interest in devising simple and sensible course regulations, and for years was our special Advisor of Studies on these matters. He served as a wise, understated yet very effective Head of Department from 2006 until 2008. He also became actively involved in the Australasian Association for Philosophy, as a member of its Council and General Secretary of its New Zealand Division (2004–2011). He was elected President of the New Zealand Division in 2011.

Some years ago the University decided that we should all have business cards bearing appropriate mottos. Colin chose Rationis Defensor as the motto for his card. This has now become the title of this volume. It is an apt title, for Colin has always been a resolute and effective defender of reason in all of his academic pursuits. The devotion to reason did not, however, extend to all aspects of his life. It did not, for example, interfere with his passionate devotion to all things Italian – Italian opera, Italian movies, Italian food, Italian weather, Italian scenery. He even on occasions extolled the virtues of Italian politics! Worse, despite (or because of?) the fact that he was born and bred in a rugby-mad nation, he developed an idiosyncratic and irrational interest in 'Australian Rules Football', played only in Sydney and Melbourne. For no apparent reason he became a devoted fan of a club called 'Carlton', and would regularly regale his colleagues at morning tea with its latest result. Reason has inevitable limitations, even for committed defenders of it such as he.

The contributions to this volume come from a wide variety of philosophical fields, indicating the breadth of Colin's own interests and contributions. What unites them all is simple – an admiration for the man and his work.

University of Otago — Alan Musgrave

Preface

Colin Cheyne is a consummate analytic philosopher; curious, enthusiastic, precise, and forthright. He is widely published in epistemology, philosophical logic and philosophy of mathematics but, in truth, Colin is a philosophical omnivore with great breadth of interest and great breadth of enthusiasm. In departmental seminars, Colin always has a view. He usually speaks last and he often asks the crucial question that half the people in the room were struggling to formulate. Omnivory is a hard thing to celebrate in a world that is ever more specialised—yet, celebrate it we should. This volume brings together a diverse group of philosophers, many close friends of Colin and of the Otago department, some of whom were among Colin's excellent students. All have been eager contributors to a volume that straddles epistemology, science, metaphysics and logic, the areas of philosophy with which Colin has a particular affinity. But above all this book recognises and celebrates Colin's abiding interest, not just in solving philosophical problems, but in sorting out how we should reason about philosophical problems.

That question is asked many times and in many different ways in Rationis Defensor: How do we judge our success in the pursuit of knowledge? (Alan Musgrave's Getting over Gettier, Gregory Dawes's Justified Believing and Pavel Materna's Mathematical and Empirical Concepts); How do we assess our moral obligations? (Tim Mulgan's The Future of Utilitarianism and Andrew Moore's The Buck-Passing Stops Here); How should philosophers characterise propositions (Heather Dyke's Propositions: Truth vs. Existence and Bjørn Jespersen's Post-Fregean Thoughts on Propositional Unity), modality (Josh Parsons's Against Advanced Modalizing and Charles Pigden and Rebecca Entwisle's Spread Worlds, Plenitude and Modal Realism) or logical inference? (Martin Frické's Best-Path Theorem Proving); Where should philosophers look for evidence? (Gregory Currie's Literature and Truthfulness, Juan Gomez's The Progress of Scotland and the Experimental Method and Alberto Vanzo's Kant on Experiment); What sort of propositions can form part of philosophical inference? (Jc Beall's A Neglected Reply to Prior's Dilemma and Hannah Clark-Younger's Is Imperative Inference Impossible?); What can philosophy tell us about what there is? (Emily Gill's Defending Quine on Ontological Commitment, James Maclaurin's Universal

Darwinism: Its Scope and Limits, Kirsten Walsh's Did Newton Feign the Corpuscular Hypothesis? and Vladimír Svoboda's The Scandal of Semantic Platonism).

Colin Cheyne is truly a defender of reason. He has devoted his philosophical career to understanding how philosophy works and to applying philosophical reasoning to problems that matter. He has also spent uncountably many hours helping colleagues and students in the Department of Philosophy at the University of Otago. We are lucky to have had such a good and constant friend who will now be bristling at my sloppy use of "uncountably many" in the preceding sentence.

University of Otago James Maclaurin

Acknowledgements

My first thanks of course goes to Colin who inspired this volume. How happy to be in charge of a project in which all the contributors leap at the chance to be involved and produce wonderful essays well before the publisher's deadline. 'Colin Cheyne' truly is a powerful phrase. Special thanks goes to Jc Beall for his help with the logic chapters and to those who I consulted on the structure of the book, particularly Heather Dyke and Alan Musgrave. Thanks also to Alan for writing a foreword that is accurate, illuminating and fun. Thanks to all at Springer and particularly to Lucy Fleet who has been a pleasure to work with.

My greatest debt is owed to my diligent and uncomplaining editorial assistant, Kirsten Walsh (who is also the author of chapter 12) for making this project much easier than it might have been and for all the care she took over a manuscript with more than its fair share of typographical challenges. As always, my thanks to my remarkable colleagues and to our department's generous and inspiring students who make philosophy fun.

Finally, my special thanks to Kristen Gillespie and George Maclaurin for all their help and good humour, for picking me up when I'm down and for putting up with another year of having a philosopher for a husband and father.

University of Otago James Maclaurin

Contents

Contributors

Jc Beall is Professor of Philosophy jointly in the Departments of Philosophy at University of Connecticut (US) and at the University of Otago (NZ).

Hannah Clark-Younger is a PhD student in the Department of Philosophy at the University of Otago (NZ).

Gregory Currie is Professor of Philosophy in the Department of Philosophy and Dean of the Faculty of Arts at the University of Nottingham (UK).

Gregory W. Dawes is an Associate Professor in the Department of Philosophy and in the Department of Theology and Religion at the University of Otago (NZ).

Heather Dyke is an Associate Professor in the Department of Philosophy at the University of Otago (NZ).

Rebecca E.B. Entwisle was a graduate student in the Department of Philosophy at the University of Otago (NZ) and at Princeton (US).

Martin Frické is an Associate Professor in the School of Information Resources and Library Science at the University of Arizona (US).

Emily Gill is a PhD student in the Department of Philosophy at the University of Otago (NZ).

Juan Gomez is a PhD student in the Department of Philosophy at the University of Otago (NZ).

Bjørn Jespersen is a Post-doctoral Research Fellow in the Department of Values and Technology at the Delft University of Technology (NL).

James Maclaurin is Head of Department and an Associate Professor in the Department of Philosophy at the University of Otago (NZ).

Pavel Materna is a Professor in the Department of Philosophy at Masaryk University (CZ).

Andrew Moore is an Associate Professor in the Department of Philosophy at the University of Otago (NZ).

Tim Mulgan is Professor of Moral and Political Philosophy at the University of St. Andrews (UK).

Alan Musgrave is a Professor in the Department of Philosophy at the University of Otago (NZ).

Josh Parsons is a lecturer at Corpus Christi, Oxford (UK).

Charles R. Pigden is an Associate Professor in the Department of Philosophy at the University of Otago (NZ).

Vladimír Svoboda is a researcher at the Academy of Sciences of the Czech Republic (CZ).

Alberto Vanzo is a Post-doctoral Research Fellow in the Department of Philosophy at the University of Otago (NZ).

Kirsten Walsh is a PhD student in the Department of Philosophy at the University of Otago (NZ).

Part I
In Epistemology

Chapter 1
Getting Over Gettier

Alan Musgrave

1.1 Introduction

Lucky teachers sometimes have students who are cleverer than they are. I have been lucky have Colin Cheyne first as my student and then as my colleague. Over the years we have come to see eye-to-eye on most things, but not on all. This *Festschrift* for Colin seems an appropriate place to revisit what has perhaps been our sharpest disagreement. That concerns the so-called 'Gettier Problem'. Traditionally, knowledge was analysed as justified true belief. Then Edmund Gettier refuted that traditional analysis. Or so most philosophers, including Colin, think. I disagree.

According to the traditional view, A (a person) knows that p (a proposition) if and only if (a) A believes that p, (b) p is true, and (c) A is justified in believing that p or reasonably believes that p. It is the third justification condition that is crucial. And that third condition inherits the notorious act/content ambiguity of the term 'belief'. The term 'belief' can refer to the content of the belief, the proposition believed, as when we say that two people may have one and the same belief. Or the term 'belief' can refer to the mental act or state of believing something. So when we speak of justifying or giving a reason for a belief, do we mean justifying or giving a reason for the belief-content or for the belief-act? It is obvious that we mean the latter. After all, one person might have a good reason for believing something, and another person believe the same thing for no good reason or for no reason at all. We deem the first belief (believing) reasonable and the second belief (believing) unreasonable. Yet the belief (the proposition believed) is exactly the same.

Despite these platitudes, it is widely assumed by epistemologists as well as plain folk that *a reason or justification for believing a proposition must be a reason or justification for the proposition believed*. I call this assumption *'Justificationism'*.

A. Musgrave (✉)
Department of Philosophy, University of Otago, Dunedin, New Zealand
e-mail: alan.musgrave@otago.ac.nz

J. Maclaurin (ed.), *Rationis Defensor: Essays in Honour of Colin Cheyne*,
Studies in History and Philosophy of Science 28, DOI 10.1007/978-94-007-3983-3_1,

In support of justificationism, we can point out that, after all, to believe something is to think it true or more likely true than false. So a reason for believing something must show that what is believed is true or more likely true than false. Must it not?

Still, justificationism raises the obvious question of what a reason or justification for a *proposition* might be? The answer is equally obvious – logic tells us that. A conclusive reason for a proposition is another proposition that entails it. And an inconclusive reason for a proposition is another proposition that entails that it is more likely true than false. (We need a theory of logical probability, a generalization of deductive logic, to make good the idea that one proposition can be an inconclusive reason for another proposition.) Suppose we accept that logic tells us what reasons for propositions are. Then justificationism will drive us to what I call *logomania,* the view that only logical reasoning can provide us with a reason to believe anything. And logomania will in turn result in total irrationalism, the view that all beliefs (believings) are unreasonable.

There are always propositions that entail any given proposition. And if a theory of logical probability can be worked out, there are always propositions that 'probabilify' a given proposition, entail that it is more likely true than false. If the mere existence of a reason for a proposition provided a reason to believe it, then everybody would have a reason to believe any proposition whatever. So the mere existence of these other propositions cannot provide anybody with any reason to believe any proposition. Obviously, the person in question must *believe* the other proposition in question. What we must say is that a reason for believing P is that you believe R and infer P from R. After all, people do sometimes come to believe things by inferring them from other things that they believe.

But again, this is too easy. Everybody might have a conclusive reason for believing any P, if they waste mental energy inferring P from itself. Or if they believe some stronger R from which they infer P. And if a theory of logical probability can be worked out, everybody might have an inconclusive reason for believing any P if they believe some weaker R from which they infer that P is more likely true than false. Obviously, then, we must require not just belief in the justifying proposition, but *reasonable belief* in it. Justificationism said that a reason for believing P must be a reason for P. This has now resulted in the view that the only reason for believing P is that you have inferred it from some other belief R for which you have reason. This is *logomania*, the view that only logical reasoning can provide a reason for believing anything.

But now, as sceptics down the ages have tirelessly pointed out, an infinite regress of reasons looms. Since nobody can complete an infinite series of reasons for the propositions they believe, everybody must start with non-inferential beliefs. But logomania entails that these non-inferential beliefs are unreasonable. From which it follows that all beliefs are unreasonable. If we are to avoid total irrationalism, we must reject logomania and the justificationism that spawned it. We must think that some non-inferential beliefs are reasonable beliefs. And we must think that the reasons for them are reasons for believings, not for the propositions believed.

The two prime sources of non-inferential beliefs are, of course, sense-experience and testimony. Plain folk, asked why they believe things, often invoke sense-experience and testimony as the answer. "Why do you believe that there is a cat in the corner? – Because I see her there". "Why do you believe that Everest is the tallest mountain? – Because I read it in the encyclopaedia". The answers here invoke the causes and the reasons for acts of believing. Of course, for a logomaniac there are no reasons here at all – seeing the cat or reading the encyclopaedia are not propositions and so cannot be reasons for propositions. But logomania is mistaken, as we have seen.

What goes for beliefs acquired from sense-experience or testimony may also go for other beliefs as well. They, too, may be non-inferential, in the sense that they are not brought about by inferring them from other beliefs. They, too, may be reasonable beliefs – but the reasons for them are reasons for believings, not for the propositions believed. As I have argued elsewhere, that a hypothesis has best withstood serious criticism, or provides the best available explanation of some phenomena, are reasons for believing that hypothesis. But they are not reasons for the hypothesis itself, they do not entail that it is true or more likely true than false. This is how I have defended Popper's critical rationalism and so-called 'inference to the best explanation'. But all that is controversial, and I will say no more about it now. It is high time that I turned to Gettier.

1.2 The Gettier Problem

In 1963 Edmund Gettier published the three most important pages in the entire history of philosophy – at least if we measure importance by the volume of literature they spawned (Gettier 1963). Gettier produced cases (now called 'Gettier Cases') that are generally supposed to have refuted the justified true belief account of knowledge. One such case will give the flavour of all of them. A is in a room with B, who he knows well, and C, who he does not know from Adam. A is justified in believing that B owns a Ford. He has heaps of evidence and/or argument for this, the details of which need not concern us. A validly deduces from "B owns a Ford" that someone in the room owns a Ford, and so is justified in believing this as well. In fact, B does not own a Ford – he sold his Ford that very morning. However, as luck would have it, C does own a Ford. A justifiably believes that B owns a Ford, but he does not know it because it is false. A also justifiably believes that someone in the room owns a Ford, and as luck would have it this is true. But does A know it? The widespread intuition is that A does not know that someone in the room owns a Ford, despite the fact that he believes it, it is true, and he is justified in believing it. The widespread intuition is, in other words, that this case and others like it refute the justified true belief account of knowledge.

But do they? Is the widespread intuition correct? It encapsulates a kind of epistemic Puritanism. We can have epistemic bad luck, and lack knowledge even when we have justified belief – as A in the story unluckily does not know that B owns a Ford. But we cannot have epistemic good luck. In the story A behaves impeccably from an 'internalist' point of view – he validly derives a conclusion from a premise that he is justified in believing. Despite this, A does not know that someone in the room owns a Ford either, because he is lucky that this belief is true. We may unluckily lack knowledge – but we cannot luckily possess it.

It is unclear how far this Puritanical ban on epistemic good luck is supposed to go. Suppose that all the blackbirds in some town are black, except for a single albino. Notice, by the way, that "All blackbirds are black" is not analytic, despite the fact that its predicate is contained in its subject. (Sorry – I cannot resist anti-Kantian jokes, even when they are bad ones!) Suppose that all the people in the town form the false belief that all blackbirds are black, except for one person who luckily happens to see the albino blackbird. The folk do not know that all blackbirds are black, for this is false. Does the person who saw the lone albino also not know that some blackbirds are not black, because he was lucky to see it? Anticipating, his true belief was caused by the fact that makes it true (the fact that there is an albino blackbird). Is this kind of epistemic good luck acceptable?

The 'false barns' scenario says not. Imagine a district where all the barns are 'false barns' so that all perceptual beliefs of the form "There is a (true) barn" are all false. EXCEPT that there is one true barn left in the district and an unsuspecting traveller is lucky enough to see it and form the belief "There is a barn". This belief is caused by the fact that makes it true, the lone true barn in the district. Yet some puritans say that it does not count as knowledge, presumably because it was sheer good luck that the traveller saw the one true barn and formed the belief. Other puritans soften their puritanism by distinguishing evidential good luck, which does not disqualify you from knowing, from veritic good luck, which does.

1.3 Externalism

Suppose we accept the widespread intuition that, despite behaving impeccably from an internal epistemic point of view, A does not know that someone in the room owns a Ford. Then we must add a fourth external condition which tells us which justified true beliefs count as knowledge and which do not. What shall this fourth condition be?

The first thought is that the trouble with A's belief that someone in the room owns a Ford is that it was not caused by the fact that makes it true, the fact that C owns a Ford. This is how it differs from the lone villager's fortuitously-true belief that there is a non-black blackbird. This yields the strong condition that for a justified true belief to count as knowledge, the belief must be caused by the fact that makes it true. But the trouble with this strong condition is that few, if any, beliefs are caused by the facts that make them true. The only remotely plausible cases are perceptual beliefs. But no perceptual belief is caused solely by the fact that makes it true, and many perceptual beliefs are false.

Still, we might suppose that most perceptual beliefs are true, that perception is a reliable though not an infallible way of acquiring beliefs. This leads to the less demanding view that for a justified true belief to count as knowledge it must be caused by a reliable process of acquiring beliefs, a process that produces more true beliefs than false ones. Thus we arrive at the view that knowledge is reliably-caused justified true belief. Reliable causation is an 'externalist' condition because it is not assumed that the epistemic agent in question knows or believes anything about what causes her beliefs.

Does the addition of this externalist or reliabilist condition solve the Gettier problem? Does it show, for example, that A's justified true belief that someone in the room owns a Ford does not count as knowledge? To show this, the reliabilist must make out that this belief is NOT caused by a reliable process, a process that produces more true beliefs than false ones. What did cause A's belief? It was caused by A's validly deducing it from his justified false belief that B owns a Ford. So the reliabilist must be assuming that acquiring beliefs by deduction from justified false beliefs yields more false beliefs than true ones. How does the reliabilist know this? Every proposition has infinitely many consequences, infinitely many of which will be true, and infinitely many of which will be false. It is not obvious that false premises yield more false conclusions than true ones. It is even less obvious that justifiably believed false premises yield more false conclusions than true ones. (The problem here is the same problem that dogged Popper's attempt to define verisimilitude or closeness to the truth in terms of true/false ratios within consequence classes.) So it is not at all obvious that adding an externalist or reliabilist constraint to the JTB account will deliver the intended result. But never mind this – suppose that A's belief that someone in the room owns a Ford is not caused by a reliable process. There are deeper problems.

The first problem is that even reliably-caused justified true beliefs may be only fortuitously true – in other words, reliabilism can itself be 'Gettierised'. Colin Cheyne has a Gettier case to illustrate the point (Cheyne 2001, p. 51). Suppose we form beliefs about the temperature by reading an extremely reliable thermometer. The thermometer is not totally reliable because it is designed with a self-checking mechanism that operates for 10 seconds every 10 hours (or days, or years), and when this mechanism is operating it displays random numbers. When the self-checking mechanism is operating we acquire false beliefs about the temperature by reading the thermometer. EXCEPT for the rare occasions when the random number displayed happens to be the correct temperature! THEN we acquire a fortuitously true belief through a process that is as reliable as you like. Reliabilism can always be 'Gettierised' in this way. Any process that rarely produces false beliefs may even more rarely produce true beliefs that are only fortuitously true. So reliabilism does not solve the problem it was supposed to solve, does not rule out 'fortuitous knowledge'.

Despite this, reliabilism or externalism became flavour of the month, and has come to dominate epistemology. So far I have presented it merely as supplementing the traditional account of knowledge with a fourth condition, so that knowledge is now thought of as reliably-caused justified true belief (an RJTB account). No one should quarrel with this or any other definition. One can wonder whether, on this definition, anybody ever knows that they know anything. Do we know, of any belief-producing process, that it is reliable? I shall come back to this.

However, externalists or reliabilists go further than this. They say that the traditional JTB account is not to be supplemented but rather is to be replaced by a simpler account whereby knowledge is simply defined as reliably-caused true belief (RTB). Or they say that knowledge is still justified true belief, but that we need an externalist or reliabilist view of what justification amounts to (JTB with J=R). Either way, I think, a crucial issue is being set aside. And ultimately what is at work in setting it aside is justificationism, the conflation of reasons or justifications for acts of believing with reasons or justifications for the things believed. Reliabilism is the offspring of justificationism. That a believing is caused by the fact that makes its content true shows its content to be true. That a believing is caused by a reliable process shows its content to be more likely true than false. But neither shows anything about whether the believing is reasonable.

If we simply define knowledge as reliably caused true belief then, obviously, we say nothing whatever about the reasonableness or justifiability of the act of believing. An act of believing that is totally unreasonable or irrational from an 'internalist' point of view might count as an item of knowledge. The same applies if we say instead that we are presenting an externalist or reliabilist view of the justification or reasonableness or acts of believing. Again, a totally unreasonable or irrational or irresponsible act of believing counts as justified. The believing counts as justified because the belief-content is justified – as justificationism maintains.

Here, again by courtesy of Colin Cheyne (2001, p. 49), is an imaginary scenario to illustrate the point. Alvin wakes up each morning with a new belief. Sometimes Alvin's new beliefs concern completely inaccessible matters of historical fact, sometimes unsolved problems of mathematics, and so forth. Alvin has no idea how he comes by these beliefs, and can produce no evidence or argument for any of them. But he cannot help having them. He persists in them however unlikely they seem to be, and he is impervious to counter-evidence or counter-argument. Cash out the 'ethics of belief' how you like, Alvin ought not to have these beliefs, it is not reasonable for him to have them. However, the facts are these. God exists and has singled Alvin out for special attention in the way of belief. God somehow zaps true beliefs into Alvin while he sleeps. There could be no more reliable way of acquiring beliefs than this one. According to the externalist view of knowledge, Alvin knows all these things. According to the externalist view of justification, Alvin is perfectly justified in believing all these things.

This scenario is fanciful, of course – but not completely so. Some atheists and agnostics said that, whether or not God exists, it is unreasonable to believe in God. To defeat this epistemic objection to theism, Alvin Plantinga proposed an externalist theory of knowledge. He argues that if God exists he will have zapped religious belief into the chosen, not exactly as they sleep, but by some more convoluted process. Specifically, by designing us so that we will acquire the belief by the proper exercise of our cognitive machinery. (Presumably, those who lack religious belief are guilty of improper exercise of their God-given cognitive machinery. "Ignorance of the Lord is sin", just as Descartes said.) So if God exists, belief in Him is not only true but also justified or 'warranted'. And if God does not exist, belief in Him is not only false but also unjustified or 'unwarranted'.

Theism and atheism are epistemically on a par. Epistemic objections to theism are question-begging, in that they assume atheism. Thus was externalism a God-send to God-botherers like Plantinga.

The epistemic objection to religious belief was that it is unreasonable, whether or not it is true. You do not answer this objection by saying that if it is true it is reliably caused – unless you go in for an unsatisfactory reliabilist theory of rationality or reasonableness. Nor do you answer it by going in for a reliabilist theory of knowledge whereby knowledge is just reliably caused true belief. For a belief that counts as knowledge under this view may be completely unreasonable.

There is another very obvious objection to reliabilist views of justification or reasonableness. Suppose an act of believing is justified because it is produced through a reliable process. Suppose that this process is not completely reliable and that this particular belief happens to be false. Finally, and crucially, suppose that the believer comes to see this, comes to find evidence and/or argument for the falsity of the belief. The belief was justified – does it remain so? On reliabilist views it seems that it does. After all, the belief was produced by a reliable process and that fact can hardly change. On internalist views it does not – it was, perhaps, reasonably believed to begin with, but it does not remain so. It may be unreasonable to believe today what we believed yesterday. Generally, what determines whether or not a belief is reasonable is not how we acquire it (reliably or not), but what we do with it after we have acquired it.

I have resisted externalism or reliabilism because it either ignores or gives an unsatisfactory account of the reasonableness of acts of believing. This does not mean that I think the question of which methods of acquiring beliefs are reliable and which not is an unimportant or uninteresting question. Quite the contrary. If we seek truth, we have an obvious interest in the question of whether some methods of forming beliefs are more reliable than others. This is a factual or scientific question about our methods of inquiry. As such, it is subject to all the usual sceptical and/or fallibilist constraints. We cannot certainly know that any belief-producing mechanism is reliable, for all the usual reasons. Even if we could show that some mechanism has in the past produced more true beliefs than false ones, it does not follow that it will continue to do so – that is just the problem of induction again.

If there is some reliable belief-producing mechanism, then this is a contingent fact. Thus it is possible that there are no reliable belief-producing mechanisms. Should we make contingent facts about our cognitive machinery, something we seek knowledge about, part of an externalist definition of what knowledge is? Could we come to know, by the exercise of our cognitive machinery, that our cognitive machinery can deliver us no knowledge? Might externalism be self-refuting?

I think that it is reasonable to believe the so-called 'evidence of the senses' – unless and until you find a specific reason not to. I do not know whether perception is a reliable belief-producing mechanism. I do not know whether most of the perceptual beliefs I have acquired are true – let alone whether most of the perceptual beliefs that other folk have acquired, down the ages, have been true. Again, I think it reasonable to believe what other folk tell you – unless and until you find a specific reason not to. I do not know whether testimony is a reliable belief-producing mechanism. I do not know whether most of the things folk have told me are true – let alone whether most of the things other folk have told other folk, down the ages, have been are true.

What I do know, I think, is that sceptics have not shown that perception and testimony are unreliable belief-producing mechanisms. How could they show this, given the nature of sceptical arguments themselves? Nor, of course, can anybody show that reading tea-leaves is an unreliable belief-producing mechanism – that it has produced more false beliefs than true ones up until now does not show that it will continue to do so. The best we can do, in these abstract regions as in more concrete ones, is show that one hypothesis withstands criticism better than another. That is a reason to adopt the hypothesis, but not a reason for the hypothesis itself, for it does not show that it is true or more likely true than false. That is the basis on which sense-experience and testimony are to be preferred to reading tea-leaves. We have to start somewhere, with non-inferential beliefs. Sense-experience and testimony are better places to start than reading tea-leaves, because they have withstood criticism better than reading tea-leaves has. But, to repeat, what determines whether a particular belief is reasonable is not so much where we got it from, but what we do with it after we get it.

To sum up. Knowledge is justified true belief after all. But what is justified is the act of believing, not the content of the belief. So the Gettier cases are cases of knowledge after all – we can luckily know as well as unluckily fail to know. Indeed, having had Colin Cheyne as a friend and colleague has brought me a good deal of good luck, epistemic and otherwise. Which is where I came in.

References

Cheyne, C. 2001. *Knowledge, cause, and abstract objects: Causal objections to platonism.* Dordrecht/Boston/London: Kluwer Academic Publishers.

Gettier, E. 1963. Is justified true belief knowledge? *Analysis* 23: 121–3.

Chapter 2
Justified Believing: Avoiding the Paradox

Gregory W. Dawes

2.1 Introduction

Colin Cheyne has recently outlined what appears to be a *reductio ad absurdum* of a traditional account of justified belief. He begins from three apparently uncontroversial ideas, which he sets out as follows (Cheyne 2009, p. 278).

1. There are circumstances in which an agent is justified in believing a falsehood.
2. There are circumstances in which an agent is justified in believing a principle of epistemic justification.
3. Beliefs acquired in compliance with a justifiably-believed epistemic principle are justified.

Cheyne shows that there are circumstances in which these three ideas can lead to a paradox, in which a given belief is simultaneously justified and unjustified. In what circumstances does the paradox arise and can we avoid it?

2.2 Cheyne's Alleged Paradox

The paradox arises when, in accord with principle (1), an agent forms a justified false belief. But this belief is not merely a false belief; in accord with principle (2), it is a belief in a false principle of epistemic justification. Cheyne's example is that of Albert, who forms a justified but false belief "that reading tea-leaves is a reliable method for acquiring true beliefs" (Cheyne 2009, p. 287). Albert then uses this false epistemic principle to form new beliefs, which I shall call "derivative

G.W. Dawes (✉)
Department of Philosophy and Department of Theology and Religion,
University of Otago, Dunedin, New Zealand
e-mail: gregory.dawes@otago.ac.nz

J. Maclaurin (ed.), *Rationis Defensor: Essays in Honour of Colin Cheyne*,
Studies in History and Philosophy of Science 28, DOI 10.1007/978-94-007-3983-3_2,

beliefs." He forms these new beliefs by reading tea-leaves. Principle (3) – that of "justification-transference" – says that these derivative beliefs will themselves be justified (Cheyne 2009. p. 285). But since the tea-leaf-reading principle is a false principle of epistemic justification, these same beliefs will be unjustified. Hence the paradox, which seems to be a *reductio ad absurdum* of at least one traditional idea of justified belief.

Let me attempt to spell this out a little more formally. In the scenario Cheyne posits Albert justifiably believes that for every act of believing, if it is performed on the basis of reading tea-leaves, it is a justified act of believing. More generally,

1. A subject *S* justifiably believes that for every act of believing *p* performed under circumstances *C*, that act of believing is justified.

Having justifiably believed in the tea-leaf-reading epistemic principle, Albert then forms a derivative belief: one formed by reading tea-leaves. Let's say it is a belief that he will inherit money next week. We have already accepted, for the sake of the argument, a second principle, that of justification-transference (Cheyne 2009, p. 2):

2. Beliefs acquired in compliance with a justifiably-believed epistemic principle are justified.

From principles (1) and (2), it follows that the subject's belief that he will inherit money next week is justified.

In fact, however, Albert has justifiably but mistakenly formed this belief in the doxastic value of reading tea-leaves. Acts of believing performed on the basis of reading tea-leaves are not justified, even though Albert justifiably believes that they are. More generally,

3. For every act of believing *p* performed under circumstances *C*, that act of believing is unjustified.[1]

From principle (3), it follows that Albert's belief that he will inherit money is unjustified. So the three principles together entail that his belief is both justified and unjustified. This is, as I understand it, Cheyne's paradox.[2]

---

[1] The assumption here is that justification has just two possible truth-values, so that what is not justified is unjustified. Alvin Goodman (1993, pp. 274–275) suggests the possibility of a third, neutral category – that of non-justified but not necessarily unjustified beliefs – but I shall not pursue that possibility here.

[2] Throughout this paper, I have understood the falsity of the principle of epistemic justification to be *objective* falsity. In a recent conversation, Cheyne has suggested that the falsity in his scenario should be thought of as falsity *according to the theory of justification being employed*, rather than falsity *simpliciter*. But on this rather different assumption, my argument would still go through. To anticipate, what I am arguing is that an agent could be subjectively justified in accepting beliefs on the basis of a false principle of epistemic justification principle, provided that she does not realize that it is false. But this can be easily reworded to accommodate Colin's suggestion: an agent could be subjectively justified in accepting beliefs on the basis of a principle of epistemic justification that is inconsistent with another principle she holds, provided that she does not realize that it is inconsistent. In neither case does a paradox arise.

Cheyne's argument suggests that in order to avoid this paradox we need to abandon one of his three initial principles. The obvious principle to abandon is the principle of justification-transference. The *externalist*, he argues, might be able to abandon this principle, since for the externalist what is happening in the agent's mind is irrelevant to the question of justification. It is the reliability or unreliability of the process giving rise to the belief that matters. On an externalist view, therefore, "the unreliability of tea-leaf reading trumps Albert's justified belief that tea-leaf reading is reliable" (Cheyne 2009, p. 287). But it seems that no similar way out is available to the *internalist*, at least one who holds to a deontological account of justification. If justification is a matter of "fulfilling one's epistemic duties or responsibilities" (Cheyne 2009, p. 289), then Albert would be acting irresponsibly if he did not regard his derivative beliefs as justified. After all, it would seem irresponsible to form a justified belief in an epistemic principle – even a tea-leaf-reading principle – and then ignore it when forming one's beliefs. So it seems we must abandon deontological accounts of justification.

Can we avoid this conclusion? There is, as one might expect, nothing wrong with Cheyne's logic here. But what I want to argue is that there is at least one version of an internalist account of justified believing that avoids this paradox. Moreover, the account that can avoid the paradox is precisely the one that Cheyne believes needs to be abandoned, namely a deontological account. In the case of a non-deontological account, the paradox can be avoided (as Cheyne suggests) by abandoning the principle of justification-transference. But this is, I shall argue, a conclusion that any internalist could accept.

2.3 Two Internalist Conceptions of Justification

I am understanding internalism to be the view that the essential factors that make a belief justified or unjustified have to do with the mental state of the agent. Let me begin by offering, for the sake of the argument, one version of such a view. I shall begin by describing the mental states relevant to justification as the agent's reasons for believing.[3] I wish to leave open what form those reasons might take. They may consist, for instance, of other beliefs from which the content of her new belief can be inferred. Or they may consist of particular doxastic experiences, such as those of memory or sense perception. The internalist, on my view, holds that for an act of believing to be justified, these two conditions must be fulfilled.

[3] By "reasons for believing" I mean those factors that are the basis on which the agent believes. It is not easy to define this basing relation (see Korcz 2010), but very often a causal account seems appropriate. An agent's reasons for believing are those factors that are bringing about (or sustaining) her belief.

(a) The agent's reasons for believing must be matters of which she is aware.
(b) It must be the case that she would, on reflection, consider these to be adequate grounds for her belief.[4]

The key question then is: Is the state of affairs represented by these two conditions a sufficient or merely a necessary condition of justification (Kvanvig 1984, p. 72)? Each of these options will give rise to a distinctive sense of what it means for a belief to be justified: the first subjective and the second objective (Pollock 1979, pp. 109–110).

2.3.1 *Subjectively Justified Acts of Believing*

A subjective account of justified believing holds that conditions (1) and (2) are *sufficient* for justification. An act of believing is justified if the agent has (a) done all that she ought to do to examine the grounds of her belief and (b) would (on reflection) consider these to be adequate grounds.

The "ought" here indicates that this is what is customarily called a deontological view. But the term "deontological" may be misleading, since the "ought" here need not be thought of as ethical. Richard Feldman, for instance, suggests it is best regarded as a "role ought" (Feldman 2000, p. 676). ("If you are a competent pianist, you ought to be able to play Beethoven's Moonlight Sonata.") On this view, there are ways in which you ought to act, in your role as a knower. Alternatively, one could regard it as the kind of "ought" that is characteristic of practical reason. ("If you want to become a competent pianist, you ought to practise daily.") On this view, if you want to obtain knowledge, this is how you ought to act in order to attain your goal. But in precisely what sense subjective justification is deontological is not a question I need discuss here.

This kind of justification is subjective in several senses. One of these is that what it entails is person- and context-relative. If I hear rumours about possible side-effects of a vaccine, how I ought to respond will depend on who I am. If I am a doctor, it may be that I ought to examine the evidence more carefully than if I am merely a curious member of the public. (The "ought" in this context does have an

[4] Any attempt to articulate an internalist view will be controversial. My view assumes that for justified believing the agent must be aware of those factors – other beliefs or doxastic experiences – that are her reasons for believing. But she need not have formed any belief about (a) the fact that they are her reasons for believing or (b) their adequacy as reasons. What is required for justification is that she would, on reflection, consider them to be adequate reasons. (The "adequacy" here is, of course, epistemic: an adequate reason is such that if her belief that *p* is true, she can be said to know *p*.) By taking this line, I hope to avoid the notorious infinite regress problem (Ruloff 2009, pp. 144–145). But it also seems to me that a person who has long held a justified belief could say, quite intelligibly, "Yes, my belief that *p* is the reason I believe that *q*, and it seems to me a good reason, even though I've never reflected on its evidential force before."

ethical dimension.) What subjective justification entails is also person- and context-relative in another sense: whether I am acting rationally in forming this belief will depend on the methods of reasoning that are available to me. A person trained in medicine will be better placed to appraise the evidence regarding the vaccine than a person without such training, and would be falling short of her epistemic obligations if she did not make use of these skills. And a person living in the twenty-first century will have methods of assessment available to her that were not available in medieval Europe.

But the key point here is that an act of believing is subjectively justified if, at the end of this process, the agent would (on reflection) consider her reasons for believing to be adequate reasons. This judgement does not need to be correct. Her belief regarding the available reasons would be sufficient, on this view, for justification. It would remain so even if she has made some honest error of reasoning or if the mechanism that gives rise to her belief is not, in fact, as reliable as she believes it to be. The intuition behind such a notion is that one does not cease to be a rational agent as a result of making a (non-obvious) mistake regarding the available evidence. To accommodate such an intuition, some notion of subjective justification seems required.

2.3.2 Objectively Justified Acts of Believing

There are times, however, when our (internalist) question about justification is a different one. We might agree that a person is subjectively justified in forming a belief, but still want to know if her judgements regarding her reasons for belief are correct. In other words, we want to know if she has formed her beliefs in ways that are objectively appropriate (Henderson and Horgan 2001, p. 227). This is not the same as asking if what she believes is true, since it is (as we shall see) possible to make a correct judgement on the basis of the available evidence and still form a false belief (see Cheyne 2009, p. 12). Nor is it a deontological question, since it is not answered by the observation that the agent has acted as she ought to in forming the belief. She may, for instance, have examined the evidence to the best of her ability and still made an error of judgement. The question regarding objective justification has to do with what the evidence available to the agent actually warrants.

It follows that the answer to this question will *not* be person- and context-relative. Admittedly, our starting point in answering it *will* be person- and context-relative, since we will start with the evidence that was available to this agent at the time she formed her beliefs. But given that evidence, the justifying relation, however that is conceived, will be entirely objective. If the evidence available to the agent really does support the act of believing – whether by deductive, inductive, or abductive reasoning, or in some other way – the act of believing is objectively justified.

Because this sense of "justified believing" has to do with the objective conditions under which something is believed, it may appear to be an externalist notion. After all, the validity or soundness of an argument is an objective fact: it does not depend on what the subject believes about it. The similarity with externalism will be particularly striking when we are dealing with a belief that arises not from a process of reasoning, but spontaneously, as a result of some mechanism (such as sense perception). The reliability or unreliability of the mechanism is also a fact that is independent of what we believe about it. But even in these circumstances, objective justification can still be considered an internalist idea. It is not, for example, the mere fact that the mechanism happens to be reliable that makes the belief justified. The agent must be aware of this mechanism and her other beliefs must be such that she would, on reflection, consider it to be reliable. To ask if her confidence in this mechanism is objectively justified is to ask if her other beliefs really do support her conviction that it is reliable.

Here's another way of making the point. I suggested earlier than on an internalist account, there are two conditions that need to be met for a justified act of believing.

(a) The agent's reasons for believing must be matters of which she is aware.
(b) It must be the case that she would, on reflection, consider these to be adequate grounds for her belief.

If what we are interested in is *subjective* justification, then these conditions are sufficient for justification. But if what we are interested in is *objective* justification, then they are merely necessary conditions. An agent's act of believing could meet both of these conditions, but still fail to be objectively justified, since the evidence on the basis of which she is believing does not, in fact, support her belief, even though she would (on reflection) consider it to do so.

Here are a couple of examples. Galileo Galilei may have been acting rationally – he may have been subjectively justified – in holding to his particular theory of the tides. But he might also have made some poor judgements in this respect: it may be that even the evidence available to him did not in fact support his theory. A contemporary Christian may be subjectively justified in holding to her Christian faith, if she has done all she ought to do to examine the evidence and her belief seems to her to be warranted (Plantinga 2000, pp. 252–253). But she may still be mistaken in making these judgements. Even if she is subjectively justified, we can still ask if the facts or experiences that she takes to be evidence of the truth of her faith really do constitute adequate reasons for believing.[5]

[5] This is a question that Alvin Plantinga, in developing his Reformed Epistemology, entirely neglects.

2.3.3 Related Distinctions

Before continuing my argument, I should note two distinctions that could be confused with the one I have just made. While related to my distinction, they are not identical with it.

The first is one that Cheyne himself employs, distinguishing (as we have seen) between the act of believing and the content of belief. His argument chooses (rightly) to focus on the act of believing (Cheyne 2009, p. 279). But the distinction I have just made – between a subjective and an objective sense of "justified believing" – is not identical with this. I, too, wish to focus on the act of believing, the question being (on an internalist account) whether an agent has adequate reasons for a particular act of believing. But this question, as I hope my discussion has shown, can be formulated either subjectively or objectively. It can be a question about whether the agent has reasons that *she would (on reflection) consider* adequate reasons or a question about whether *she has what really are* adequate reasons. She may believe, for instance, that inference to the best explanation (IBE) warrants belief, and thus be subjectively justified in believing the best available explanation of some fact. Or she may believe that sense perception, in these particular circumstances, is reliable, and thus be subjectively justified in accepting the evidence of her senses. But we can still ask if IBE really does warrant belief or if sense perception really is reliable in these circumstances. So it is possible that the subject's subjectively justified act of believing may not be objectively justified.

A second distinction, which is closer to mine but still not identical with it, is that between propositional and doxastic justification. In a recent article, Colin Ruloff sets it out as follows:

> Propositional justification obtains when an agent *S*'s total available evidence makes a proposition *p* likely to be true – even if *S* does not believe *p* or *S* believes *p* for the wrong reason. Doxastic justification, by contrast, obtains when a proposition *p* has propositional justification for *S* and when *S* believes *p* on the basis of justifying evidence or grounds. Doxastic justification ..., thus, requires that *S*'s belief that *p* be appropriately related to *S*'s evidence or grounds. (Ruloff 2009, p. 134)

Propositional justification resembles objectively justified acts of believing since it, too, has to do with an objective relation between evidence and belief. But in common with Cheyne, I am focusing not on the proposition believed, but on the act of believing, so the justification with which I am concerned is doxastic justification. My distinction between subjectively and objectively justified acts of believing has to do with different ways of conceptualising the "appropriate relation" between an agent's act of believing *p* and the grounds on which she believes *p*. On a subjective account, an agent's act of believing is justified if it would (on reflection) seem to her that an appropriate relation exists. On an objective account, an agent's act of believing is justified only if that relation really is appropriate: if the reasons on the basis of which she believes are, in fact, adequate reasons.

2.4 Internalism and the Paradox

Armed with this distinction, I can now return to Cheyne's paradox. What I hope to show is that no paradox arises in the case of a *subjective* (or deontological) accounts of justification. A paradox might arise in the case of an *objective* account of justification, but this is a fact which an internalist could easily accept.

2.4.1 Subjective (Deontological) Justification

Let's begin with an example that may seem a little more plausible than Cheyne's tea-leaf reading example. Let's say that Mary believes in a principle (let's call it *R*) which states that one is justified in believing whatever the Pope, as Bishop of *R*ome, teaches. Mary would, on reflection, consider that the reasons on the basis of which she believes this principle are adequate reasons. She is, therefore, subjectively justified in believing the following proposition:

(R) For any proposition *p*, simply by virtue of the fact that the Pope has declared *p* to be true, then one is justified in believing *p*.

Mary would then consider that she had adequate reasons to believe a particular papal teaching – such as the idea that the Blessed Virgin Mary was assumed bodily into heaven – the adequate reason being precisely principle *R*. So her belief in the doctrine taught by the Pope is also subjectively justified. Given a principle of justification-transference (*JT*), this seems correct.

(JT) Beliefs acquired in compliance with a justifiably-believed epistemic principle are justified.

Let's now say, for the sake of the argument, that *R* is false.

(not-R) For any proposition *p*, it is not the case that simply by virtue of the fact that the Pope has declared *p* to be true, then one is justified in believing *p*.

Cheyne's argument, as I understand it, is that this gives rise to a paradox. If Mary is subjectively justified in believing *R*, even though *R* is false, then by *JT* she is also justified in believing *p*, a belief taught by the Pope, simply by virtue of the fact that he teaches it. But according to *not-R*, she is not justified in believing *p*, simply by virtue of the fact that the Pope teaches it.

The problem with this argument is that it fails to distinguish the two senses of justified believing that I have attempted to spell out. Given that Mary is *subjectively* justified in accepting *R* then – given *JT* – she is also *subjectively* justified in accepting a particular papal teaching. What *not-R* shows is that she is not *objectively* justified in accepting that teaching. But there seems no reason to assume that a person cannot be both subjectively unjustified and objectively unjustified. She has simply made a mistake about the proper way to form beliefs.

2.4.2 *Objective Justification*

But what about objective justification? We have seen that the falsity of a principle of justification leads to no paradox in the situation in which I am *subjectively* justified in believing that principle. But would the paradox arise in the circumstances in which I am *objectively* justified in believing a false principle of justification?

Could this situation ever arise? It's difficult to know. Let's look again at the situation of Mary. Mary has been reading a learned Roman Catholic apologist. This apologist has produced valid arguments for principle *R*. Those arguments might state, for example, that Jesus appointed St Peter as the rock on which the church was to be built (Matt 16:18) and that the Bishops of Rome inherited his office. From these two premises he deduces that in matters of faith the Popes will never be in error and that principle *R* is therefore justified. Papal authority is, in effect, a reliable mechanism producing true beliefs. Let's grant, for the moment, the logic of the apologist's argument. But let's also assume that one of the premises is false. It may be, for instance, that the words attributed to Jesus in the Gospel of Matthew are a fiction or that the Popes did not in fact inherit the office given to St Peter. Let's assume, too, as we did before, that principle *R* is false. Would Mary be objectively justified in believing *R*?

The answer one gives to this question will depend on how one answers a second question: Could a false belief taken as evidence constitute objective justification for an act of believing? If, for example, Mary falsely believes that the premises of the apologist's arguments are true, does the validity of his argument mean that her belief is objectively justified? It may be that a false belief cannot *objectively* justify another belief (Lehrer 1974, pp. 213–214), even though it could *subjectively* justify it. This seems plausible. If that which (objectively) justifies a belief is the fact that one could construct a deductive argument leading from evidence to proposition believed, then it is not sufficient that such an argument be valid. It must also be sound: its premises must be true. If this is the case, then Mary's belief in *R* is not objectively justified, no matter what she believes about it. But if Mary's belief in *R* is not objectively justified, then no paradox arises.

Are there other circumstances in which I might be objectively justified in believing a false principle of justification? I have suggested that my act of believing cannot be objectively justified when the evidence on which I base my belief is false. But what if the evidence on which I base my belief is true but incomplete? In these circumstances, it seems, I *could* be objectively justified in holding a false belief. Given the evidence available to me today, I am objectively justified in believing that the butler was responsible for the murder. But new evidence might turn up tomorrow that exonerates the butler and makes it clear the maid was responsible. In a similar way, it is conceivable that the arguments in favour of a particular belief-forming policy (such as *R*) may be sound. But if they are inductive or abductive arguments – arguments in which the premises can be true but the conclusion false – then new evidence could still indicate that the conclusion they supported was the wrong one.

Only in these circumstances – in which I am objectively justified in accepting a false principle of justification – would an internalist need to confront Cheyne's paradox.[6] Let's say that Mary was objectively justified in believing the (false) principle *R*.

(R) For any proposition *p*, simply by virtue of the fact that the Pope has declared *p* to be true, then one is justified in believing *p*.

Let's accept, too, the principle of justification transference, *JT*.

(JT) Beliefs acquired in compliance with a justifiably-believed epistemic principle are justified.

Mary is then objectively justified in forming a belief that *p*, on the grounds that the Pope declares *p* to be true. But *ex hypothesi*, R is false.

(not-R) For any proposition *p*, it is not the case that simply by virtue of the fact that the Pope has declared *p* to be true, then one is justified in believing *p*.

It follows that Mary's belief that *p* is objectively unjustified. So the same belief would be both objectively justified and objectively unjustified, *quod est absurdum*.

In these circumstances, and in these circumstances alone, Cheyne would be right. When faced with an objectively justified false belief, we cannot maintain the principle of justification-transference. Even if Mary is objectively justified in believing the false principle *R*, this does not mean that the beliefs she forms on the basis of *R* are objectively justified. I see no reason, however, why an internalist could not accept this conclusion, since the subjective (or deontological) sense of justified believing has emerged from this discussion unscathed.

2.5 Conclusion

At first sight, Cheyne's paradox looks like a *reductio ad absurdum* of a popular view of justified belief. It demands the surrender of a principle – that of justification-transference – that seems vital to a deontological view of epistemic justification. But on closer examination, we find that it is precisely the deontological view of justified believing that escapes the paradox and allows us to retain that principle. The reason

[6] In an earlier draft of this paper, I distinguished four cases (rather than merely two): (a) subjectively justified belief in a false principle of subjective justification, (b) subjectively justified belief in a false principle of objective justification, (c) objectively justified belief in a false principle of subjective justification, and (d) objectively justified belief in a false principle of objective justification. While this did exhaust the alternatives, I now believe it to have been unnecessarily confusing. But for the record, the case I am here considering is most accurately characterized as an example of (d). None of the others could rise to a paradox, since there is no reason why a belief cannot be subjectively justified but objectively unjustified or objectively justified but subjectively unjustified.

for this turns out to be a simple one. It is that the falsity of a principle of epistemic justification does not render an act of believing that relies on this principle *subjectively* unjustified. It would do so only if the agent realized that the principle was, indeed, false, but *ex hypothesi* she does not. It follows that no paradox arises in the case of the subjective (or deontological) sense of justified belief.

On the other hand, it may be that the falsity of a principle of epistemic justification renders an act of believing that relies on this principle *objectively* unjustified. If so, then the objective sense of justified believing will give rise to a paradox in the conditions that Cheyne outlines. It would follow that when we speak of justified believing in the objective sense, we will have to abandon the principle of justification-transference. But this is not a conclusion that should bother internalists, given that the deontological (subjective) use of this principle has been saved.

My conclusion, then, is that a deontological view of justified believing can survive Cheyne's criticism. This is not, however, to say that it can survive all criticisms. There remain powerful arguments against internalist accounts of justified believing. I have made no attempt to address such arguments, since this paper does not pretend to be a defence of epistemic internalism *tout court*. All I have attempted to do is to defeat one objection to an internalist view, understood in deontological terms.

References

Cheyne, C. 2009. A paradox of justified believing. *Ratio NS* 12: 278–290.

Feldman, R. 2000. The ethics of belief. *Philosophy and Phenomenological Research* 60(3): 667–695.

Goodman, A. 1993. Epistemic folkways and scientific epistemology. *Philosophical Issues* 3: 271–285.

Henderson, D., and T.E. Horgan. 2001. Practicing safe epistemology. *Philosophical Studies* 102: 227–258.

Korcz, K.A. 2010. The epistemic basing relation. *The Stanford encyclopedia of philosophy (Spring 2010 Edition)*. Retrieved from http://plato.stanford.edu/archives/spr2010/entries/basing-epistemic/.

Kvanvig, J.L. 1984. Subjective justification. *Mind* 93: 71–84.

Lehrer, K. 1974. *Knowledge*. Oxford: Clarendon.

Plantinga, A. 2000. *Warranted Christian belief*. New York: Oxford University Press.

Pollock, J.L. 1979. A plethora of epistemological theories. In *Justification and knowledge: New studies in epistemology*, ed. G.S. Pappas, 93–113. Dordrecht: Reidel.

Ruloff, C. 2009. Epistemic supervenience and internalism: A trilemma. *Theoria* 75: 129–151.

Chapter 3
Literature and Truthfulness*

Gregory Currie

There is a kind of character in thy life
That to th' observer doth thy history
Fully unfold.

Measure for Measure 1.1

"Shakespeare was inconceivably wise" said Emerson, expressing a traditional view of the reverence due to great literature. In philosophical circles this view has survived the ravages of the deconstructive turn, with Martha Nussbaum's recent advocacy strikingly reminiscent of Trilling half a century earlier; Robert Pippin and Jenefer Robinson take the philosophical arm of the tradition into the new century (Nussbaum 1994; Robinson 2005; Trilling 1951). Even the current literary world is not always hostile: Jonathan Bate notes a "welcome return. . . to the Johnsonian idea that [Shakespeare's] plays illuminate not just the mentalities of their own age but rather. . . 'the human condition'"; David Bevington, in Johnsonian mood, tells us that *Hamlet* is "able to speak to persons and societies of all nations and all ages who have turned to it for a better understanding of themselves" (Bate 2011; Bevington 2011, p. vii).

Views of this kind have many forms, but a common thought is that works of great literature are not merely worthy objects of aesthetic attention; they are the products of exceptional minds, sensitive to aspects of the real world, especially the human world of decision, thought and feeling. They are works which have the

* This essay is warmly dedicated to Colin Cheyne, for many years my colleague at Otago. It is part of the work I have been able to do over the last year with the support of the Arts and Humanities Research Council and St John's College, Oxford. I'm also grateful for discussion of these issues with the many helpful, patient and critical colleagues involved in these projects, among them Terence Cave, Matthew Kieran, Aaron Meskin, Margaret Moore, Jon Robson.

G. Currie (✉)
Department of Philosophy, Faculty of Arts, University of Nottingham, Nottingham, UK
e-mail: gregory.currie@nottingham.ac.uk

J. Maclaurin (ed.), *Rationis Defensor: Essays in Honour of Colin Cheyne*, Studies in History and Philosophy of Science 28, DOI 10.1007/978-94-007-3983-3_3,

capacity to deepen our understanding of ourselves, our predicament and our capacities, especially in relation to choices concerning co-operation, commitment and competition, and the kinds of emotions that inform and sometimes derail these choices: love and liking, hatred and envy, solidarity and suspicion. Call that the positive view.

The positive view generally combines an interest in the psychological and the moral dimensions of narrative, through the idea that the best kinds of fictions are those which focus on ways in which characters negotiate or fail to negotiate moral problems and dilemmas. Works such as Conrad's *Lord Jim* are valued partly for their vivid portrayal of the stresses which produce moral failure, and the psychological aftermath of such failure. Thus the focus tends to be on literary representations of decision and action, and the roles of emotion, weakness of will, ignorance, self-deception, and partiality of view in complicating (and occasionally in enabling) such action. There is also a special emphasis in the novel on the long terms psychological and moral effects of decisions and actions, and their tendency to produce unpredicted and ironic consequences for the agents concerned: something that Shakespeare, with little time to spare, gives us in a great rush at the end of *Measure for Measure*.

Some recent opposition to the positive view within literary theory and criticism derives from a general suspicion about notions of truth hospitable to the idea that there is a reality independent of its various representations. These objections have received detailed rebuttals and I do not wish to revisit this issue here.[1] Besides, denying integrity to truth, etc. would make it impossible for any advocate of the Positive View to argue that truth is a value in literary representations, thereby cutting off much of the debate at the start. Wanting to give my opponents at least a chance to get started, I avoid the nuclear option and accuse no one of incoherence when the advocate of the Positive View claims that literary representations sometimes convey to us important truths. I'm doubtful about this claim, but my doubts are not based on philosophical principles; I simply think, rather boringly, that the idea will turn out to be, in large measure, unsupported by the relevant evidence.

Some advocates of the Positive View will themselves be hesitant about deploying the notion of truth in the service of their cause. It is a mistake, they say, to think that literature gives us truths, for truth ties us to knowing-that, and the kind of knowing in question is knowing-how. Nussbaum emphasises the extent to which literature helps to tune us to the exquisite particularity of ethical judgment, helping us to become those people on whom, in Henry James' words, "nothing is lost". Coming from a completely different perspective, Keith Oatley and his colleagues at Toronto have recently argued that fictions educate, not by standing in for lectures, by providing simulations not unlike those we use to train pilots.[2]

[1] See e.g. Devitt and Sterelny (1999), also Lamarque and Olsen (1994), about which I say more later.

[2] See e.g. Mar et al. (2006, 2008), Djikic et al. (2009).

It's not clear, in the end, how far such an approach will take us away from truth and truthfulness. For a start, a number of people have argued that the supposed distinction between knowing that and knowing how is bogus, with the latter actually a special case of the former.[3] But even if we think of knowing how as non-propositional, the case against truth is not settled. Nussbaum often seems to think of propositional knowledge as co-extensive with knowledge of simple, stateable, general truths. But some truths fall into none of these categories. Take the idea of direct reference, which allows an object to be part of the proposition in question. And just as I may refer directly to Smith, so may I refer directly to Smith's way of dancing, as when I say "Smith dances like this", where "this" refers directly to a sample of dancing that I now produce for you. By means of this construction I express a proposition of which a sample of dancing is a constituent. Grasping the referent of "this" requires a perceptual confrontation with this sample. This proposition and many like it is not knowable by me simply on the basis of hearing the sentence just quoted, or indeed any sentence likely to be formulable in a human language. Once we untie propositional knowledge from the cat-is-on-the-mat model of expressibility, it is more inviting to say that literature does teach us propositions.

Suppose now that the Oatley model of fictional learning is the right one, and the know-how gained that way has no propositional basis. Trainee pilots pick up know-how in simulators, but the know-how they pick up helps them fly real planes only if the simulations provided match in relevant ways the conditions of flight they are being trained for. "Truth" might not be the right word for a reliably flight-matching simulator, but we certainly need some notion of correspondence between the two. Presumably the same holds of fiction's relation to the world, if its insights are to be understood in Oatley's terms.

The same observation holds for a somewhat different approach to knowing-how suggested by the phenomenon of artificial grammar learning (AGL), which has been described as follows:

> In a typical AGL study, participants are first exposed to a set of strings, derived from an artificial grammar, under the pretext of a memory or pattern learning experiment. Participants are then told of the rule-based nature of the strings and asked to classify a new set of strings—some of which are generated by the same grammar and some of which are not—according to whether these novel strings follow the same rules as the ones they saw earlier. Classification performance on such AGL tasks is typically above chance, despite participants' general lack of overt knowledge of the underlying regularities (Christiansen et al. 2010).

Subjects in these experiments do not seem to have propositional knowledge of the rules governing these artificial grammars, though they are able, to a degree, to distinguish grammatical from ungrammatical strings. Such subjects may, as a result of training, come to represent, at a sub-personal level, grammatical rules corresponding more or less to those of the target grammar. In that case their competency is explained in terms of correspondence (or lack of it) between the

[3] See Stanley and Williamson (2001). For criticism of this approach, see Devitt (2011).

sub-personally represented rule and the rule of the grammar. I conclude that truth, or some generalised notion of correspondence, will continue to be very useful to us in thinking about what can and cannot be learned from literature.

Even so, disputing the truth of the Positive View is a messy business, for no advocate of that view claims that *all* literature conveys important and insightful truth. And no wise opponent of it should argue that *no* work of literature ever does. The argument ought to be about the aptness of literature, or certain kinds of literature, to convey insight. Fortune tellers are surely right some of the time, but that does not make them useful resources for those who want to know their futures. Their methods, their institutions, their goals, are not apt to convey insight.[4]

I said I would ignore those who deny coherence to the idea that literature is apt to convey important and insightful truths. But there is another claim which threatens to derail the debate before it begins, and this one is more interesting. I have in mind the position of Lamarque and Olsen, who argue that, while literature may speak to serious human concerns, it "is the *content* of the proposition [expressed], what it is about, not its truth as such, that confers interest ..." (Lamarque and Olsen 1994, p. 330). If correct, this would allow us to celebrate literature's bearing on humans concerns while avoiding the argument about the truthfulness of its claims altogether. Nice work, as David Armstrong would say.

The passage just quoted may sound odd, because it seems to deny that there might be any interest in finding out whether an interesting claim is true or not. But L&O do not think this. Their claim is that, *so far as literary criticism is concerned*, the question of truth is of no interest. They may think, and probably do, that truth is always of interest from some point of view—but not from this one. That is enough to set them apart from Trilling and the others I referred to earlier, who think insightfulness is a literary value: it is part of what makes literature valuable as literature. And one can imagine the response of F. R. Leavis, that stern humanist, to the proposition that literary values run independently of whether literature tells us anything important about ourselves and our predicament. Undeterred, Lamarque and Olsen say that there is no place in criticism for "debate about the truth or falsity of general statements about human life or the human condition" (Lamarque and Olsen 1994, p. 332).

Now this prohibition is part of their more general project of ruling out questions about the truth value of statements made or implied in literature, a position for which Lamarque and Olsen have a number of arguments, none of which I am inclined to accept. But one thing they additionally claim and which, if true, would support their position, is that, as a matter of fact, "debate about the truth or falsity of the propositions implied by a literary work is absent from literary criticism". If this is true, someone who claims that such debate has a rightful place in criticism would have some explaining to do; they would be claiming that something is an intrinsic part of criticism but has never shown up in the history of critical practice.

[4] As I argue in *The Mind in Literature*.

That would certainly be strange. I'll spend the rest of this paper considering this remarkable claim. While I don't think it gets L&O where they want to go, there is something to the idea.

One person who has been struck by the power of this idea is Peter Kivy, who agrees that "there is, most of the time, no true-false evaluation in [criticism] of the general thematic statements [in a literary work]" (Kivy 1997, p. 123). But Kivy is not on the side of L&O. He goes on to argue that, while evaluation is not visible in criticisms, it does not follow that it is irrelevant to the evaluation of narrative fictions. Evaluation, he claims, is part of appreciation, and appreciation is the job of the reader, not of the critic. The job of the critic is to make available to the reader "whatever hypotheses the fictional work may, directly or indirectly, propose" (Kivy 1997, p. 125). It is then up to the reader to evaluate those hypotheses.

As responses to L&O go, this strikes me as a weak one. Kivy's distinction between criticism, which makes the content of the work available, and appreciation, which involves evaluation of that content, surely does not correspond to anything we find in critical practice. Thus critics regularly have appraised literary fictions on various grounds, especially moral ones. For example—there is a vast array of such examples, and picking one is rather like illustrating the colour of grass by holding up one blade of it—James Gould Cozzens novel *By Love Possessed* was widely criticised on its publication by critics such as Irving Howe, Frederick Crews, and Dwight Macdonald for racism, snobbery and general conservativism. Macdonald's review in particular was responsible for the almost total destruction of Cozzens' literary reputation.

This example and many others are not, it will be said, examples of criticism which focus on psychological truthfulness; they are cases of ethical criticism. But such cases usually involve, at least implicitly, appraisal as to psychological verisimilitude: the troubling implication which critics claimed to find in the treatment of Jews, black people and Catholics in *By Love Possessed* was that such people were morally inferior in virtue of certain traits of character or disposition which are attributed to characters in the work and which—the critics suspected—we are invited to "export" to the real world.[5] And the critics are presumably rejecting those factual claims, at least implicitly. True, these critics did not then set about *arguing* that Jews, black people and Catholics did not have any of the status-reducing characteristics which they took Cozzens to be implying they did have; they simply took it as evident that they did not.[6] But their objections to the novel would have little force unless it was an at least implicit part of their critique that the novel was based on, and expressed, false assumptions about these groups. And much of the fictional literature of morality is an investigation of the fit, or lack of it, between moral precepts and psychological reality as the author sees it; Jane Austen and Henry James are good examples here.

---

[5] On exporting, see Gendler (2010, ch. 12), Stanley and Williamson (2001); also Snowden (2003). For criticism of this approach, see Devitt (2011).

[6] This is an important issue to which I will return.

Anyway, it strikes me that the question of what is the proper domain of the critic as opposed to that of the private reader is so contested and contestable, that the idea that we would find general adherence in the history of critical practice to one view or another about this—as Kivy implies we do—is very implausible indeed.

Kivy could, I suppose, argue that cases where critics depend on the truth of psychological claims are ones where they overstep the boundary of their responsibilities as critics and make pronouncements which they either should refrain from making or which they make when speaking as readers rather than as critics. In that case Kivy could not claim to find the distinction he makes between readers and critics in the historical record; he would be imposing on the record a distinction which he thinks people have failed to be aware of; critics certainly do not preface their psychological remarks by saying "I am now speaking as a reader rather than as a critic". One would not want to rule out such a strategy a priori, but it would need a good deal of arguing for. Of the many arguments against proceeding in this way, I will mention here one.

The argument's central premise is that much of the best and most celebrated criticism of the last 300 years could not be understood without assuming that the project of criticism is, in part, the project of assessing the work and its characters against standards of presumed psychological truthfulness. The most obvious examples here come from the literature on Shakespearean drama, much of which is concerned with the psychological plausibility of Leontes, Othello, Lear and, most notably, Hamlet. In all these cases, a great deal of speculation and debate goes on about the extent to which these characters are motivated from within—by plausibly human models of deliberation and feeling—or from without, by dramatic necessity, or poetic inclination, or the need to make the always difficult fourth act work, or whatever. It is only if we find Hamlet's delay psychologically unmotivated that we invoke some dramatic reason for slowing the progress of the plot; it is only if we think that Hamlet hasn't really got a good reason—a reason that makes sense within the boundaries of the plot—to go to England that we say that Shakespeare was following the ur-Hamlet too closely. A. C. Bradley argued that Iago acts as he does because plotting enhances his sense of power; Empson agrees, supporting the claim with the observation that this is a common explanation in real life for "apparently meaningless petty cruelties". Bradley's explanation, he says, is to that extent more plausible than Coleridge's motiveless malignity, which he, Empson, thinks of as a literary conceit (Empson 1951). Psychological credibility is the first test, and if a character's behaviour passes that, there is no more explaining to do.

For this reason, it seems to me that ideas of truth and truthfulness are woven into the fabric of a kind of criticism that is widespread now and comes with a long and distinguished history. A critic may—and some critics do—reject the idea that truth plays a role in criticism; she may pursue such strategies as deconstruction. I shan't argue that such people are not true critics; all I ask is that they and others not argue the same about those who are guided in their criticism by notions of truthfulness.

Psychological plausibility is of course not the only value visible in critical practice. It is highly defeasible, and sometimes is given little weight because of the perceived anti-naturalism of the work in question, or simply because a sacrifice of naturalism here makes for a gain in some other value. Take for example Max Beerbohm's *Zukeila Dobson*, an ironic fantasy of Oxford undergraduate life a 100 or so years ago. The characters' speech and motives are absurd in about equal measure, but one would be missing the point to hold this against the work.[7] For the author is plainly not seeking psychological verisimilitude. But it is very rare for it to be given no weight at all; if characters behave in ways that have absolutely nothing to do with the ways we think real people behave (and what would such behaviour be like?), a critical discussion is likely to ensue about what might justify such a dramatic deviation from the path of truthfulness.

So I reject entirely Lamarque and Olsen's claim about the irrelevance of truth to critical practice, and Kivy's attempt to rescue truth by making its determination wholly the responsibility of the reader. And yet: Lamarque, Olsen and Kivy have noticed something important. There *is* something odd in the practice of humanistic critics who appeal to notions of truthful representation in their criticism. Critics who make quite startling claims about the psychological truthfulness of authors hardly ever argue for these claims. Often they present them as rhetorical flourishes. To look no further than my latest edition of the TLS: in Barry Lyndon "the narrowing of the gap between voice and ventriloquist allows Thackeray to achieve a profound truthfulness about his protagonist's inner life" (Taylor 2011, p. 15). I'm not sure this makes any sense at all, but certainly a psychologist would be glad to hear that profound truthfulness about someone's inner life could be gained simply by an adjustment to your mode of narration.

Perhaps it is this absence of reason-giving which prompted Kivy to say that in criticism there are no true-false "evaluations". If that is what he meant I agree with it, though it does nothing to support his contention that critics merely make explicit "whatever hypotheses the fictional work may, directly or indirectly, propose", leaving it to readers to evaluate those hypotheses. For if critics ascribe truth values to these hypotheses they cannot be said to be leaving evaluation up to readers, whether or not they give reasons for assigning the truth-values they do assign. And Lamarque and Olsen are right to say that "debate about the truth or falsity of the propositions implied by a literary work is absent from literary criticism." They are merely wrong to infer from this that truth is not a value in literary criticism.

Of course readers may reject the truth-value ascriptions of critics, or simply not know about them, and do their own truth-evaluation. But on the whole I suspect that readers do no more arguing than critics do. Critics and readers alike simply form judgments, more or less spontaneously, concerning the truthfulness of the implicit psychological claims of a literary work.

On what basis do they do that? Why did Johnson say of Shakespeare "he has not only shewn human nature as it acts in real exigences, but as it would be found in trials, to which it cannot be exposed."[8] Perhaps Johnson knew from experience how

[7] The interest of the work is essentially that of a *tour de force*: how long can the author retain our interest while so consciously eschewing psychological plausibility?

[8] *Preface to Shakespeare*.

human nature acts in real exigencies; it is less clear how he could be confident that people would act this or that way in circumstances he admits no one, including himself, has experienced—or even could experience. And could Johnson have been rationally confident that Shakespeare has shown how human nature acts in real exigencies, when he, Johnson, carried out no surveys, no carefully structured experiments, to find out whether it really was so? Johnson was delightfully confident in his opinions of many things, and rated himself a great observer of his fellow creatures, but the last 50 years of psychological investigation has shown how often we are wrong about our own motives and actions, and those of others, and how little mere penetrating intellect and common sense can help us overcome our ignorance.[9] When Leavis says, rather grudgingly, that *Hard Times* does not give "a misleading representation of human nature" (Leavis 1948, p. 233) it is tempting to ask how he could possibly know something that not even the greatest psychologist would think of claiming: what human nature is.

This tendency of critics to imagine themselves able to survey human nature as well as its literary representation is the understandable core of the post-modern revulsion against humanistic criticism. But there are other ways to respond: one is to think seriously about what human nature is (the answer will be very complicated, conditional and uncertain) and to think seriously about the extent to which it is fairly represented in literature. We are a long way from that now.

References

Bate, J. 2011. Review of Lupton: Thinking with Shakespeare. *Times Higher Education*, June 16. Retrieved 24 November 2011, from http://www.timeshighereducation.co.uk/story.asp?storycode=416504.

Bevington, D. 2011. *Murder most foul: Hamlet through the ages*. Oxford: Oxford University Press.

Christiansen, M.H., M.L. Kelly, R.C. Shillcock, and K. Greenfield. 2010. Impaired artificial grammar learning in agrammatism. *Cognition* 116: 382–393.

Devitt, M. 2011. Methodology and the nature of knowing how. *Journal of Philosophy* 108: 205–218.

Devitt, M., and K. Sterelny. 1999. *Language and reality*. Oxford: Blackwell.

Djikic, M., K. Oatley, S. Zoeterman, and J.B. Peterson. 2009. On being moved by art: How reading fiction transforms the self. *Creativity Research Journal* 21(1): 24–29.

Doris, J. 2002. *Lack of character: Personality and moral behavior*. New York: Cambridge University Press.

Empson, W. 1951. *The structure of complex words*. London: Chatto & Windus.

Gendler, T.S. 2010. Chapter 12, Imaginative contagion. In *Intuition, imagination and philosophical methodology*. Oxford: Oxford University Press.

Kivy, P. 1997. *Philosophies of arts: An essay in differences*. New York: Cambridge University Press.

Lamarque, P., and S. Olsen. 1994. *Truth, fiction and literature*. New York: Oxford University Press.

[9] See e.g. Doris (2002), Wilson (2002).

Leavis, F.R. 1948. *The great tradition: George Eliot, Henry James, Joseph Conrad*. London: Chatto & Windus.

Mar, R.A., K. Oatley, J. Hirsh, J. dela Paz, and J.B. Peterson. 2006. Bookworms versus nerds: Exposure to fiction versus non-fiction, divergent associations with social ability, and the simulation of fictional social worlds. *Journal of Research in Personality* 40: 694–712.

Mar, R.A., M. Djikic, and K. Oatley. 2008. Effects of reading on knowledge, social abilities, and selfhood. In *Directions in empirical studies in literature: In honor of Willie van Peer*, ed. S. Zyngier, M. Bortolussi, A. Chesnokova, and J. Auracher, 127–137. Amsterdam: Benjamins.

Nussbaum, M. 1994. *Love's knowledge*. New York: Oxford University Press.

Robinson, J. 2005. *Deeper than reason: Emotion and its role in literature, music, and art*. Oxford: Oxford University Press.

Snowden, P. 2003. Knowing how and knowing that: A distinction reconsidered. *Proceedings of the Aristotelian Society* 104: 1–29.

Stanley, J., and T. Williamson. 2001. Knowing how. *Journal of Philosophy* 98(8): 411–444.

Taylor, D.J. 2011. Goosequill days. *Times Literary Supplement*, July 15, p. 15.

Trilling, L. 1951. *The liberal imagination*. London: Secker and Warburg.

Wilson, T. 2002. *Strangers to ourselves: Discovering the adaptive unconscious*. Cambridge: Belknap Press of Harvard University Press.

Chapter 4
The Buck-Passing Stops Here

Andrew Moore

Thomas Scanlon influentially argues that in the provision of reasons to act or respond, goodness and value 'pass the buck' to other properties. Section 4.1 of this paper outlines Scanlon's buck-passing view and his arguments for it. Section 4.2 *extends* Scanlon's arguments: if they show that goodness and value pass the reason-providing buck, then relevantly analogous arguments show, contrary to Scanlon, that duty and wrongness pass the reason-providing buck too. Section 4.3 *reverses* the direction of Scanlon's buck-passing arguments: if they show that goodness and value pass the reason-providing buck, then relevantly analogous arguments show that *reasons* likewise pass the buck for generating goodness, value, duty, and wrongness. Section 4.4 extends and reverses the buck-passing arguments further still, in application to two additional buck-passing claims that Scanlon makes. Section 4.5 considers the implications for Scanlon of this whole cluster of buck-passing arguments, and argues that buck-passing considerations fail to offer any support to Scanlon's reasons-centred approach to ethical theory. Section 4.6 steps beyond buck-passing arguments, and beyond Scanlon, to consider some options for all of us on the broad issue of how evaluative or ethical considerations such as goodness, value, wrongness and reasons are related to one another, and to non-evaluative or non-ethical properties.

4.1 Scanlon's Buck-Passing Arguments

In matters of US federal politics, Harry S. Truman has famously declared: "The buck stops here". On the ethical matter of providing reasons to agents, Thomas Scanlon has influentially argued that the goodness or value of things 'passes the buck' to other properties or facts (Scanlon 1998, pp. 96–8). Scanlon makes three

A. Moore (✉)
Department of Philosophy, University of Otago, Dunedin, New Zealand
e-mail: andrew.moore@otago.ac.nz

J. Maclaurin (ed.), *Rationis Defensor: Essays in Honour of Colin Cheyne*, Studies in History and Philosophy of Science 28, DOI 10.1007/978-94-007-3983-3_4,

different buck-passing claims, though it is not clear whether he intends these to be distinct claims. His 'reason-provision thesis' is about *goodness or value*; it claims that: "it is not goodness or value itself that provides reasons but rather other properties that do so". His 'judgment thesis' is about *judgments* of goodness or value; it claims that: "to call something valuable is to say that it has other properties that provide reasons for behaving in certain ways with regard to it". His 'property thesis' is again about *goodness or value*; it claims that: "to be good or valuable is to have other properties that constitute ... reasons" (Scanlon 1998, pp. 96–7). This paper focuses primarily on Scanlon's reason-provision thesis, and then also applies essentially the same lines of argument to his judgment thesis and his property thesis.

How are properties such as the pleasantness of a resort or the insightfulness of a scientific discovery related to goodness and value, and to the reasons we have for acting or for responding to valuable things? Scanlon presents two intuitions in support of his buck-passing answer to this question. His first intuition is that reasons are provided by the properties that make a thing good or valuable in each particular case, and that it is unclear how goodness or value themselves could provide reasons. Call this his 'case-based intuition'. The fact that a resort is pleasant provides a reason to visit it or to recommend it to a friend, for example, and the fact that a discovery casts light on the causes of cancer gives us a reason to applaud it and to support further research of that kind. Scanlon's second intuition is that many different things are good or valuable, the grounds for this vary widely, and there is no single, reason-providing property common to all these cases (Scanlon 1998, pp. 97–8). Call this his 'pluralist intuition'.

Scanlon implicitly takes his case-based intuition to generalise to *all* cases in which there is goodness or value, and in which reasons are provided. This wide scope is assumed also below. His pluralist intuition is more problematical; the most that might follow from it is that neither goodness nor value provides reasons in *all* cases in which it is present. This leaves open the possibility that goodness or value provides reasons in *some* cases. Still, Scanlon does also argue (Scanlon 1998, pp. 72–6 and 366–73) for the universality of reasons, and from this it might be argued that if goodness or value fails to provide reasons in some cases in which it is present, it fails to do so in all cases. All of this is granted below. A final preliminary point is this: Scanlon's case-based intuition and his pluralist intuition are the only arguments he has so far offered for his buck-passing view. The discussion below consequently focuses on the significance of these two intuitions.

4.2 Extensions of Scanlon's Arguments

As is noted above, Scanlon argues for the reason-provision thesis about goodness and value: it is not goodness or value itself that provides reasons but rather other properties that do so. Others have endorsed this claim, including: Dancy (2000, p. 164), Parfit (2001, p. 20), Stratton-Lake (2002, p. 12), Olson (2004a, p. 192), and Crisp (2005, p. 84). This section *extends* Scanlon's arguments for his reason-provision thesis, beyond goodness and value, to generate parallel reason-provision theses about rightness, wrongness, obligation, duty, virtue and ought. These extensions are presented below as though they are in the voice of Scanlon.

It is not rightness, wrongness, obligation, duty, virtue, or ought that provide reasons, but rather other properties that do so. In each particular case, the reasons are provided by the properties that make an action right, wrong, obligatory, a duty, or something that ought to be done, or that make a character state a virtue. So, for example, the fact that she is hurting and I am at hand makes it right for me to help her and provides me with a reason to do so, the fact that you promised gives you an obligation to deliver on your promise and provides a reason for me to applaud your doing so, the fact that he is highly vulnerable makes it wrong of you to press him just now when you could easily wait and it gives you a reason not to press him, and her being considerate makes this a virtue of hers and gives her a reason to express this character state of hers. It is unclear how rightness, wrongness, obligation, virtue, duty, or ought could provide any additional reasons in such cases. Furthermore, many different actions are right or obligatory, are wrong, are duties, or are things that ought to be done, and many different character states are virtues. Within each of these sorts of ethical consideration – ought, right, wrong, virtue, and so on – the grounds for this vary widely. For none of these ethical considerations does there seem to be a single, reason-providing property that is common to all its instances. I therefore accept a buck-passing account: right, wrong, duty, obligation, ought, and virtue do not generate reasons; other properties do so.

Discussing goodness and rightness, Dancy (2000, p. 172) notes that: "a challenge might be mounted in favour of consistent buck-passing". The previous paragraph does not present any such challenge to Scanlon. Nor does it present an argument for a buck-passing account of wrongness, or for a buck-passing account of any other ethical consideration. Such arguments are presented, for example, by Dancy (2000, pp. 165–7) and Stratton-Lake (2003). The previous paragraph instead presents an argument for the following *conditional* claim: if Scanlon's buck-passing intuitions support his reason-provision thesis about goodness and value, then extensions of these intuitions equally support the reason-provision theses that it is not duty, nor wrongness, nor any of the other ethical considerations discussed above, that provides reasons, but it is instead other properties that do so.

4.3 Reversals of Scanlon's Arguments

This section *reverses* the direction of Scanlon's buck-passing intuitions. It does this first for Scanlon's intuitions about goodness and value. Then it does the same for the intuitions that were extended in the previous section to other ethical considerations. The next two paragraphs present these reversal arguments as if in Scanlon's own voice. They share a common structure, but they are independent of one another.

It is not reasons that generate goodness or value but other properties that do so. The fact that a resort is pleasant gives it value, for example, and the fact that a discovery casts light on the causes of cancer makes it good. Those properties generate the goodness or value of these things. It is unclear how reasons could generate additional goodness or value in such cases. Furthermore, many different things provide reasons, and the grounds for this vary widely. There is no single, goodness-generating or value-generating property common to all these cases. I therefore accept a buck-passing account: reasons do not generate goodness or value; other properties do so.

It is not reasons that generate rightness, wrongness, obligation, virtue, duty, or ought, but other properties that do so. The fact that he breaches his partner's trust if he has an affair generates the wrongness of this action of his and a reason for him not to do it. The fact that appropriately facing fear is second nature to you makes it a virtue of yours and a reason for you to respond in this way. Your young child's need for you gives you a duty or obligation to care for him, supplies the 'ought' and the rightness of this caring action of yours, and gives you related reasons for action. It is unclear how reasons could provide additional rightness, wrongness, obligation, virtue, duty, or 'oughtness' in such cases. Furthermore, many different things are right, wrong, obligatory, virtues, duties, or ought to be done, and the grounds for this vary widely within each of these sorts of ethical consideration. There is no single property that generates rightness, wrongness, obligation, virtue, duty, or 'oughtness', and that is common to all the relevant cases. I therefore accept a buck-passing account: reasons do not generate rightness, wrongness, obligation, virtue, duty, or 'oughtness'; other properties do so.

4.4 Further Extensions and Reversals

This paper has so far focused on Scanlon's reason-provision thesis, and it has both extended and reversed that thesis. This section widens the scope of the arguments further still, to encompass Scanlon's judgment thesis and his property thesis, the other two constituents of his buck-passing view. It considers the significance of Scanlon's case-based and pluralist intuitions for these two further theses; and the significance of related intuitions for extensions and reversals of these two further theses.

Scanlon's judgment thesis is that: "to call something valuable is to say that it has other properties that provide reasons for behaving in certain ways with regard to it" (Scanlon 1998, p. 96). His claim here is that there are equivalences between *judgments* of goodness or value and *judgments* of reasons. This is a claim about intentional states or acts of these sorts.

Scanlon's property thesis is that: "to be good or valuable is to have other properties that constitute... reasons" (Scanlon 1998, pp. 96–7). His claim here is that there are equivalences between goodness or value and other properties. This is a claim about these *objects* of goodness or value judgments.

Some, including Dancy (2000), Hooker and Stratton-Lake (2006), Olson (2004a), argue that Scanlon overlooks various important rivals to his judgment thesis and his property thesis. The present discussion sets all such matters aside, to focus instead on extension and reversal of Scanlon's judgment thesis and of his property thesis.

Insofar as Scanlon's judgment thesis that "to call something valuable is to say that it has other properties that provide reasons for behaving in certain ways with regard to it" is supported by his case-based and pluralist intuitions, extensions and reversals of these intuitions also offer this same level of support for extension and reversal of his judgment thesis. These extensions and reversals include the following: to call something a reason is to say it has other properties that provide goodness or value; to say an act is wrong is to say it has other properties that provide reasons not to do it; to call something a reason not to do something is to say it has other properties that make it wrong to do that thing. And things are likewise, by extension and reversal, for relations between judgments of reasons and judgments of many other ethical considerations, including rightness, ought, virtue, and duty.

Insofar as Scanlon's property thesis that "to be good or valuable is to have other properties that constitute... reasons" is supported by his case-based and pluralist intuitions, extensions and reversals of these intuitions equally support extension and reversal of Scanlon's property thesis. These extensions and reversals include the following: to have reason to perform an action is to have other properties in virtue of which that act ought to be performed; to have reason not to perform an action is to have other properties in virtue of which it would be wrong to perform that action; to have reasons is to have other properties that constitute goodness or value; and so on. The parallel points extend also to relations between reasons and other ethical or evaluative matters, including rightness, obligation, virtue and duty.

4.5 Options for Scanlon

Suppose the arguments of the previous sections all go through. What then for Scanlon? Finding that these implications would see significant revision of Scanlon's views, this section also briefly considers his options for resisting and pushing back.

One response to the previous sections is that *all* of the buck-passing arguments go through – both Scanlon's arguments and all of the extension and reversal arguments. This response is consistent with, but is not implied by, the conjunction of all of the buck-passing arguments of previous sections. From this response, it would follow that no ethical or evaluative difference is sufficient for any other ethical or evaluative difference. This would endorse Scanlon's buck-passing argument for his claim that no mere difference of value or goodness is sufficient for any difference of reasons. But it would also *contradict* two of Scanlon's claims: "the fact that something is... required by a duty... can count as a reason for doing it" (1998, p. 69); and "the fact that an action would be wrong is itself a strong, normally decisive, reason against doing it" (2003, p. 436; see also 1998, p. 11).

A different response to the earlier sections is to say that *none* of the buck-passing arguments goes through – none of Scanlon's buck-passing arguments, and none of the extension or reversal arguments either. This response too is consistent with, but is not implied by, the conjunction of the all of the above buck-passing arguments. It would follow that, for all that these buck-passing arguments shows, there might well be *mere* differences of value or goodness or wrongness that make for some difference of reasons, there might well be *mere* differences of reasons that make for some difference of value or goodness or wrongness, and so on. This would be consistent with Scanlon's claims that being required by a duty can count as a reason, and that for an action to be wrong is for there to be strong reason against doing it. But it would also *contradict* Scanlon's claim that: "it is not goodness or value itself that provides reasons but rather other properties that do so" (1998, p. 97).

Suppose that the arguments so far in this paper all go through. The overall implication for Scanlon's buck-passing arguments is this. Either Scanlon's buck-passing arguments fail to provide any backing for anything, in which case they fail to provide any backing for his project of using "the notion of a reason… to provide a general characterization of… the idea of value" (Scanlon 1998, p. 78); or Scanlon's buck-passing arguments do provide some backing for something, in which case they provide backing for extension and reversal arguments that *undermine* his project of accounting for value and goodness in terms of reasons, and that also undermine his wider project of establishing a reasons-centred approach to ethical theory. In short, Scanlon's buck-passing arguments either fail to support his reasons-centred approach to ethical theory, or they undermine that approach.

Claims about goodness or value are prominent in ethical theory, especially in the utilitarian, intrinsic value, natural law, and Aristotelian traditions. Scanlon's buck-passing arguments are sometimes thought to show that 'reasons' is a central ethical-theoretic concept and that 'goodness' and 'value' are not. But if the arguments of the present paper are sound, then Scanlon's buck-passing arguments at best fail to support any such conclusion. They make no case that goodness and value *in particular* pass the buck to other properties, and they make no case that goodness or value properties or judgments have any lesser a place in ethical theory than is had by reasons-judgments or reasons-properties.

In short summary, the foregoing arguments imply significant revision of Scanlon's views. How might he resist this, and even push back? In simple terms, he needs to find a well-located failure of generalisation in the foregoing extensions and reversals of his buck-passing arguments. A suitable failure of *extension* would allow his buck-passing arguments to go through for value and goodness, but would see the closely parallel buck-passing arguments fail to go through for duty and for wrongness. Similarly, a suitable failure of *reversal* would allow him to keep his buck-passing case against value-centred and goodness-centred accounts of ethical theory, while also ensuring that the closely parallel buck-passing case fails to go through against his own reasons-centred account of ethical theory. Actual delivery on these Scanlonesque aspirations is work for another day, and perhaps for another author.

4.6 Wide Issues

This final section considers the broad and basic issue of how ethical or evaluative considerations such as value and wrongness and reasons are related to one another and to non-evaluative or non-ethical properties or considerations.

Scanlon writes (1998, p. 97): it is "lower-order properties that provide reasons". Let us set aside Scanlon's further view that duty and wrongness are two exceptions to this. Consider instead the view itself in generalized form: for every ethical or evaluative difference, there is some lower-order difference that is sufficient for it, and there is no *other* ethical or evaluative difference that is necessary for it. Call this the what-lies-beneath view. It is widespread in ethics, and so are parallel claims in many other philosophical domains. Its ruling metaphor is that it is the 'underlying' matters, 'base' properties, 'grounding' items, 'deeper' considerations, 'subvening' features, 'lower-level' qualities, or the like, that are or even that must be sufficient unto the day. The what-lies-beneath view is not necessarily a naturalist or physicalist view. As Scanlon remarks in the present case, for example: "My thesis was that goodness is not itself a property that provides reasons, not that the underlying properties that do this are always natural properties" (Scanlon 2002, p. 513; see also Scanlon 2003, p. 428).

The what-lies-beneath view is often little more than a ruling metaphor. It tends to sit influential but unexamined in the views of philosophers. To my knowledge, for example, Scanlon does not examine or offer any argument for his variant of it. And this is the usual case. To note just one other example amongst many, the what-lies-beneath metaphor is pervasive, influential and unexamined in Mark Timmons's widely consulted *Moral Theory: An Introduction* (Timmons 2002).

The what-lies-beneath view is not the only option. After all, things lie in many places in addition to beneath. Here is one rival view: for every ethical or evaluative difference, there is some *other* ethical or evaluative difference that is necessary for it, and there is no lower-order difference that is sufficient for it. One might call this the what-lies-alongside view. The what-lies-alongside view is consistent with the claim that the buck-passing arguments entertained in the present paper all go through. It is also consistent with the claim that none of them go through. And it is consistent with the supervenience thesis, which some claim to be a conceptual truth, that for every ethical or evaluative difference, there is some lower-order difference that is necessary for it ('no ethical difference without some non-ethical difference'). The idea of the what-lies-alongside view, by contrast, can be roughly expressed in this slogan: to account for any ethical consideration we need to appeal to another ethical consideration. A plausible and wide-scope thought behind the what-lies-alongside view in ethics and elsewhere is that if we wish to account for something, then we should expect to do this at least in part in terms of something else of that same sort – in terms of what lies alongside it. In ethics, such views are often stigmatized as 'non-naturalist' views. One push-back response to this is that if any ethical matter is instead theorized about wholly in terms of mere lower-order matters, then we can be justifiably confident that that account will be a failure.

There is of course much else that lies alongside and beneath the what-lies-beneath and what-lies-alongside views. This is for another day, and perhaps the same author.

Acknowledgement Conversations with Kent Hurtig started me on this paper. For comments on earlier drafts, I thank Martin Wilkinson, Jonas Olson, and Heather Dyke. Responses from participants at the Otago Philosophy Department seminar helped me to improve the paper. I thank James Maclaurin for leading the book project that this paper sits within. Above all, I thank Colin Cheyne – excellent philosopher, fine colleague and good friend.

References

Crisp, R. 2005. Value, reasons, and the structure of justification: How to avoid passing the buck. *Analysis* 65(1): 80–85.

Dancy, J. 2000. Should we pass the buck? In *The good, the true and the beautiful*, ed. A. O'Hear, 159–173. Cambridge: Cambridge University Press.

Hooker, B., and P. Stratton-Lake. 2006. Scanlon versus Moore on goodness. In *Metaethics after Moore*, ed. M. Timmons. Oxford: Oxford University Press.

Olson, J. 2004a. G.E. Moore on goodness and reasons. In *Patterns of value: Essays on formal axiology and value analysis*, Lund philosophy reports, vol. 2, ed. W. Rabinowicz and T. Ronnow-Rasmussen, 190–200. Lund: Department of Philosophy, Lund University.

Parfit, D. 2001. Rationality and reasons. In *Exploring practical philosophy: From action to values*, ed. D. Egonnson, J. Josefsson, B. Petersson, and T. Ronnow-Rasmussen, 17–39. Aldershot: Ashgate.

Scanlon, T. 1998. *What we owe to each other*. Harvard: Belknap.

Scanlon, T. 2002. Reasons, responsibility, and reliance: Replies to Wallace, Dworkin, and Deigh. *Ethics* 112: 507–528.

Scanlon, T. 2003. Replies. *Ratio (new series) XVI* (4 December): 424–439.

Stratton-Lake, P. 2002. Introduction. In *Ethical intuitionism: Re-evaluations*, ed. P. Stratton-Lake. Oxford: Clarendon.

Stratton-Lake, P. 2003. Scanlon's contractualism and the redundancy objection. *Analysis* 63(1): 70–76.

Timmons, M. 2002. *Moral theory: An introduction*. Lanham/Oxford: Rowman & Littlefield.

Part II
In Science

Chapter 5
Universal Darwinism: Its Scope and Limits

James Maclaurin

5.1 Introduction

Richard Lewontin (1970) sets out three "Darwinian Principles" that must be true of any population undergoing natural selection:

- Members of the population must vary from one another.
- That variation must be heritable.
- That variation must have effects on fitness.

These principles form the hard core of the Darwinian research program. Abandoning any of them is abandoning Darwinian science. However, the simplicity of the Darwinian principles is a two edged sword. On the one hand, they provide enormous scope for the application of the theory, hence the vitality of the Darwinian research programme. On the other hand, taking the Darwinian principles at face value implies that natural selection acts in a variety of systems (Nelson 2007) that Darwin did not envisage and that many modern-day Darwinians think controversial. It is this ambiguity in the theory that has led to the great variety in general accounts of natural selection often going under the name of "Universal Darwinism". These include: Williams' Adaptation and Natural Selection (1966), Lewontin's The Units of Selection (1970), Dawkins' The Selfish Gene (1976) and The Extended Replicator (1982), Cavalli-Sforza and Feldman's Cultural Transmission (1981), Oyama's The Ontogeny of Information (1985), Darden and Cain's Selection Type Theories (1989), Griffiths and Gray's Developmental Systems and Evolutionary Explanation (1994), Hull, Langman and Glenn's A general account of selection: Biology, immunology and behaviour (2001) and Sterelny's Niche Construction, Developmental Systems, and the Extended Replicator (2001) and more recently Nelson's Universal Darwinism and Evolutionary Social Science (2007).

J. Maclaurin (✉)
Department of Philosophy, University of Otago, Dunedin, New Zealand
e-mail: james.maclaurin@otago.ac.nz

J. Maclaurin (ed.), *Rationis Defensor: Essays in Honour of Colin Cheyne*, Studies in History and Philosophy of Science 28, DOI 10.1007/978-94-007-3983-3_5,

The purpose of this paper is to explain why it has proved so difficult to produce a general account of natural selection. That is an account that settles, for every case, whether that case is an instance of natural selection. I begin my argument by demonstrating the latitude of evolutionary explanation afforded by the Darwinian principles. In this first part, I focus on a variety of interpretations of the requirement that Darwinian evolution rests on inheritance. I will leave to the reader the task of generalising this argument to cover the other two Darwinian principles. I will argue that the Darwinian principles are necessary but not sufficient in a general account of natural selection. Neither this fact nor the protracted disagreement about the nature and scope of natural selection shows a general account to be impossible. However, I will employ an argument from philosophy of mind to show that adding to or qualifying the Darwinian principles to produce a determinate general account of natural selection, must inevitably fail. I argue that it is a consequence of the functional nature of evolutionary explanation that natural selection is not a determinately bounded process.

5.2 Part One: The Paradox of Selection

Within biology, inheritance is usually interpreted as genetic inheritance. As Mateo Mameli puts it "The received view is that nongenetic inheritance is either something that does not exist of something without evolutionary significance" (Mameli 2004, p. 35). It is not surprising that most of those analysing evolution should restrict the scope of the term "inheritance", for outside classical evolutionary theory "inheritance" is a broad and ill-defined term. In some circumstances all the following are accounted as things one might inherit: English, Philosophy, modern sanitation, antique furniture, one's name, an oxygen rich atmosphere, the rule of law, and so on. Thus, in the non-technical sense inheritance is very easy to come by. Indeed, not only are the above items broadly heritable, they also vary in most populations. I could have inherited a different language, other furniture, a name that was easier to spell, a lawless society, and so on. Furthermore, at least most of the items on the above list are apparently fitness-enhancing (the language, the sanitation, the oxygen, etc.).

So prima facie, all the listed phenomena (and presumably many similar) fulfil the Darwinian principles. All have evolved over time. Therefore, seemingly we should conclude that all are products of natural selection. But that would be a bold claim indeed and one that would be resisted by most evolutionary theorists. My name, my furniture, my plumbing etc. simply don't seem to stand in need of the explanatory power of the theory of natural selection.

The issue is made all the more stark by the fact that outside mainstream evolutionary biology, natural selection is employed to explain a wide variety of phenomena very unlike the central biological cases (genes, organisms and groups of organisms). Recent science has given us Darwinian theories of neural development (Edelman 1987; Fernando et al. 2010), computation (Ashlock 2006), economics

(Nelson and Winter 1982), epistemology (Popper 1972), etc. All have been put forward as real instances of natural selection and in all these cases the claim is controversial. I use the term "Darwinian hard cases" to pick out all these non-standard cases that are nonetheless instances of the three Darwinian principles.

This then is the paradox of selection. Inference from the Darwinian principles indicates that all sorts of mundane items are actually products of natural selection, and yet we do not employ natural selection to explain their characteristics or existence.

There are two solutions to the paradox. A single phenomenon may be the product of many different processes. Thus while natural selection is partly responsible for various non-biological phenomena, it may not figure in explanations of those phenomena because it is not the most important or interesting process responsible for their existence. But if we are to dissolve the paradox in this way, we will need an explanation of why these are not interesting instances of natural selection or perhaps of why natural selection has not been an important cause of such non-biological phenomena.

The existence of so many Darwinian hard cases has caused many philosophers and biologists (listed in the introduction above) to opt for a second solution to the paradox of selection. This is the project of developing some supplementation of the Darwinian principles which it is hoped will rule out inappropriate application of evolutionary theory. It is this strategy that I will focus on in the general account of natural selection that I propose in this paper.

5.2.1 A Red Herring

Before I go on, I must address a red herring. There is a clear and acknowledged sense in which the theory of natural selection is indeterminate in its scope.

An obvious reason that it is so easy to join the natural selection club is that the prerequisites for membership are not discrete properties. Inheritance, variation and differential fitness are all matters of degree (i.e. continuous properties). Inheritance (as I have broadly construed it thus far) is just a causal connection, giving rise to similarity. Because similarity is a matter of degree, so must inheritance be. Conversely, variation is just dissimilarity between the members of a population and therefore it too must be a matter of degree. Differences in fitness are usually accounted as differences in the likelihood that you will leave offspring or differences in your propensity to do so. Again these are matters of degree.

So, all three of the Darwinian Principles refer to continuous properties. For this reason alone we ought to expect difficulty in making decisions about whether some phenomena count as instances of natural selection. It is this fact that Richard Dawkins seeks to address by his injunction that entities subject to natural selection ought to possess fidelity, fecundity and longevity (Dawkins 1976). These prerequisites are inferred from the observation that natural selection is really cumulative selection. However, for present purposes, they merely relocate the ambiguity leaving us wondering how much accumulation of change over how many generations and to what effect ought to be required to justify the hallmark of true natural selection.

This is an important issue, if only because those employing the theory differ in their responses to it. However, it is not my focus here. For I think there is a deeper ambiguity in evolutionary theory—one with potentially more complex effects. I will begin by setting out some of this complexity in Part 2. I will argue for the existence of the deeper ambiguity in Part 3.

5.3 Part Two: A Profusion of Evolutionary Analyses

Ambiguities inherent in the Darwinian principles make the study of evolution a difficult and controversial business. There are many facets of evolutionary theory that demonstrate this, but I will concentrate just on the analysis of non-genetic inheritance (i.e. instances where inheritance does not proceed via the genome). In this part, I will list some of the many strategies available to those who would analyse non-genetic inheritance in accord with the Darwinian Principles. I begin with a little scene setting.

5.3.1 The Problem of Non-genetic Inheritance

G. C. Williams (1966), David Hull (1981), Richard Dawkins (1976) and many who followed have analysed natural selection in terms of the constituents of systems undergoing that process. They have focused in particular on the distinction between replicators and interactors. Both are functional kinds. A replicator is anything that is copied (not that copies itself) in the course of natural selection. It is replicators that Dawkins insists must possess fidelity, fecundity and longevity (that is – they must be capable of being copied very accurately over long periods of time). Interactors (Hull's term) are entities constructed by replicators to enhance their prospects for representation in future generations. Both replicators and interactors are subject to natural selection. Both play a part in the process. However, it is replicators that evolve in the sense that they and only they can form long flourishing lineages. It is this fact that led Dawkins (somewhat more so than Hull) to conclude that replicators are genes and thus that all instances of natural selection are profitably thought of in terms of changes to genes, genomes and gene pools.

To put it in the language I have used so far, Dawkins supplements the Darwinian Principles with his injunction that natural selection requires fidelity, fecundity and longevity. He takes it that this gives a determinate answer to the question "What sort of entities can experience natural selection?" and that answer is "genes". Hence, this view is called "gene selectionism" and, while plausible, it leads to an obvious and important question—What is the gene selectionist to make of the obvious fact of non-genetic inheritance? A bird will have inherited many fitness-enhancing genes but it might also have inherited an existing nest in which to raise its young or a learned song with which to attract its mate. What does the gene selectionist say about such non-genetic bequests? Are changes in such things only instances of natural selection to the extent that they can be derived from facts about genes? Or might we employ the Darwinian principles more liberally? These are the fundamental questions behind the problem of non-genetic inheritance. The following five approaches are not intended as an exhaustive list of possible solutions to the problem.

5.3.2 Approach One: The Extended Phenotype

Richard Dawkins (1976, 1982) makes two responses to the problem of non-genetic inheritance. Firstly, abiotic elements of an organism's environment can be very important in the course of its evolution and development, but the mere fact that such elements are not alive need not imply that they stand outside the replicator interactor picture.

Physiologically, spider's webs, termite mounds and bird's nests are not parts of spiders, termites or birds, but they are nonetheless products of them and crucially, products of their genes. For this reason, Dawkins argues that for the purposes of evolutionary analysis, we should treat these abiotic elements as if they were parts of the phenotypes of the organisms that produce them. Dawkins calls this phenomenon 'extended phenotype' and it explains away some apparent instances of non-genetic inheritance by proclaiming them to be phenotype and thus actually parasitic on genetic inheritance. But he also accepts that this response will not cover all cases.

5.3.3 Approach Two: Memes

Dawkins second answer is to remind us that 'replicator' really is a functional term and its scope need not be limited to genes and their effects. Wearing his non-gene-selectionist hat, Dawkins proposes the idea of a meme. Memes are copiable elements of culture, variously defined and controversial from the outset. Some see them as ideas, others as memory traces. Some see them as inevitably stemming from imitation (Blackmore 1999). Others propose a battery of non-genetic processes by which inherited behaviours could span the generations. There is no universally accepted definition, but there are plenty of putative examples of memes. Dawkins suggests: tunes, ideas, catch-phrases, clothes, fashions, ways of making pots or of building arches (Dawkins 1976, p. 206). Daniel Dennett (1995, p. 344), nearly 20 years later and rather more ambitiously, includes the ideas of: vendetta, alphabet, the Odyssey, chess, perspective drawing, impressionism and deconstructionism.

The meme promises much, but clearly suffers from a great deal of ambiguity. This has led both to popular enthusiasm and to academic scepticism. But even here memes are making something of a comeback due to the advocacy of Dan Dennett in Darwin's Dangerous Idea, Susan Blackmore in The Meme Machine and a rather more even-handed treatment in Robert Aunger's collection Darwinizing Culture (2000). Even those inclined to be more sceptical are taking the idea seriously enough to explore the scope of, and reasons for, their scepticism. Amongst these I include Robert Brandon (1999), Bill Wimmsatt (1999), Kim Sterelny (2001), and Peter Godfrey Smith (2000). The jury is still out. Some conclude that natural selection has little part to play in the analysis of culture. Others take a more positive stance towards that project, but don't see memes as useful in its resolution.

Crucial for present purposes memetic evolution involves inheritance, variation and differential fitness. Granted, the inheritance may be rather low fidelity and the variation, although sometimes random, may also be the product of purposeful human invention. Nonetheless, nothing in memetic theory violates the Darwinian Principles.

5.3.4 Approach Three: Dual Inheritance

According to dual inheritance theory, natural selection acts directly on culture which co-evolves with biology (Boyd and Richerson 1985, 2005). It "assimilates cultural inheritance into population genetic models, developing mathematical models that incorporate both biological and cultural evolution simultaneously and interactively..." (Mesoudi et al. 2004). Unlike memetics, it does not require the cultural equivalent of a gene. Rather, it requires only that cultural change affects the environment in which genetic selection operates and that natural selection acts specifically on cultural variants as well as on processes such as transmission biases (Henrich and McElreath 2003). Crucially it treats cultural and biological evolution as occurring within the same adaptive landscape in a way that, arguably, memetics does not. However, unlike other theories addressed here (such as Dawkins' extended phenotype), it acknowledges that natural selection can act via horizontal as well as vertical transmission of information.

5.3.5 Approach Four: Developmental Systems Theory

Richard Dawkins has always slightly distanced himself from the meme idea due in part to its initial reception from those who took him to be promoting a new form of sociobiology. But in the 1990s Dawkins central claim that replicators are first and foremost genes came under withering fire from several quarters.

Developmental systems theorists such as Paul Griffiths and Russell Gray (1994) as well as Susan Oyama (1985) sought to widen the scope of interest in non-genetic inheritance. It is not merely elements of culture that are transmitted from one generation to the next. The development of any biological individual relies upon the existence of a host of non-genetic factors and in one sense those 'developmental resources' are just as important as its genetic developmental resources. Were they not present, it would not develop at all. Furthermore, it turns out to be very difficult to demonstrate (in terms of causation or information) what it is that genes have that non-genetic inheritance could not have (Maclaurin 1998). Thus developmental systems theorists are in favour of treating all developmentally salient inheritance as 'on a par'.[1]

Hence, they argue, what is inherited from one generation to the next is an entire developmental system—a whole set of developmental resources that is reliably reconstructed generation after generation. In the current context this view is distinctive for two reasons. First, it is holistic, treating the evolving entity as a conglomerate of biotic and abiotic factors. Secondly it is very liberal as to the sort of things that form the conglomerate. It is perhaps the most liberal interpretation of the Darwinian principles, counting even interactions with sunlight and gravity as part of the evolving entity (Griffiths and Gray 1997, p. 482).

Not surprisingly, many have sought a middle ground between the opposing poles of gene selectionism and developmental systems theory.

5.3.6 Approach Five: Extended Replicator Theory

Kim Sterleny, Kelly Smith and Mike Dickerson (1996) argue that, rather than, in effect, treating the whole developmental system as a replicator, we might instead analyse its elements as replicators. So we can include non-genetic inheritance provided that the elements in question really are subject to natural selection. This is the extended replicator picture of natural selection. In a sense, this is treating all non-genetic inheritance in the way that memetic theory treats memes. Indeed, memes can get into this extended replicator picture except that according to Sterleny, they probably don't. For the other beef between developmental systems theory and the extended replicator view is that Sterelny et al. dislike the suggestion that we ought to treat all inherited developmental resources as being on a par. Sterelny in particular, still holds a special place for genetic replication and is in general conservative about bestowing the status of replicator on non-genetic resources (2001).

[1] The point is rather more nuanced than is often recognised. Oyama (2000) argues against its being interpreted as requiring that the contributions of all developmental resources be evaluated in any biological study. Rather it stipulates (contra gene selectionism) that there is no general sense in which genes are to be thought of as especially important contributors to evolution and development.

5.3.7 *Why Are There So Many Approaches?*

In summary then, the analysis of non-genetic inheritance in evolution presents a complex picture. There are many reasons for this theoretical diversity, but clearly one of them is the fact that the Darwinian principles can be interpreted in many different ways.

Some think the principles imply the replicator/interactor distinction (Dawkins, Hull, Sterelny, Smith and Dickerson). But this casts doubt on memetic theory, so meme theorists are more likely to think the principles imply the existence of replicators but not necessarily interactors. Other's think even replicators are optional, e.g. Dual Inheritance theorists.

Many think that as biological natural selection paradigmatically acts cumulatively over very long periods of time, we should infer that inheritance should be very strong and variation should be very weak. Others apply the principles much more liberally. Any change in a system that can be explained by inheritance, variation and different fitness counts as an instance of natural selection. Never mind if variation is rife or inheritance is weak. If we can explain what's going on in terms of the Darwinian principles, then it's natural selection.

The analysis of non-genetic inheritance in evolution is littered with powerful theoretical tools. But there is no agreement about how to referee disputes over which perspectives are viable and which ought to be used in particular circumstances. I've represented those differences as disagreements about the proper scope of Darwinian theory. It now falls to me to substantiate my earlier claim that there is an important reason why such disagreements cannot be settled.

5.4 Part Three: Natural Selection Meets Functionalism

In this part, I draw an analogy between evolutionary theory and philosophy of mind. Much of philosophy of mind addresses the issue of characterising minds and mental states. One common school of thought, Functionalism, is based on the idea that minds are functional systems. Thus Functionalists draw a distinction between minds and the physical systems in which they are instantiated that is broadly analogous to the distinction between computer software and computer hardware. However, an important qualification to the basic Functionalist idea was developed at length by William Lycan (1987). Specifically, Lycan warns us against, "two-levelism". This is the idea that you can view a system in two ways, either in terms of its structure or in terms of its function.

Lycan's counter-argument is that functional abstraction is a matter of degree. Complex systems (including computational ones) can, in general, be analysed at many different functional levels. Owners of cars think of them in very abstract terms as expensive but convenient possessions for getting from A to B. This is a high-level functional abstraction. Much lower-level functional abstractions are employed by mechanics who think of cars as collections of systems performing functions to do with braking, suspension, combustion and so on. Even lower level abstractions will have to settle on physical details of the system in question, such as whether the suspension is pneumatic or mechanical. These are still of course functional abstractions as there are, for example, many ways in which you can achieve mechanical suspension.

In philosophy of mind, the Language of Thought hypothesis and connectionism are both functional claims. They could be instantiated in a variety of physical systems, but they are clearly tied to particular classes of cognitive architectures. We can though make claims about cognition that are at a much higher level of functional abstraction. The Turing test characterises the mental without making any claims about fine detail of functional architecture. Indeed, you don't need to be a fan of the Turing test to recognise that psychology is full of high-level functional talk.

Furthermore, cognitive scientists and philosophers of mind are adept at inferring from one level to another. For example, Jerry Fodor's (1975) claims about productivity and systematicity in thought employ high-level facts about cognition as justifications for lower-level hypothesis about mental representation.

This has much to tell us about evolutionary theory, for Darwinians also employ functional abstraction, as when we characterise natural selection in terms of "replicators" and "interactors".

In the remainder of this paper, I shall argue for the following claims:

- Two-levelism is no better in evolutionary theory than it is in philosophy of mind. Darwinian functional abstraction is a matter of degree and there is therefore a hierarchy of functional abstractions available in the characterisation of such systems.
- Just as in philosophy of mind, evolutionary theorists can make inferences from one level of functional abstraction to another. Indeed, such inferences are often employed in arguments concerning Darwinian hard cases.
- Multi-levelism in evolutionary theory provides an in-principle objection to the idea that natural selection is a determinately bounded process. Thus, I claim, there can be no fact of the matter as to which Darwinian hard cases are genuine instances of natural selection.

5.4.1 *Evolution's Turing Test*

There are two reasons why two-levelism is a bad idea in evolutionary theory. The first is that the current level of functional analysis of choice, that of replicators and interactors, might well be optional. The argument is based on a thought experiment from Peter Godfrey Smith's The Replicator in Retrospect (2000).

Godfrey Smith starts form Lewontin's dictum that what matters in evolution is inheritance, not how you get it. He then develops a thought experiment in which an organism has the ability to reverse transcribe DNA from protein (the converse of the familiar biological process). This organism therefore doesn't need to maintain a germline. It doesn't need to keep a store of sex cells because it can just synthesise DNA when it wants to reproduce. But, despite the biochemical eccentricity of such a species, nothing here violates the Darwinian principles. In such a species, therefore, we would have natural selection in the absence of replication. Of course, the details of this thought experiment are biologically farfetched, but this is beside the point in a discussion of Darwinian hard cases not all of which are biological. The important point being made here is that "the importance of replicators in evolution here on earth is a contingent matter" (Godfrey Smith 2000, p. 413).

If Godfrey Smith is right then we cannot assume that natural selection requires replication. This has two important consequences. Firstly, if natural selection is multiply realisable, there must be some other non-replicator-based (but still functional) characterisation of that process. If this new characterisation subsumes both replicator-based and non-replicator-based natural selection, then it must be a higher level functional abstraction than the familiar replicator/interactor picture. Therefore, two-levelism in evolutionary theory fails. From this, flows the second important conclusion, viz., that we therefore cannot rely on any replicator/interactor supplementation of the Darwinian principles as a means of adjudicating Darwinian hard cases.

However, I think two-levelism in evolutionary theory is unworkable even without recognising the possibility of non-replicator-based natural selection. This is because, even those whose central task is the analysis of replication, seem bound to employ inferences between different levels of functional abstraction as a means of making precise our notion of Darwinian replication. I will give two examples.

In Niche Construction, Developmental Systems and the Extended Replicator (2001) Kim argues (contra developmental systems theory) that accepting the existence of non-genetic replicators does not imply accepting the further idea that all replicators of any sort are somehow on a par with one another. Here is how Sterelny draws the crucial distinction.

He asks—If you wanted to produce a planet that had a "rich, complex, well-adapted, and varied biota—biota in many ways like life here" (Sterelny 2001, p. 339) and you wanted to grow this from seed, just to put one seed of life on this planet and let it evolve, what sort of replicator would that seed have to have. Omitting the technical detail, the answer is that you need a system that is substantially like genetic inheritance systems with which we're familiar. As I understand it, this is an inference from two high level functional properties[2] of systems undergoing natural selection (diversity and complex adaptation) to a lower level functional characterisation of such systems in terms of replicators and interactors.

Sterelny's argument is sound but it is premised upon his particular choice of high-level functional properties and there were others he could have chosen. Not all instances of (even uncontroversial biological) natural selection lead to complex

[2] Note: I call diversity and complex adaptation high level functional properties precisely because they do not specify details of the replication system.

adaptations. Also, the extent to which natural selection is a producer of diversity is famously controversial (Gould 1989). What Sterelny provides is an analysis of the sort of natural selection that will maximise diversity and complexity of adaptation. But this gets us no closer to the adjudication of Darwinian hard cases precisely because we could have chosen different high level functional properties, and indeed, in other contexts other theorists have done just this.

Bill Wimmsatt's argument in Genes, Memes and Cultural Heredity (1999) focuses on two very different high-level functional outputs of natural selection. One is linkage—the idea that genes close to each other tend to get passed on together. Wimmsatt argues that cultural traits also seem to get passed on in clumps rather than just as a random assortment.

The other high-level functional output is generative entrenchment. This is the evolutionary rigidity that results form natural selection building trait atop trait atop trait. In the end Wimmsatt settles on a characterisation of natural selection in terms of the Darwinian principles plus generative entrenchment. In Wimmsatt's terms—"Generative systems would occur and be pivotal in any world – biological, psychological, scientific, technological, or cultural – where evolution is possible."(1999, p. 295)

Specifically Wimmsatt requires that, along with the Darwinian principles, evolving systems contain:

1. Structures which are generated over time so they have a developmental history (generativity); and
2. Some elements which have larger or more pervasive effects than others in that production.

But why might we go for this supplementation of the Darwinian principles? Wimmsatt's arguments are characteristically cogent. Generative Entrenchment is a natural outcome of systems that obey the rest of the Darwinian principles. Generative entrenchment has genuine predictive and explanatory power. Furthermore, it is theoretically fertile. It causes us to ask questions about natural selection outside the central biological cases we might otherwise not have noticed. In this context Wimmsatt (1999, p. 289) quotes with approval Kenneth Boulding's claim that: "A car is just an organism with an exceedingly complicated sex life."

But all this brings us to an apparent impasse, for Sterelny's characterisation of natural selection is not like Wimmsatt's. They are not inherently incompatible. We could join them together, but that would make for a characterisation that is much less Liberal than Wimmsatt's, for example. I think the answer to this impasse lies in the fact that Sterleny and Wimmsatt are focussing on different high level functional characteristics of evolving systems... Furthermore, the discussion so far has only scratched the surface of the variety of high level functional characteristics which might be employed in general characterisations of natural selection and so might form the basis of inferences about its workings. Other possibilities include:

- The production of vestigial traits – initially adaptive but no longer
- The evolution of evolvability – segmentation

- The maintenance of an inverse relation between variation and longevity
- Modularity
- Speciation
- Arms races
- Co-evolution
- Evolutionary convergence

5.5 Conclusions

This brings me back to my central argument about Darwinian Hard Cases and the scope of the theory of natural selection. Natural selection is under-characterised by the Darwinian principles. This leads to much room for theoretical manoeuvre. Earlier I presented a number of ways in which we can analyse non-genetic inheritance. In just this single parameter of evolutionary analysis, there was a good deal of variety, but what all these accounts had in common was that they were all functional. Furthermore, when we look at the theoretical justifications for such theories, we find inferences from facts at a higher level of functional description. Furthermore such inferences seem inevitable, for a low-level characterisation of natural selection is only of value if it secures for us those characteristics of evolving systems with which we are familiar (adaptation, generativity, speciation, evolutionary convergence. . .).

The theory of natural selection will have a determinate scope only if there is a fact of the matter as to which high-level functional characteristics are a necessary part of that process. As I see no argument in favour of such a fact of the matter, then I take natural selection to be an indeterminately bounded process.

References

Ashlock, D. 2006. *Evolutionary computation for modeling and optimization*. New York: Springer.

Aunger, R. (ed.). 2000. *Darwinizing culture: The status of mimentics as a science*. Oxford: Oxford University Press.

Boyd, R., and P. Richerson. 1985. *Culture and the evolutionary process*. Chicago: University of Chicago Press.

Boyd, R., and P. Richerson. 2005. *Not by genes alone: How culture transformed human evolution*. Chicago: University of Chicago Press.

Blackmore, S. 1999. *The meme machine*. Oxford: Oxford University Press.

Brandon, R. (1999). The Units of Selection Revisited: The Modules of Selection. *Biology and Philosophy* 14: 167–180.

Cavalli-Sforza, L., and M. Feldman. 1981. *Cultural transmission and evolution: A quantitative approach*. Princeton: Princeton University Press.

Darden, L., and A.J. Cain. 1989. Selection type theories. *Philosophy of Science* 56: 106–129.

Dawkins, R. 1976. *The selfish gene*. Oxford: Oxford University Press.

Dawkins, R. 1982. *The extended phenotype*. Oxford: Oxford University Press.

Dennett, D.C. 1995. *Darwin's dangerous idea*. New York: Simon and Schuster.

Edelman, G. 1987. *Neural Darwinism: The theory of neuronal group selection*. New York: Basic Books.

Fernando, C., R. Goldstein, and E. Szathmáry. 2010. The neuronal replicator hypothesis. *Neural Computation* 22(11): 2809–2857.

Fodor, J. 1975. *The language of thought*. New York: Crowell.

Godfrey-Smith, P. 2000. The replicator in retrospect. *Biology and Philosophy* 15: 403–423.

Gould, S.J. 1989. *Wonderful life*. New York: W. W. Norton.

Griffiths, P., and R. Gray. 1994. Developmental systems and evolutionary explanation. *Journal of Philosophy* XCI(6): 277–304.

Griffiths, P., and R. Gray. 1997. Replicator II – Judgement day. *Biology and Philosophy* 12: 471–492.

Henrich, J., and R. McElreath. 2003. The evolution of cultural evolution. *Evolutionary Anthropology* 12: 123–135.

Hull, D. 1981. Units of evolution: A metaphysical essay. In *The philosophy of evolution*, ed. R. Jensen and R. Harre. Brighton: Harvester.

Hull, D., R. Langman, and S. Glenn. 2001. A general account of selection: Biology, immunology, and behaviour. *The Behavioral and Brain Sciences* 24: 511–573.

Lewontin, R. 1970. The units of selection. *Annual Review of Ecology and Systematics* 1: 1–18.

Lycan, W.G. 1987. *Consciousness*. Cambridge, MA: MIT Press.

Maclaurin, J. 1998. Reinventing molecular Weismannism: Information in evolution. *Biology and Philosophy* 13: 37–59.

Mameli, M. 2004. Nongenetic selection and nongenetic inheritance. *The British Journal for the Philosophy of Science* 55: 35–71.

Mesoudi, A., K. Laland, and A. Whiten. 2004. Is human cultural evolution Darwinian evidence reviewed from the perspective of the origin of species. *Evolution* 58: 1–11.

Nelson, R.R. 2007. Universal Darwinism and evolutionary social science. *Biology and Philosophy* 22: 73–94.

Nelson, R.R., and S.G. Winter. 1982. *An evolutionary theory of economic change*. Cambridge: Harvard University Press.

Oyama, S. 1985. *The ontogeny of information*. New York: Cambridge University Press.

Oyama, S. 2000. Causal democracy and causal contributions in developmental systems theory. *Philosophy of Science* 67(Suppl. 3): 332–347.

Popper, K.R. 1972. *Objective knowledge: An evolutionary approach*. Oxford: Oxford University Press.

Sterelny, K. 2001. Niche construction, developmental systems, and the extended replicator. In *Cycles of contingency*, ed. S. Oyama, R.D. Gray, and P. Griffiths, 333–349. Cambridge, MA: MIT Press.

Sterelny, K., K. Smith, and M. Dickison. 1996. The extended replicator. *Biology and Philosophy* 11: 377–403.

Williams, G.C. 1966. *Adaptation and natural selection: A critique of some current evolutionary thought*. Princeton: Princeton University Press.

Wimsatt, W.C. 1999. Genes, memes and cultural heredity. *Biology and Philosophy* 14: 279–310.

Chapter 6
The Future of Utilitarianism*

Tim Mulgan

6.1 Introduction

My recent work has focused on the demands of utilitarianism, and our obligations to future people (Mulgan 2001, 2006, 2007). In my current work, I draw on that earlier work, and ask how utilitarians might deal with the ethical challenges of climate change (Mulgan 2010a, 2011, forthcoming a).[1] Climate change has obvious practical implications. It will kill millions of people, wipe out thousands of species, and so on. My question in this paper is much narrower. How might climate change impact on moral theory – and especially on the debate between utilitarians and their non-utilitarian rivals?

6.2 The Broken World

I am especially interested in the philosophical implications of the possibility that dangerous human-induced climate change may produce a *broken world*, where resources are insufficient to meet everyone's basic needs, where a chaotic climate

---

* While a French translation of this paper has appeared in *The tocqueville Review/La Revue Tocqueville* (vol. XXXII, no. 1–2011), it appears here for the first time in English. I dedicate it to Colin Cheyne – colleague, friend, and cosmopolitan Dunedinite.

[1] I have also presented work-in-progress on this research project to audiences at the University of St Andrews, the Universite de Rennes, the Open University, the University of Edinburgh, the University of Auckland, the University of Otago, the British Society for Ethical Theory, and the Centre for Human Values at the University of Princeton. I am grateful to all these audiences for helpful comments.

T. Mulgan (✉)
Moral and Political Philosophy, University of St Andrews, Scotland, UK
e-mail: tpm6@st-andrews.ac.uk

J. Maclaurin (ed.), *Rationis Defensor: Essays in Honour of Colin Cheyne*,
Studies in History and Philosophy of Science 28, DOI 10.1007/978-94-007-3983-3_6,

makes life precarious, where each generation is worse-off than the last, and where our affluent way of life is no longer an option. This is not our world. Humanity currently has the resources to meet everyone's needs. But nor is the broken world merely imaginary. It is one possible future.

Everything about climate change is controversial in public debate. If the internet teaches us nothing else, it does remind us that every fact is denied by someone. So I want to stress the modesty of my empirical assumptions. I claim only that past and present human behaviour may produce something like a broken world at some point in the future. This modest claim is sufficient to motivate our discussion; and no-one can reasonably be confident that it is false.

My current research project on ethics for a broken world has two dimensions. The first re-imagines ethics *within* a broken world. To make the thought experiment vivid, I imagine a history of philosophy class *in* the broken world, studying classic texts from a past age of affluence (Mulgan 2011a, 2011b, Forthcoming a, b,). This highlights the *contingency* of our moral and political ideals – as when we study *past* political philosophers in *their* historical context.

The second dimension of my project studies the impact of a broken future *on us*. If we consider only present people and their needs, then our world is not broken. But, on any plausible moral theory, the well-being of future people matters as much as our own. So the needs of 'our world' include the needs of future people, and 'our resources' include its future resources. If, on this wider definition, our resources are insufficient to meet all our needs, if we must choose between present and future needs, then our world is *already broken*.

This commitment to *temporal impartiality* is a striking feature of contemporary utilitarian moral philosophy. It is also very relevant to the public debate over climate change. As the scientific evidence that humans are causing climate change has become overwhelming, climate change 'sceptics' have shifted from science to economics. One very common sceptical argument is that the *future* benefit of preventing climate change is not worth the *present* cost. Central to this 'cost-benefit analysis' is a *social discount rate*, where future costs or benefits are discounted. Even a modest discount of 5% per annum makes it 'uneconomic' to spend even one dollar today to avert a global catastrophe in 500 years time.[2]

Everyone agrees that it is reasonable to discount as a proxy for uncertainty, or to accommodate the remote possibility that there will be no future people. (Humanity might be wiped out by an asteroid strike, of instance.) It would also make ethical sense to discount if you reasonably believed that future people will be richer than present people. (Of course, *this* argument may be *reversed* if we accept the possibility of a broken future where people are worse-off than today.)

However, many economists go further, and apply a *pure time preference*. Future happiness counts for less simply *because* it lies in the future. This pure discounting

[2] To justify the expenditure of a dollar today, the catastrophe has to cost $39,323,267,827 at that future date.

is justified on the grounds that this is how people actually behave. We *do* discount future benefits, both to ourselves and to others.

While they may accept discounting for other reasons, utilitarian moral philosophers reject any pure time preference. Utilitarians reject the inference from the descriptive to the normative. The fact that people do discount does not imply that they should. Utilitarians embrace temporal impartiality. Human beings are equally valuable, no matter when they live. Present happiness is not intrinsically more valuable than future happiness. The challenge for utilitarians is that, in a world with a broken future, this principled refusal to discount *at all* threatens to overwhelm all other moral concerns (Mulgan 2011a, part two).

My question in this paper is how utilitarianism can adjust to a broken future. This might seem to be merely a question in applied ethics – just a matter of adjusting our preferred moral theories or principles to new circumstances. I shall argue instead that a broken world has implications both for the content of moral theories, and for the choice between them. This is because every moral theory rests on (often implicit) background assumptions that may no longer hold in a broken world (Mulgan forthcoming a). At the extreme, this means that some moral theories or principles simply fail to apply at all.

Some failures are obvious. Consider the standard libertarian view that justice is exhausted by absolute property rights, and that such rights can only be justly acquired if one leaves 'enough and as good for others'.[3] In a broken world, where this requirement can never be met, libertarian principles of justice have no application. There are no legitimate property rights over external objects. No-one owns anything, and the state has no legitimate business at all.

Another telling example is John Rawls's special conception of justice – the heart of his famous liberal egalitarian theory of justice (Rawls 1971, pp. 54–55).[4] Rawls's special conception explicitly applies only in *favourable conditions* – where all basic needs can be met without compromising basic liberties. By definition, a broken world lacks favourable conditions. In such a world, Rawls would have to fall back on his much less-developed *general* conception of justice. This has many implications for Rawls's liberal egalitarianism. In particular, Rawls himself conceded that, if they ever did conflict, basic needs would trump even basic liberties (Rawls 1993, pp. 297 and 325, Rawls 1999, p. 65, Rawls 2001, p. 44, footnote 7). So the famous Rawlsian commitment to the priority of liberty would be in doubt in a broken world.

Other philosophical consequences of the broken world might be more surprising. Here is one speculative example drawn from contemporary meta-ethics. Some ethical naturalists identify moral facts with the end-points of moral inquiry.[5]

[3] This is the famous Lockean proviso, borrowed from Locke's original discussion. For further discussion, and references, see Mulgan (2011a, part one).

[4] For further discussion and references, see Mulgan (2011a, part three).

[5] One prominent example is Frank Jackson (1999), who identifies moral facts with the 'mature folk morality' of some future community of rational inquirers.

But it might turn out that our current methods of moral inquiry are all inextricably linked to an unsustainable way of life. Inquiry *as we know it* simply cannot continue indefinitely. If this is so, then it may not be possible for moral inquiry to converge in any possible worlds that is relevantly similar to the actual world. Would these ethical naturalists then have to accept that there are no moral facts?

Failures as dramatic as these will probably be the exception. Most moral theories will be able to adapt to a broken world. However, the broken world still has a significant moral impact. In particular, some moral theories will cope with the transition to a broken world better than others. No moral theory copes perfectly with all situations, or perfectly captures all considered moral intuitions or judgements. The choice between different theories is often driven largely by our pre-theoretical moral *priorities*. When two or more intuitions clash, we must decide which is most important (Mulgan forthcoming a).

Here is a loose analogy. Philosophers often test competing theories of personhood against imaginary cases involving sentient aliens or conscious computers. We begin with the intuition that a certain imaginary extra-terrestrial or machine would (or would not) be a person, and then we ask whether a given analysis of personhood respects that intuition. At present, these debates are a theoretical curiosity. (Or perhaps a desperate attempt to make our subject appealing to teenagers.) But suppose we really did encounter sentient or rational aliens, or we did develop computers with plausible claims to personhood. We would then have to decide whether or not to treat these new beings *as persons*. In such a situation, it really *would* matter what our moral theory said about those boundary cases. And a moral theory that could not give plausible advice in these cases would no longer be a plausible contender.

The broken world has a similar impact. Certain types of situation that we currently regard as uncommon or peripheral will become much more significant. Two obvious examples are emergency situations and actions that impact adversely on future people. In a broken world, moral principles that currently apply only in rare moral emergencies will loom large. When developing moral principles to govern their everyday lives, people in a broken world may find that our current philosophical debates over scarce and expensive medical treatment are more relevant than those areas of contemporary moral and political philosophy that implicitly assume an abundance of resources. (They may even find the harsh infanticide or euthanasia choices of the Inuit or the Ancient Greeks more instructive than our no-longer-attainable moral standards.) And these future broken world dwellers will reject as irrelevant any theory that can only work by setting scarcity aside (Mulgan 2011a, forthcoming a).

I address emergency cases elsewhere.[6] The present paper focuses on future people. The broken world raises the significance of our moral relations with distant future people by removing three key reasons for setting those people aside: the belief

[6] Mulgan, 'Theory and intuition in a broken world'; Mulgan, *Ethics for a broken world.*

that we cannot affect them, the belief that there is no real conflict between their interests and ours, and the optimistic assumption that future people will inevitably be better-off than ourselves.[7]

Of course, no-one believes that our actions have no impact on the future. But much contemporary moral and political philosophy proceeds on the assumption that, because the foreseeable long-term impacts of different actions cancel one another out, we can safely set distant future people aside. This complacency collapses once we recognize the significance of long-term environmental effects, and the possibility of a broken world.

Rawlsian optimists argue that future people are easily accommodated. We do best for future people if we leave behind stable democratic institutions and a thriving economy; and these good things outweigh any accompanying environmental costs. Our best interests do not conflict with those of distant future people, and we can rest assured that they will be better-off than us. Rawls himself makes these assumptions (often implicitly) in his original discussion of intergenerational justice (Rawls 1971, section 44). Subsequent debate within liberalism has largely taken them from granted.

Rawls's optimism is now very controversial, and perhaps even untenable. In this paper, I shall assume, if only for the sake of argument, that Rawls is wrong. In other words, I assume that climate change produces a broken future. In that broken future, of course, no-one will share Rawls's optimism, and moral theories that are able to accommodate obligations to distant future people (*and* to offer advice about how to deal with intergenerational *conflict*) will have a very significant prima facie advantage over theories that cannot even admit the possibility of such conflict.

There is an obvious link here to the second dimension of my project. Ethics for a future broken world cannot easily be separated from ethics for a present world with a (possibly) broken future. The same considerations that will force future people in a broken world to place their future-oriented obligations at the top of the ethical agenda should also force us to give our obligations to them a similar weight.

In the rest of this paper, I explore in more detail two places where the prospect of a broken future affects present-day moral theory: the debate between consequentialist and social contract accounts of intergenerational justice; and the debate between moderate and extreme forms of consequentialism. I shall argue that a broken future favours consequentialism over its rivals, but it also leads consequentialism in new and uncomfortable directions. (A terminological note before we proceed to the main argument. Analytic moral philosophers tend to prefer the term 'consequentialism' to 'utilitarianism'. The former is more general, as it covers any moral theory where moral rightness is linked to the promotion of value. On this usage, utilitarianism is a specific case of consequentialism, where the value to be promoted is aggregate human well-being.)[8]

[7] For more discussion of our obligations to future people, see Mulgan (2006, 2007, chapter 9). For further references, see Mulgan (2011b).

[8] For further references to current debate, see Mulgan (2011b).

6.3 Two Models of Intergenerational Justice

Two approaches dominate discussion of intergenerational justice within current analytic philosophy: consequentialism and social contract. A broken future strengthens familiar objections to social contract and thereby strengthens consequentialism.

In the social contract tradition, justice is modelled as a bargain or agreement between rational individuals. The difficulty of contracting with distant future people is obvious (Mulgan 2006, chapter 2, Mulgan 2011a, part three). Due to a complete lack of reciprocal interaction, we cannot bargain, negotiate, or cooperate with those who will live long after us. In technical terms, intergenerational relations are not a game. In a game, each player's payoff depends on what all players do. In intergenerational relations, our payoff depends only on what we do. This is why attempts to model climate change as, for instance, an inter-generational prisoners' dilemma are doomed to fail (Gardiner 2004, 2011).

These problems are compounded by Derek Parfit's *non-identity problem*, where the people who exist in different possible futures are not numerically identical to one another. Following Parfit, many philosophers have argued that, in such cases, common moral principles often break down (Mulgan 2006, chapter one; Parfit 1984, Part Four; Roberts and Wasserman 2009). Consider a simple tale introduced by Parfit (1984, p. 371). We must choose between two energy policies. The first is completely safe. The second is cheaper, but riskier – burying nuclear waste where there is no earthquake risk for several centuries, but a significant risk in the distant future. Suppose we choose this risky policy. Many centuries later, an earthquake releases radiation, killing thousands of people.

Our choice seems clearly wrong. But why? Intuitively, we do wrong because we harm those who die. But suppose the two energy policies lead to radically different futures – with different patterns of migration and social interaction. Now take any particular individual killed by the catastrophe. Suppose the precise chain of events leading to her existence would not have occurred if we had chosen differently – her parents would not have met, and might not even have existed themselves. But now it appears we have harmed no-one. For how can someone be harmed by an action without which she would not exist? And, if we harm no-one, how can our choice be wrong?

The non-identity problem is especially difficult for social contract theorists. How can we begin to imagine contracts, bargains, or cooperative schemes involving future people whose existence and identity depend upon what we decide? Contractualists as diverse as Kant, Rawls, Gauthier, and Scanlon all face serious difficulties here (Mulgan 2006, chapters 1 and 2; Gosseries and Meyer 2009).

For consequentialism, by contrast, future people are theoretically unproblematic. Faced with two possible futures, we simply choose the one with happier people. Moral obligation flows from the needs of sentient beings. Future people will be such beings. End of story. Of course, the content and scope of our intergenerational obligations are controversial – as we'll see in the remaining sections of this paper.

But these are problems within consequentialism, rather than attacks on its very coherence.

Confronted by the incoherence of a contract with distant future people, some social contract theorists 'bite the bullet'. They accept that there are no obligations to future people. If morality and justice require existence and interaction, then we have no moral obligations (and no obligations *of justice*) to future people. We may happen to care about them, we may choose to take their interests into account, but we owe them nothing.

Most social contract theorists do not take this route. They offer positive accounts of intergenerational justice. These are based on a bewildering array of theoretical devices or assumptions: ad hoc motivational assumptions, contracts between overlapping generations that somehow reach indefinitely into the future, the appointment of trustees or ombudsmen for the future, intergenerational original positions where the parties know neither when nor whether they exist, or (in Kant's case) God.

These intergenerational extensions of the social contract are controversial. Many critics think they don't work even in a world that enjoys favourable conditions in both the present and the future. Setting that debate aside, I shall now highlight two respects in which, whatever their fate under favourable conditions, intergenerational contracts fare *much worse* in a broken world.

The first problem is that *specific* extensions of the social contract typically presuppose Rawls's two assumptions of favourable conditions and intergenerational optimism. I will briefly discuss two examples. The first is Rawls himself (Mulgan 2006, chapter 2, Mulgan 2011a, part three). Rawls's discussion of intergenerational justice is very limited. He focuses on only two problems: just savings (or 'accumulation') and stability. The just savings problem only arises because we assume that future people will be better-off, and Rawls only addresses the question of stability under the same assumption. Rawls ignores population policy, thereby side-stepping non-identity issues altogether. He also ignores the possibility of environmental threats to the long-term stability of liberal democratic institutions.[9] Rawls's assumptions affect how he frames his two problems, as well as the details of his solutions. Most importantly, Rawls's assumptions make those problems seem much more tractable. For instance, it is much easier to ensure the stability of a political system if each generation knows it will be better-off than the last. If future people might be worse-off, then Rawls's own theoretical commitments would force him to give the just savings principle a much more prominent position in his theory.

My second example of an intergenerational contract is David Gauthier's contractarianism (Gauthier 1986; Mulgan 2006, chapter 2). Gauthier's intergenerational contract only works if we confine ourselves to adjacent generations, assume that we are dealing with self-sufficient rational individuals, and build-in a sufficient abundance of natural resources that Gauthier's eccentric reformulation of Nozick's version of Locke's proviso on acquisition can be satisfied in perpetuity. None of these

[9] In his *Law of Peoples* (Rawls 1999), Rawls also optimistically ignores the impact of climate-related scarcity on international relations.

assumptions is remotely plausible in a broken world. (Actually, Gauthier's solution still doesn't work even if we grant all his assumptions – but that is another story).

Of course, the piecemeal rebuttal of extensions of specific intergenerational contracts is a never-ending business. There are always new contracts, new extensions, new epicycles. Life is too short. It is more satisfactory to simply undermine the whole enterprise by questioning its underlining motivation. I will now argue that, once future people move centre stage, it is no longer credible to see intergenerational justice as an *extension* of 'normal justice' between contemporaries.

Suppose you really did believe that justice *is* cooperation for mutual advantage. Once you realise that future people cannot interact with us, wouldn't you simply lose interest in this contradictory topic called 'intergenerational justice'? Of course, we can imagine an intergenerational original position. Philosophers, after all, can imagine pretty much anything. But why would you bother? Rawls's original 'original position' models the requirements of fairness in the distribution of the products of social cooperation between contemporaries. What exactly could any intergenerational original position model?

A philosopher who really is committed to the social contract should thus simply ignore intergenerational justice entirely. (Or, at most, they should regard it as an afterthought: an optional extension of the basic theory of justice between contemporaries.) Justice is really about cooperation between contemporaries. This would fit with a very plausible *sociological* hypothesis about the motivations that draw people to this intellectual tradition. People are not drawn to social contract because they think it has a compelling account of our obligations to distant future people. Rather, they are drawn to it for other reasons. Perhaps they believe that social contract copes comparatively well with other (more pressing) moral problems, or that it offers a better account of other (more significant) foundational issues, or that it is more realistic in what it assumes about (or requires of) human moral agents. (Of course, the consequentialist thinks these beliefs about the advantages of the social contract are mistaken. But that is another story.) Having been attracted to social contract for these reasons, its proponents then continue to accept it *despite* its obvious inability to account for the possibility of intergenerational justice. After all, no theory is perfect.

In a broken world, this strategy becomes untenable. No plausible theory of justice can set distant future people aside. If you start with the interests of people – and then notice that most of the people whose interests matter most are distant future people – then the very idea of modelling morality as a contract simply makes no sense.

This argument may seem circular. Consequentialism first tells us that distant future people are significant, and then uses this result against its opponents. Why not simply deny that we have any obligations to distant future people? Perhaps social contract *does* offer the best account of intergenerational justice: there is no such thing.

My reply is that the intuitive force of the claim that we should, at least, leave distant future people with a habitable planet, is not a theoretical artefact. It is not only consequentialists who worry about climate change. Few social contract theorists do, in fact, bite the bullet. Even Gauthier, who is not otherwise worried by the counter-intuitive (some would say morally repellent) consequences of his own theory, feels compelled to avoid this one.

6.4 Towards Moderate Consequentialism

I argued in the previous section that the shift from a world enjoying favourable conditions to one that has a broken future favours consequentialism over contractualism. The application of consequentialism to a broken world may seem straightforward. We simply choose whatever actions, rules, or institutions will best promote human well-being. It may be practically impossible to perform the necessary calculations, but that is not a new problem. There is nothing *theoretically* puzzling about the broken world.

Unfortunately, things are more complicated.[10] The broken world has different implications for different versions of consequentialism. *Act consequentialists* say that every agent should always perform the action that produces the best consequences. For them, things remain theoretically simple – even in a broken world. But act consequentialism is notoriously demanding even in our own world. Consider a very simple example. You have ten dollars in your pocket. You could buy a book, see a movie, or give the money to a reliable charity who will use it to restore someone's sight. It's pretty clear which produces more happiness. So you make the donation, and go to the cash machine to get money to go to the movies. But now you have ten dollars in your pocket. What should you do? You can see where this is going.... No movies for you.

Moderate consequentialists seek less demanding alternatives. One popular strategy is to shift attention from the consequences of individual actions to the collective consequences of codes of rules.[11] The result is *rule consequentialism.* Act consequentialists image a single consequentialist agent, heroically maximising human happiness in a non-consequentialist world. Unsurprisingly, her life is demanding, alienating, and unattractive. Rule consequentialists picture morality as a task given, not to each individual agent, but to a *community* of *human beings*. We imagine ourselves choosing a moral code to govern our community – deciding what code to teach the next generation. For rule consequentialism, the fundamental moral questions are: 'What if *we* did that?', and 'How should *we* live?' We first seek

---

[10] For more on consequentialism and a broken world, see Mulgan (2011a, part two). For discussion of consequentialism in general, see Mulgan (2001, 2006). For further references, see Mulgan (2011b).

[11] Another common strategy for moderate consequentialists is to *directly* incorporate *agent-centred prerogatives* that allow agents to give disproportionate weight to their own interests. This strategy is exemplified by Samuel Scheffler's hybrid view (Scheffler 1982; Mulgan 2001, chapter 6, Mulgan 2006, chapter 4).

an *ideal moral code*. Acts are then assessed *indirectly*: the right act is the act called for by the ideal code. One leading proponent offers this succinct formulation.

6.4.1 Hooker's Rule Consequentialism

An act is wrong if and only if it is forbidden by the code of rules whose internalisation by the overwhelming majority of everyone everywhere in each new generation has maximum expected value in terms of well-being (with some priority for the worst-off). The calculation of a code's expected value includes all costs of getting the code internalised. If in terms of expected value two or more codes are better than the rest but equal to one another, the one closest to conventional morality determines what acts are wrong (Hooker 2000, p. 32).

In this paper, I focus on rule consequentialism as an exemplar of moderate consequentialism. On this interpretation, rule consequentialism emerges as a moderate, humane alternative to the extreme demands of act consequentialism. And its ability to deliver *moderate* rules is thus an essential feature of rule consequentialism.

We saw earlier that a broken world presents no new *theoretical* difficulties for act consequentialism. But act consequentialism would be even more demanding in that new situation. This would raise the comparative appeal of moderate consequentialism. Unfortunately, the transition to a broken world is *not* simple for rule consequentialism. In particular, it threatens to undermine its moderate credentials altogether. As we are considering rule consequentialism as a moderate *alternative* to act consequentialism, this is a potentially fatal objection.

6.5 The Lexical Threshold

To see why the broken future poses this challenge to rule consequentialism, we next explore a moral phenomenon that I call the *lexical threshold*.[12] A pervasive feature of our moral experience is a tension between moral concepts that posit sharp dividing lines, and the continuous natural properties on which they seem to supervene. One example is the fraught boundaries of the concept 'person' that plague debates over abortion, euthanasia, or animal ethics. In all these debates, commonsense morality seems to put great weight on concepts with very fuzzy boundaries.

My second example of a lexical threshold comes from an infamous puzzle in contemporary consequentialist value theory – Derek Parfit's *repugnant conclusion*. Imagine a series of possible human populations. In A, ten billion people enjoy wonderful lives. B contains twice as many people, each more than half as happy as

[12] This section presents a brief sketch of material that I explore at much greater length in Mulgan (2006, especially chapters 3 and 5–9). See also Mulgan (2005).

the A-people. B has more total happiness than A. Repeat this process until you reach Z, where a vast population have lives barely worth living. As each step increases total happiness, overcrowded Z is *better* than flourishing A. Parfit finds this conclusion 'repugnant', and argues that any acceptable theory must avoid it (Parfit 1984, p. 388).

Parfit's simple example has generated an enormous literature (Mulgan 2006, chapter 3; Ryberg and Tannsjo 2005).[13] Some reject Parfit's intuition, and defend the total utilitarian view that the best outcome contains the greatest total happiness. Others embrace the average utilitarian view that the best outcome contains the highest average happiness. But this view has problems of its own. Many moral philosophers reject both total and average utilitarianism. One popular alternative is the *lexical view*. We can introduce this view using a simple tale developed by Roger Crisp, who presents it as a counter-example to Bentham's own hedonism (Crisp 1997, p. 24).

6.5.1 Ollie and the Oyster

The good fairy offers Ollie two choices. He can live a flourishing human life for 100 years; or be a happy oyster with very simple pleasures. The oyster life lasts as long as Ollie wants – even up to millions of years.

One common objection to hedonism is that, because it equates human well-being with aggregate pleasure, it must allow *any* human life to be trumped by a large enough number of oyster years. Many people find this intuitively unappealing.[14] They believe instead that Ollie should choose a flourishing human life over an indefinitely long oyster life. This doesn't necessarily mean that oysters have no value. Rather, human lives are *incomparably* better than oyster pleasures. No amount of oyster pleasure could outweigh the value of a good human life. On this view, flourishing human lives are *lexically superior* to oyster lives.

In exactly the same way, some philosophers avoid Parfit's repugnant conclusion by positing a lexical gap between flourishing human lives and those that are barely worth living. A is better than Z because A-lives are lexically superior to Z-lives. The obvious problem for any lexical view is where to draw the line. Do flourishing and destitute humans differ in the same way as humans and oysters? Or do different human lives differ in *degree*, rather than in *kind*? Parfit himself regards the lexical view as the most promising solution to his repugnant conclusion, but he also presents a seemingly devastating objection to that view (Parfit 1986). Parfit asks his readers to imagine a continuum of possible lives, going from the best to the worst.

---

[13] For further references, see Mulgan (2011b).

[14] This has certainly been my experience in teaching such examples over many years.

Where on that continuum would we locate our lexical gap? Any location seems arbitrary and ad hoc. Why draw the line *there*?

The simplest response to Parfit' continuum objection is to abandon lexical thresholds altogether, and adopt a morality of continuous graded properties. This is probably the most appealing solution *for the act consequentialist*, who is happy to embrace an austere and counterintuitive moral theory. However, I argue elsewhere that those rule consequentialists who seek to avoid the relentless demands of act consequentialism *must* posit lexical thresholds (Mulgan 2006, chapter 3).

The solution I develop instead is to treat the lexical threshold, not as part of our theory of comparative value, but as part of our theory of moral obligation. The lexical threshold is not an objective feature of the world. Instead, it is something we adopt in particular deliberative contexts.

I shall now illustrate this solution using Parfit's repugnant conclusion. I believe that the vast literature on the repugnant conclusion actually addresses the wrong question. Our strongest intuitions in this area concern, not the comparative values of possible futures, but our obligations to future people. We are most concerned, not with what is good, but with what we are obliged (or permitted) *to do*. Those who support Parfit's rejection of the repugnant conclusion are often motivated by an objection, not to the initial claim that Z is better than A, but to the inference that A-people have an obligation to transform their world into Z. For the act consequentialist, this inference is automatic. If Z is better than A, then anyone who has a choice between the two possible futures *must* opt for Z over A. Act consequentialists who wish to avoid the obligation must deny the value claim. But rule consequentialism severs this tight connection between value and obligation. It thus opens the possibility of a more nuanced response to Parfit's repugnant conclusion – one where we *agree* that Z is better than A, but then *deny* that anyone has an obligation to turn an A-world into a Z-world.

In my own recent work, I expand the notion of a lexical threshold. I use that notion to unite a broad range of seemingly disparate departures from act consequentialism. On this broader usage, the lexical threshold structures the many commonsense prohibitions and permissions of the rule consequentialist ideal code.[15] The lexical threshold marks out a protected moral sphere – a private space where each individual is both morally and practically free to concentrate on her own projects, goals, and relationships even at the expense of aggregate wellbeing. When deliberating, I can legitimately discount options that would take me below my lexical threshold. I argue that the lexical threshold can thus evolve from a single point on a scale of wellbeing to a richer notion of the essential components of a flourishing human life.

This is only a very brief sketch of controversial work-in-progress. In the rest of this paper, I focus on the significance of a broken world for any moral theory based on my extended notion of a lexical threshold. An essential feature of my account is

[15] In my combined consequentialism, I also use the lexical threshold to set the boundaries of the agent-centred prerogative (Mulgan 2006, chapter 11).

that the lexical threshold is context-dependent. Over time, as our notion of the worthwhile life has evolved, our lexical threshold has risen. We now take for granted a broader range of goals, a much longer life span, and a far greater level of security than previous generations would have thought possible. This context-dependence immediately raises a very significant set of moral questions. When we encounter people whose situation differs markedly from our own – when we look to the past, to the future, to other lands, or even to less fortunate areas of our own society – we find different lexical thresholds. When my actions impact on other people, and I am deliberating about what I should do, should I use *my* interpretation of the lexical threshold or *theirs*?

Consider a simple example. Like all moderate moral theorists, rule consequentialists believe both that I am obliged to ensure that others have worthwhile lives, but also that I am not thereby obliged to sacrifice a worthwhile life for myself. There is a vast literature within analytic moral philosophy devoted to the demands of beneficence (Mulgan 2001).[16] That whole literature boils down to a simple question: Does 'worthwhile life' have the same referent in both the other-regarding obligation and the self-regarding permission? Am I obliged to bring myself down to the level to which I raise others? Can I insist for myself on what cannot be guaranteed to all? Within the framework I have developed in this section, this question becomes: Can I legitimately employ two distinct lexical thresholds in the same deliberative context?

The act consequentialist has a very definite answer: 'No'. Each is to count for one, and none for more than one. If we are to employ lexical thresholds at all, then we must always employ the same threshold. I argue elsewhere that, to avoid extreme demands, moderate consequentialism must allow us, not only to give greater weight to our own interests, but also to privilege our own lexical threshold – claiming for ourselves what cannot be guaranteed for all.

We are familiar with cases where the lexical level rises through time. We expect goods and guarantees that our ancestors could not even have imagined. The true significance of the broken world lies in the fact that it introduces a hitherto unacknowledged phenomenon: the possibility of a *lowering* of the lexical threshold.

6.6 Lexical Thresholds in a Broken World

To see why the broken world is so unsettling for moderate consequentialism, we first ask how the people dwelling in a broken future might determine their lexical threshold. Suppose they begin with *our* moral ideals. Unfortunately, they realise that our affluent ideals are simply unattainable in their new situation. How might they adapt our lexical threshold? Some changes are obvious. If our

[16] For further references, see Mulgan (2011b).

broad range of resource-intensive goals, and our long lifespan, are not widely available, then no-one can reasonably insist on them. So these elements can no longer be built into anyone's interpretation of the lexical threshold.

Other revisions might be more drastic. One role that the lexical threshold plays in my account of rule consequentialism is that it provides a notion of a worthwhile life that is *guaranteed* to everyone. (For instance, a rule consequentialist will insist that just institutions must ensure that everyone has enough to eat, and that their basic rights are respected.) In a world where it is not possible for everyone to survive, if we still seek a common lexical threshold for all, then we can no longer think of it as a guarantee of a worthwhile life or even as a guarantee of survival. Instead, we might think of the lexical threshold as a fair and equal *chance* of surviving (or living a worthwhile life) – and then insist that *this* fair chance is what must be guaranteed to all. Take one simple case: the allocation of water under conditions of scarcity. If an equal share of water is insufficient for survival, then it makes no sense to give everyone an equal inadequate share rather than an equal *chance* of an adequate share.

Today, perhaps with good consequentialist reasons, we regard the violation of basic human rights as unthinkable – and we build inviolable rights into our lexical threshold. But, in a future broken world, we must rethink both the content and the strength of rights. Consequentialist political institutions may shift from securing everyone's survival to managing a fair distribution of chances to survive. In this context, lotteries may emerge as the paradigm of utilitarian justice.

Act consequentialists will insist that, once we acknowledge a broken future, and see our world as already broken, we must adopt this lower lexical threshold *now*. Take a concrete example. Suppose we discover that, if we *insist* on 70 years of good health for ourselves, and insist on the necessary investment in medical technology, then our descendants can only hope for a reasonable chance of 50 years of moderate health. Can we still regard a lifespan of 70 years as a *right*? If so, why is it a right for us and not for future people? Or suppose we discover that, if we guarantee our basic needs, then our descendants will have to institute a survival lottery – a social decision procedure to determine who lives and who dies. Can we insist on guaranteed survival for ourselves, or should we move in their direction – operating a survival lottery across the generations? (And, if we did so, what might that lottery look like?)

Moderate consequentialists will seek to avoid these demanding questions (or at least to avoid the extreme demands of act consequentialism), just as they do in the parallel case of our obligations to present people in distant lands. This may seem to be a straightforward task. Why not treat this as just another case where two groups have different lexical thresholds? After all, I argued in the previous section that, in other similar cases, moderate consequentialism does allow us to privilege our own interpretation of the lexical threshold.

The problem is that, as *consequentialists*, rule consequentialists cannot just help themselves to different lexical thresholds. They need a *justification* that acknowledges that other people's well-being is just as important as our own. Moderate consequentialists must build partiality on an impartial foundation. And, unfortunately, the tricks that moderate consequentialists use in the case of obligations to distant present people are unlikely to work for the distant future.

6.7 Three Moderate Consequentialist Tricks

In this final section, I briefly examine three tricks commonly used by moderate consequentialists to avoid extreme demands, and ask whether they would still work for our obligations to future people in a broken world.

6.7.1 First Trick. A Background of Innocence

Many moderate moral philosophers implicitly treat the external circumstances that enable some groups to adopt more ambitious lexical thresholds than other groups as a *morally neutral* background, and *not* as something for which the better-off groups are themselves *responsible*. One reason why we have only a modest obligation to assist people who are starving in distant countries is that their comparative destitution is not our fault. Of course, many philosophers – especially consequentialists – question this assumed lack of responsibility in the case of relations between contemporary people in the developed and developing worlds (Pogge 2002; Singer 1972). In this paper, we set that particular debate aside. What *is* clear is that this move could not reasonably be made in relation to the broken future. We are imagining a broken future caused by *human-induced* climate change. Ex hypothesi, well-off people in the developed world – people like us – are (collectively) responsible for *that*.

6.7.2 Second Trick. A Background of Entitlement

A second common trick is to frame the demandingness debate against an implicit background of legitimate resource entitlements (Pogge 2002; Murphy 2000). The moral question becomes: 'When may I legitimately use *my* resources for myself?' Unfortunately, as I argue at length elsewhere, all standard attempts to ground property rights *against future people* – rights to use resources that they might want or need – break down in a broken world (Mulgan 2011a).

The appeal to legitimate property rights takes two distinct forms. Some moral pluralists would combine a consequentialist principle of beneficence with a non-consequentialist account of property. However, non-consequentialist accounts typically require either that we leave as much and as good for future people, or that our property regime benefits future people, or something on similar lines. And no such condition is met in a broken world. Ex hypothesi, if our actions lead to a broken world, then we have failed to either leave enough and as good for our descendents, or to leave them better-off.

Of course, a thoroughgoing consequentialist will not take this route. Instead, they will seek to develop a *consequentialist* justification of property rules themselves. As we are focusing on rule consequentialism, we will consider its account of property rights. This account, in turn, is part of the general defence of a liberal ideal code. That defence will be our third and final moderate consequentialist trick.

6.7.3 Third Trick. A Liberal Ideal Code

Rule consequentialists, drawing especially on the liberal utilitarian tradition of J. S. Mill (and also on the earlier arguments of Bentham), argue that human well-being is maximised if we recognise a range of familiar rights and freedoms – including property rights, individual liberties, agent-centred prerogatives and restrictions, and the freedom to privilege our own lexical threshold (Mulgan 2006, chapters 5 and 6, Mulgan 2007, chapters 2 and 6).

The consequentialist case for liberalism borrows from other currents in contemporary moral thought. These include Scheffler's hybrid view (briefly mentioned in footnote 26 above) and the social contact tradition discussed in Sect. 6.2. From Scheffler, rule consequentialism borrows the claim that each individual's agent-centred prerogatives make a contribution to her own well-being. From the social contract theorist, rule consequentialism borrows the claim that the bargains that rational agents will make include familiar liberal principles. On the assumptions that a prerogative that promotes each individual's well-being is good for all, and that rational agents will agree to what is best for all, rule consequentialism transforms these claims of other theories into claims about what would best promote aggregate human wellbeing.

In an affluent *world* – one that enjoys favourable conditions and has a bright and unbroken future – it is reasonable to argue that these three rival traditions (rule consequentialism, Scheffler's hybrid view, and contractualism) will support the same general moral principles. (And, in particular, that they will support liberal principles.) Some moral philosophers – myself included – have attempted to get a lot of mileage out of this coincidence (Mulgan 2001, chapter 10, Mulgan 2006, chapter 11). Unfortunately, the different traditions may well come apart in a world with a broken future (Mulgan 2011a, part two).

We must first note that any consequentialist case for liberalism is *empirical*. Specifically, it rests *both* on optimistic claims about our present social and environmental context, *and* on optimistic claims about the impact of human freedom in that particular context. All these claims are borrowed from the social contract tradition. We first assume that we are dealing with a situation where contemporaries can reciprocally interact to their mutual advantage; and *then* we make optimistic assumptions about the human response to freedom, the productivity of liberal capitalism, and so on. In *that* context, on *those* assumptions, it is not unreasonable to conjecture that freedom will promote wellbeing. Universal freedom benefits everyone.

Unfortunately, all these borrowed optimistic assumptions are out-of-step with a broken world. When we are dealing with our obligations to *future* people, reciprocity is impossible. For future people, liberal democracy has all the defects of despotism. Present people are free to arbitrarily impose their will on future people. Furthermore, once we face up to the possibility of a *broken* future, pessimistic assumptions are surely more reasonable than optimistic ones. As we saw in Sect. 6.2, the transition to a broken world is fatal for the social contract. But it is also deeply worrying for any liberal consequentialist. Given the ways we might exercise it, how could we reasonably believe that *our* freedom will benefit future people? If our freedom leads to a broken future, then it seems clear that it does not maximise well-being across time. Mill could feel confident that utilitarianism and liberalism would coincide. But we have no grounds for similar confidence.

It is important not to get carried away. Many rule consequentialist arguments will translate to any future – however broken – that contains recognisable human agents. Some logically possible codes of rules will remain forever too demanding, alienating, or intricate to be effectively taught to any human population – whatever the benefits of doing so. A broken future doesn't collapse rule consequentialism back into act consequentialism. But it does remove our confidence that the set of rules that maximises human wellbeing into the future will resemble commonsense morality – or that it will safeguard agent-centred prerogatives or coincide with contemporary contractualism.

The broken world thus provides a decisive intuitive test for moderate consequentialists. Is our commitment to moderation an external constraint on consequentialist reasoning, or is it a contingent output of empirical reasoning? Is moderation something we impose on the broken world come what may, or is it rather something that we might jettison in the harsh light of a broken future? I think the most consistent response for the moderate *consequentialist* is to reject an external constraint as unjustified. Therefore, my current broader research project takes the second option – asking what a genuinely consequentialist moral theory might look like without the presumption of moderation. Unfortunately, many of our moral intuitions are saturated with an implicit commitment to moderation. Therefore, if we are to locate a new lexical threshold for a broken world, we must turn from intuition to utilitarian moral theory. We must place less reliance on our own considered moral judgements, and perhaps return to the more empirically-minded arguments characteristic of Bentham and Mill.

References

Crisp, R. 1997. *Mill: On utilitarianism*. New York: Routledge.
Gardiner, S. 2004. Ethics and global climate change. *Ethics* 114: 555–600.
Gardiner, S. 2011. *A perfect moral storm*. New York: Oxford University Press.

Gauthier, D. 1986. *Morals by agreement*. New York: Oxford University Press.
Gosseries, A., and L. Meyer (eds.). 2009. *Intergenerational justice*. New York: Oxford University Press.
Hooker, B. 2000. *Ideal code, real world: A rule-consequentialist theory of morality*. New York: Oxford University Press.
Jackson, F. 1999. *From metaphysics to ethics*. New York: Oxford University Press.
Mulgan, T. 2001. *The demands of consequentialism*. New York: Oxford University Press.
Mulgan, T. 2005. Two parfit puzzles. In *The repugnant conclusion: Essays on population ethics*, ed. J. Ryberg and T. Tannsjo, 23–45. Dordrecht: Springer.
Mulgan, T. 2006. *Future people*. New York: Oxford University Press.
Mulgan, T. 2007. *Understanding utilitarianism*. Stocksfield: Acumen.
Mulgan, T. 2010. Utilitarianism for a broken future. *Report to the Club of Amsterdam, November 2010*.
Mulgan, T. 2011a. *Ethics for a broken world: Reimagining philosophy after catastrophe*. Acumen Publishers.
Mulgan, T. Forthcoming a. Theory and intuition in a broken world. *La società degli individui*.
Mulgan, T. 2011b. Consequentialism. In *Oxford bibliographies online*. doi:10.1093/OBO/97801 95396577-0026.
Mulgan, T. Forthcoming b. Utilitarianism for a broken world. Plenary lecture delivered to the 'Two Centuries of Utilitarianism' conference at the Universite de Rennes II in June 2009. To be published by the Presses Universitaires de Rennes.
Murphy, L. 2000. *Moral demands in nonideal theory*. New York: Oxford University Press.
Parfit, D. 1984. *Reasons and persons*. New York: Oxford University Press.
Parfit, D. 1986. Overpopulation and the quality of life. In *Applied ethics*, ed. P. Singer, 145–164. New York: Oxford University Press.
Pogge, T. 2002. *World poverty and human rights*. Cambridge: Polity Press.
Rawls, J. 1971. *A theory of justice*. Cambridge: Harvard University Press.
Rawls, J. 1993. *Political liberalism*. New York: Columbia University Press.
Rawls, J. 1999. *The law of peoples*. Cambridge: Harvard University Press.
Rawls, J. 2001. *Justice as fairness: A restatement*. Cambridge: Harvard University Press.
Roberts, M., and D. Wasserman (eds.). 2009. *Harming future persons: Ethics, genetics and the nonidentity problem*. Dordrecht: Springer.
Ryberg, J., and T. Tannsjo (eds.). 2005. *The repugnant conclusion: Essays on population ethics*. Dordrecht: Springer.
Scheffler, S. 1982. *The rejection of consequentialism*. New York: Oxford University Press.
Singer, P. 1972. Famine, affluence and morality. *Philosophy and Public Affairs* 1: 229–243.

Chapter 7
Kant on Experiment

Alberto Vanzo

In this paper I illustrate the relations in which experiments stand to hypotheses, laws, and principles in Immanuel Kant's natural philosophical methodology. My aim is not to provide a rational reconstruction of Kant's philosophy of experiment or to assess its internal coherence, but to illustrate it by contrasting it with an alternative conception of experiment that had a widespread following in the early modern period. This is the philosophy of experiment that was first sketched by Francis Bacon and later developed by Robert Boyle and Robert Hooke.

Needless to say, Bacon, Boyle, and Hooke are neither the first, nor the only philosophers to comment on the nature and functions of experiments before Kant. However, Bacon, Boyle, and Hooke provided a fairly elaborated and extremely influential set of reflections on the nature, types, and functions of experiments. Their philosophy of experiment embodies a set of beliefs and attitudes – such as the focus on fact-gathering and the rejection of hypotheses – that were widely held in the second half of the seventeenth century, among others, by many English natural philosophers and members of the Royal Society, the Florentine Accademia del Cimento, and the early French Academy of Sciences. A similar emphasis could be found in many British and French authors throughout the eighteenth century.[1]

Sketching a history of the anti-hypotheticalism that characterizes the Bacon-Boyle-Hooke view of experiment, Larry Laudan claims that it was endorsed by "most scientists and epistemologists" from the 1720s to the end of the eighteenth century. Laudan mentions Kant as one of the authors for whom "the method of hypothesis is fraught with difficulties" (Laudan 1981, p. 10). In contrast to Laudan,

---

[1] For instance, Bacon's, Boyle's, and Hooke's attitude towards hypotheses was shared, among others, by Locke, Newton, Hume, Turnbull, d'Alembert, and Reid. See Locke (1976–89, vol. 4, pp. 563, 629), Newton (1687, pp. 939–944; 1714/15, pp. 222–24), Hume (1740, §2), Turnbull (1740, p. 2), d'Alembert (1751–77, pp. i, v, vi, xxv), and Reid (2001, p. 50).

A. Vanzo (✉)
Department of Philosophy, University of Otago, Dunedin, New Zealand
e-mail: alberto.vanzo@email.it

J. Maclaurin (ed.), *Rationis Defensor: Essays in Honour of Colin Cheyne*, Studies in History and Philosophy of Science 28, DOI 10.1007/978-94-007-3983-3_7,

I will argue that Kant's view on the relations between experiments on the one hand and hypotheses, laws, and theories on the other hand, is best seen as an alternative to the aversion to hypotheses and sharp contrast between experiments and speculations that characterizes the Bacon-Hoyle-Hooke view of experiment.

Kant often emphasizes the importance of experiments for natural science. Experiments, together with observations, are "the single road of natural science" (Kant 1922, vol. 11, p. 142)[2] and "the source" of physics (Kant 1936–1938, vol. 22, p. 331).[3] They enable us to discover "the properties of object[s] of outer senses" (Kant 1922, vol. 11, p. 142, trans. modified) and the laws of nature (Kant 1966f, pp. $815_{7–9}$, and 898_{48}). Empirical physics entered "the highway of science" only when Galileo, Torricelli, and Stahl followed "the suggestion of the ingenious Francis Bacon" (Kant 1781/1787, Bxii; 1980, p. 107) and started performing experiments:

> When *Galileo* rolled balls of a weight chosen by himself down an inclined plane, or when *Torricelli* made the air bear a weight that he had previously thought to be equal to that of a known column of water, or when in a later time *Stahl* changed metals into calx and then changed the latter back into metal by first removing something and then putting it back again, a light dawned on all those who study nature [...] (Kant 1781/1787, Bxiii)

Galileo's, Torricelli's, and Stahl's experiments are not only the basis for what Kant regarded as a scientific revolution ("a sudden revolution in the way of thinking," Bxii). They are also the basis for a revolution in metaphysics, which Kant undertakes in the *Critique of Pure Reason*. Its method "consists in this: to seek the elements of pure reason in that *which admits of being confirmed or refuted through an experiment*" (Bxviii). Kant's "experiment of pure reason" (Bxxi n.; 1793, p. 291) aims to set metaphysics on the same "secure path of a science" (Kant 1781/1787, Bix) on which physics entered thanks to Galileo's and Torricelli's experiments. In the theoretical field, the experiment of pure reason will enable Kant to explain the nature, extent, and limits of a priori knowledge (Kant 1781/1787, B6, A3/B7, A57/B81; 1783, p. 276; Seigfried 1989). In the practical field, that experiment will clear the ground for the new foundations of morality which Kant articulates in the *Critique of Practical Reason* (Kant 1781/1787, Bxxiv–xxx; Sato 2008).

Despite his claims on the importance of experiments, Kant was not an experimenter (Adickes 1924–25, vol. 1, pp. 6–11; vol. 2, pp. 484 and 487),[4] although he was aware of many results of the experimental sciences of his day. Kant's writings do not contain any accurate experimental report or discussion of the practical aspects of experiments. The Kantian corpus does not provide any detailed, self-contained

---

[2] Translations are my own for those writings of Kant which have not been translated into English. All other translations are from the *Cambridge Edition of the Works of Immanuel Kant*. I have replaced American spelling with British spelling in quotations.

[3] See Kant (1936–1938, vol. 22, pp. $299_{1–2}$ and 329_{32}–330_{3}). The expression "study of nature" in the *Opus postumum* is often followed by the explanation: "through observation and experiment" (see e.g. Kant 1936–1938, vol. 22, pp. $322_{17–18}$, 328_{16}, 344_{15}, and $346_{25–26}$).

[4] Kant briefly describes an experiment that he performed in (1747, p. 153).

discussion of what an experiment is, except for a claim that was customary among Kant's German contemporaries (Lambert 1764, vol. 1, pp. 351–53; Erxleben 1772, §4). This is the claim that, by means of experiments, we *intervene* in nature, placing objects in states in which they would not otherwise be (Kant 1980, pp. 102–3).[5] By contrast, observations do not modify the state in which observed objects are. Accordingly, we can perform experiments on animals and the medium-sized objects that we have at hand, but not on distant stars and planets, because we cannot modify the state they are in.[6]

Despite his apparent disinterest in the practical aspects of experiments, Kant provides reflections on three important topics in the philosophy of experiment. These are the relations of experiments to hypotheses, laws of nature, and the heuristic principles that guide scientific inquiry.[7] I will reconstruct these relations in Sects. 7.2, 7.3 and 7.4 of the paper, after sketching the Bacon-Boyle-Hooke conception of experiment in Sect. 7.1.[8] Some critical remarks are provided in Sect. 7.5.[9]

7.1 Bacon, Boyle, and Hooke

According to Bacon, Boyle, and Hooke, the main function of experiments is not testing theories or enabling us to discover the laws of nature. It is providing data, factual information on the properties and behaviour of bodies in determinate circumstances.

---

[5] This text classifies several types of experience, following a schema similar to those of Hennings (1774, pp. 151–52) and Walch (1775, cols. 1083–84).

[6] Nevertheless, Kant often mentions observations and experiments in one breath. See e.g. Kant (1922, vol. 11, p. 142_{36}); *Reflexion* (henceforth *R*) 5645 (1780–88?), vol. 18, p. 288_{3}; Kant (1936–1938, vol. 21, pp. $15_{16–17}$ and $76_{28–29}$). Kant often employs the term "experiment" in a loose sense, to refer to observations, e.g. in (1798b, p. 98; 1966b, p. 611).

[7] Kant also has interesting views on the role of theoretical concepts for experimentation. I will not comment on this topic, as it is discussed in Wartenberg (1992, pp. 242–45). I will not discuss Kant's views on the role of experiments for the a priori inquiries that he develops in the first two *Critiques*. On this issue, see esp. Seigfried (1989), Gloy (1996, 2009), Sato (2008).

[8] I follow Peter Anstey (2005, unpublished).

[9] In reconstructing Kant's views, I will draw not only on the texts that Kant published, but also on his manuscript notes (the so-called *Reflexionen*) and on the notes of his lectures. These materials raise several philological problems: see Conrad (1994). In the face of those problems, when citing *Reflexionen* and lecture transcripts, I will mostly rely on statements which can be found in more than one source: several *Reflexionen* or lecture transcripts, or *Reflexionen* alongside Kant's works and letters. The following datings are assumed for the lectures on which the lecture transcripts are based. 1966c, 1966e, 1966f, 1980, and 1998b are based on lectures given from the early 1780s onwards. 1961 is based on lectures given between 1777 and 1782. 1966b and 1998a are based on lectures given in several different years, probably including pre-Critical materials. 1966a and 1966d are based on lecture given in the early 1770s.

Experiments aimed primarily to verify or falsify theories were eschewed because Bacon, Boyle, and Hooke conceived of natural philosophical inquiry as a two-stage process (Hooke 1705, p. 7). The first stage was the construction of natural histories, that is, structured collections of a large number of facts on designated topics (Boyle 1666). The second stage was the development of theories on the basis of the information collected. This second stage was to be initiated after the completion of the first stage (Boyle 1662, p. 12). Compiling natural histories was an enormous endeavor which would occupy many generations of researchers. As a consequence, theory construction was seen as a task which could only be accomplished in a distant future (Parker 1666, pp. 45–46).

To be sure, Boyle and Hooke claimed that it is useful to know the main theories which are available on a given topic in order to design experiments (Boyle 1666, p. 2; Hooke 1705, p. 19). However, they were wary of drawing general conclusions from observations and experimental results. For instance, Boyle regarded the so-called Boyle's law as a generalization, observed to be true in specific places, but he was reticent to say that it is universally valid (Boyle 1662, p. 60).[10]

Bacon, Boyle, and Hooke held that one should perform experiments in a state of mind which is as free from theoretical assumptions as possible (e.g. Hooke 1705, p. 20). Otherwise, one will easily fall prey to the prejudices (the famous Baconian *idola*: Bacon 1620, part 1, §61) which often thwart our endeavors to discover truth. Accordingly, Boyle and Hooke professed themselves adherents of experimental philosophy.[11] They counseled against the premature formulation of theories, systems, and hypotheses (e.g. Hooke 1665, sig. A4[12]). These were distinctive marks of speculative philosophy, as the adversary of experimental philosophy was called (Boyle 1662, p. 12; Hooke 1665, sig. a3, b1; Sprat 1667, p. 341; Anstey 2005; Gaukroger 2006, pp. 352–451). The most frequent examples of speculative philosophy were Aristotelian and Epicurean natural philosophy and, later, Cartesian natural philosophy (Boyle 1666, p. 2).

With their factual and experimental approach, Bacon, Boyle, and Hooke, together with the members of the Royal Society and the other adherents to the program of experimental philosophy, gave an extraordinary impulse to the natural sciences. However, their aversion to speculation and hypotheses eventually undermined the Baconian research program of constructing natural histories. In the absence of organic links between experiments and observations on the one hand, and natural philosophical theories and hypotheses on the other, the mere accumulation of facts did not lead to the substantial progress in the explanation of the newly observed phenomena that experimental philosophers were expecting. It also did not lead to the establishment of corpuscularianism which many

[10] Boyle's law states there is an inverse proportional relationship between the pressure and volume of a gas in a closed system where the temperature is constant.

[11] Bacon was posthumously recruited as the "Patriark of Experimental Philosophy" (Power 1664, p. 82).

[12] On the origin of this passage, see Birch (1756–57, vol. 1, pp. 490–91).

"new philosophers" hoped would replace the traditional Aristotelian matter theory.[13] Therefore, it is not surprising that, after the publication of Newton's Philosophiae naturalis principia mathematica in 1687, British natural philosophers were quick to embrace the new Newtonian model of natural philosophical inquiry.

Experiments and observations in the *Principia* are not aimed at the compilation of natural histories. They are aimed at the establishment of mathematical and nomological explanations of natural phenomena. By contrast, determinate nomological explanations were not central to Bacon's or Boyle's natural philosophical projects. Newton claimed that the principles and laws which are the core of natural philosophical theories are deduced or induced from the phenomena. However, Newton did not spell out in detail how this deduction or induction takes place. Moreover, Newton continued to decry hypotheses and speculation. Like Bacon, Boyle, and Hooke, Newton "did not feign hypotheses".[14] For Newton, as for Bacon and Boyle, experiments and hypotheses "were on different sides of the methodological divide" (Anstey unpublished). By contrast, according to Kant, experimental activity starts with the formulation of a hypothesis.

7.2 Experiments and Hypotheses

7.2.1 *Experiments, Hypotheses, and Preliminary Judgements*

Discussing "how we can discover the hidden qualities of natural bodies by means of experiment", Kant states:

> we must always first presuppose something here (*begin with a hypothesis*) from which to begin our course of investigation [. . .] For to venture forth blindly, trusting good luck until one stumbles over a stone and finds a piece of ore and subsequently a lode as well, is indeed bad advice for inquiry (Kant 1798a, pp. 223–24, italics added).

Unlike Bacon, Boyle, and Hooke, Kant holds that we neither can, nor should perform experiments in a theoretical void. Starting with a hypothesis is not only good experimental practice. It is also what every experimenter, more or less consciously, actually does: "[e]very man who makes experiments first makes hypotheses, in that he believes that this or that experiment will have these consequences" (Kant 1966f, p. 889).[15]

[13] Boyle's endorsement of corpuscularianism is an example of indulgence in the sort of speculative hypotheses that experimental philosophers officially eschewed. Also, it should be granted that Bacon, Boyle, and Hooke regarded some experiments as tests for theories and hypotheses. My claim is only that they did not regard hypothesis-testing as the main function of experiments.

[14] See Newton (1687, vol. 2, p. 764; 1714/15, pp. 222–24).

[15] See Kant (1998b, p. $377_{617/18}$).

Hypotheses are not only important for experiments. They are the starting point of every activity which requires one to make judgements on the basis of uncertain grounds. "Thus a doctor makes hypotheses when he cures the sick[;] he has to subsume everything under hypotheses, and see whether the consequences that he now has before his eyes follow therefrom" (Kant 1966a, p. 220). A businessman makes a hypothesis when he "finds a ground that is sufficient for undertaking something" (Kant 1966c, pp. 750–51). A general facing the enemy "must necessarily judge and decide something" on the basis of hypotheses (Kant 1966c, pp. 750–51).

Bacon or Boyle could object that hypotheses may derive from prejudices and therefore they must be eschewed. Kant would agree that hypotheses derive from prejudices. However, he would add that, in line of principle, there is nothing wrong in having pre-judices, understood in the etymological sense of preliminary judgements. On the contrary, preliminary judgements are necessary for invention and discovery:

> There has never been an inventor in the world, and there has never been anyone who invented something, who did not at the same time make a preliminary judgement concerning his invention and the invented thing. He was not certain of the thing, but *the judgement cleared the path for him to try, and to experiment* (Kant 1966a, p. 162, trans. modified and italics added).[16]

The hypotheses to be tested by means of experiments are "half-judgements" (Kant 1966f, p. 862), suppositions, and tentative assumptions about the properties and existence of objects and the laws and forces to which they are subjected.

Why should we believe that our inquiries always start with hypotheses and preliminary judgements? Kant's texts contain two sets of considerations in support of this claim. First, Kant formulates pragmatic considerations, which are independent from his transcendental philosophy. Kant sketches several brief descriptions of the activities of doctors, judges, miners, inventors, and experimenters. These descriptions are aimed to yield plausibility to the view that inquiries in all these fields are based on hypotheses and preliminary judgements.

Second, Kant's conception of the mind places a great emphasis on the active role of the human subject in shaping our experience of the world (Kant 1781/1787, A126). In Kant's view, we continuously subsume the objects that we come by under concepts, such as those of substance and causality (B161, B164). We do this by formulating judgements about those objects (B143), even though we are normally unaware of this mental activity. If one endorses this view of the mind, it will be much easier to admit that we constantly frame hypotheses and preliminary judgements than it would be for those philosophers who deny the existence of unconscious mental contents. According to Kant, the preliminary judgements and

[16] See Kant (1798a, pp. 223_{14-20} and 405_{8-11}). To be sure, Kant distinguishes between prejudices in the proper sense (*Vorurteile*) and preliminary or provisional judgements (*vorläufige Urteile*): see La Rocca (2003).

hypotheses which are at the basis of our experiments are part of the continuous flow of our spontaneous activity of judgement (Kant 1961, p. $243_{2–33,36}$; 1998b, $359_{189–90}$).

7.2.2 *Hypotheses and Induction*

Kant's concise explanations of the notion of hypothesis employ several technical terms. It is helpful to unpack them in three steps:

1. Hypotheses are judgements that we regard as true: they are "a holding to be true" [*Fürwahrhalten*].
2. We hold them to be true because they explain the reasons for given phenomena. The metaphysics treatises of Kant's eighteenth-century German predecessors provided detailed treatments of the notion of reason in connection with the principle of sufficient reason (e.g. Crusius 1747, §§139–54). Adopting their terminology, Kant states that hypotheses illustrate the *Gründe*, that is, the causes or reasons of given phenomena. Having in mind the distinction between sufficient and insufficient reasons (e.g. Crusius 1747, §143), Kant states that the reason expressed by a hypothesis must be sufficient to explain why the phenomena take place.
3. Hypotheses describe presuppositions [*Voraussetzungen*] (Kant 1966c, p. 746_5; 1998a, p. 146_{14}; 1800a, p. 84_8). They describe what must be preliminarily [*voraus*] posited [*gesetzt*], that is, exist or take place, for certain phenomena to take place.

The *Jäsche Logic* expresses all this as follows:

> *A hypothesis is a holding-to-be-true of the judgement of the truth of a ground for the sake of its sufficiency for given consequences*, or more briefly, *the holding-to-be-true of a presupposition as a ground.*
>
> All holding-to-be-true in hypotheses is thus grounded on the fact that the presupposition, as ground, is sufficient to explain other cognitions as consequences (1800a, pp. 84–85).[17]

For example, "I suppose that the earth has cavities, because on the basis of this [supposition] it is possible to explain how valleys arose; then I suppose the ground, because on the basis of it one can explain [its] consequences. This is a hypothesis" (Kant 1966d, p. 440).

The example makes clear that, when we formulate hypotheses, we follow a pattern of reasoning which leads us from certain events (the consequences) to their ground (*R* 2687 [1776–89?], vol. 16, p. 471; Kant 1966c, p. $746_{5–7}$). On the face of

[17] See e.g. *R* 2678 (1764–75?), vol. 16, p. 465; *R* 2690 (ca 1780–89), vol. 16, p. 471; *R* 2694 (1790–1804), vol. 16, p. 472.

it, this pattern of reasoning would seem to be an abduction or an inference to the best explanation. Instead, Kant's discussions of hypotheses mention induction. Hypotheses "are not taken to be true apodictically, but *per inductionem*" (Kant 1966e, p. 558).[18] "Induction" is a technical term for Kant. It designates the ascription of a property to all members of a class, on the ground that it belongs to some of them.[19]

Since Kant claims that hypotheses are the result of inductions, and the conclusions of inductions are universal judgements, Kant must hold that hypotheses are universal judgements. These are judgements of the form "all *S* are *P*". However, this claim is implausible. Kant himself makes many hypotheses on the constitution of individual objects or on the causes of just one event: for instance, the hypothesis that there is heated matter at the centre of the Earth (Kant 1800b, pp. 259–60; 1966f, p. 887_{7-8}; 1998b, p. 377). In what follows, I will ignore this difficulty and presuppose Kant's claim that hypotheses are universal judgements derived by induction.

7.2.3 Hypotheses, Certainty, and Probability

The claim that hypotheses are the conclusions of inductive inferences is the basis for Kant's further claim that hypotheses cannot be certain.[20] They do not enjoy the high epistemic status that early modern philosophers typically required for a statement to be part of the body of proper science, or *scientia*.

In Kant's vocabulary, to be certain of a statement means to know that it is necessarily true.[21] For Kant, we can never know that a hypothesis is necessarily true. For universal statements, necessary truth goes hand in hand with "true or strict" generality, "i.e., in such a way that no exception at all is allowed to be possible" (Kant 1781/1787, B3–4). However, we cannot know whether there is any exception to the state of affairs described by a hypothesis.

This is because hypotheses are universal statements established by induction on the basis of information gathered in the course of experience. However,

[18] See an addition (1770 or later) to *R* 2130, vol. 16, p. 246; 1800a, vol. 9, p. 85_{15-16}.

[19] See Kant (1966e, p. 594; 1998b, pp. 476–77; 1800a, pp. 132–33). The inductive pattern of reasoning which leads to the formulation of hypotheses is described in Kant (1781/1787, A647/B675).

[20] E.g. *R* 2681 (1776–89?), vol. 16, p. 469_{10-11}; *R* 2687 (1776–89?), vol. 16, p. 471. More precisely, hypotheses cannot be apodictically certain. On the distinction between apodictic and assertoric certainty, see Capozzi (2001, pp. 572–76). By "certainty" I will mean apodictic certainty in what follows.

[21] See Kant (1966e, p. 517): certainty "is awareness of the necessity of truth". See also Kant (1966e, p. 530; 1800a, p. 66).

> Experience never gives its judgements true or strict but only assumed and comparative *universality* (*through induction*), so properly it must be said: as far as we have perceived, there is no exception to this or that rule (Kant 1781/1787, B3, italics added; see A24, A90/B124).

Experience cannot guarantee that a statement is necessarily true either, because necessity, together with strict universality, is a "secure" indication "of an *a priori* cognition" (B4; see A823/B851).

Kant's contemporary Johann Georg Daries, among others, held that some inductions are the source of strictly universal judgements, of whose truth we can be certain. They are the inductions that are based on the enumeration of all members of a class (Daries 1776, §135). However, according to Kant, experience never gives us any guarantee that our enumerations are complete. For every given genus about which we make a generalization, there could be some members that we have not yet experienced and that are counterexamples to our generalization. The inductive employment of reason for devising hypotheses

> is not properly *constitutive*, that is, not such that if one judges in all strictness the truth of the universal rule assumed as a hypothesis thereby follows; for how is one to know all possible consequences, which would prove the universality of the assumed principle if they followed from it? (Kant 1781/1787, A647/B675)

Therefore, "[h]ypotheses always remain hypotheses, that is, presuppositions, whose complete certainty we can never attain" (Kant 1800a, p. 85).[22] They can only be more or less probable.

7.2.4 *The Three Requirements for a Good Hypothesis*

Not every hypothesis should be regarded as a possible explanation of natural phenomena. Kant details three requirements that every hypothesis must satisfy. In the first place, we must be sure that whatever state of affairs the hypothesis adduces as an explanation for given phenomena can actually take place (Kant 1781/1787, A770/B798). To this end, the hypothesis must be consistent with the body of our knowledge. "If, for example, to explain earthquakes and volcanoes we assume a subterranean fire, then such a fire must be possible, if not as a flaming body, yet as a hot one" (Kant 1800a, p. 84; see 1998b, p. 377). Hence, we should not make the hypothesis that there are flames at the centre of the earth, because combustion would be impossible due to the lack of air (Kant 1800b, pp. 259–60). However, we can make the hypothesis that the centre of the earth is composed of heated matter, because a body can be heated in absence of air.[23]

[22] Based on Kant (1966e, p. 558); see *R* 2681 (1776–89?), vol. 16, p. 469.

[23] Kant emphasizes that hypotheses must be consistent especially with that particular body of knowledge which is constituted by the "conditions of possible experience" (1781/1787, A771/B799), outlined in the first *Critique*.

In the second place, we must make sure that the events to be explained really follow from the assumed hypothesis. Otherwise, the hypothesis is "a mere chimera" (Kant 1800a, p. 85; see 1966e, p. $559_{6–7}$; 1998b, p. $378_{635–37}$).[24]

In the third place, a hypothesis must be sufficient to explain a whole set of phenomena, without the need to integrate it with further hypotheses.

> Thus *Tycho* Brahe's hypothesis, for example, did not suffice for the explanation of many appearances; hence he assumed several new hypotheses to complete it. Now here it is to be surmised that the assumed hypothesis cannot be the real ground. The Copernican system, on the other hand, is a hypothesis from which everything can be explained that ought to be explained therefrom, *so far as it has yet occurred to us*. Here we do not need any *subsidiary hypotheses* (*hypotheses subsidiarias*) (Kant 1800a, pp. 85–86; see 1781/1787, A774/B802; 1966e, p. 559).

The prohibition to use subsidiary hypotheses prevents natural scientists from framing a whole system out of hypotheses which integrate each other to form comprehensive explanations, but are not systematically related to experiments and observations. An example of such a system is Descartes' vortex theory, which for Locke and Newton was a paradigm example of an unfounded speculative hypothesis (Anstey 2005, pp. 229–31). It is occasionally targeted in the Kantian corpus as well (Kant 1966a, p. 222; 1763, p. 144).[25]

Compliance with the three criteria ensures that we do not assume "mere chimeras" (Kant 1966f, p. 888_{23}), "empty figments of the brain" (1790, p. 466_{18}), "empty fictions" (1966c, p. 746_{19}), "romances of reason" (1966a, p. 220_{33}) or "daring adventure[s] of reason" (1790, p. 419 n.) as hypotheses. These are typical expressions of the anti-hypothetical rhetoric to be found in the writings by Boyle, Hooke, or Locke (e.g. Locke 1976–89, vol. 4, p. 628). They employed those expressions in their wholesale rejection of any hypothesis from the current stage of natural philosophy.

Kant, like his German contemporaries (e.g. Erxleben 1772, §9), employs the same expressions of seventeenth century British philosophers. However, unlike his British predecessors, Kant employs those expressions within a framework which is not hostile to hypotheses *per se*. Kant states that hypotheses, like castles in the air, are fictions, but not all fictions must be rejected. The power of imagination, kept "under the strict oversight of reason" (Kant 1781/1787, A770/B798), can give rise to useful "heuristic fictions" (1966a, p. 262_{28}). The three criteria laid out above discriminate castles in the air from heuristic fictions. The task of experiments is determining which heuristic fictions portray actual states of affairs, rather than mere possibilities.

---

[24] The first *Critique* adds that we should be able to determine the consequences of a given hypothesis a priori (Kant 1781/1787, A774/B802). I take this to mean that, when we explain how certain events follow from the assumed hypothesis, we should show that they follow from it on the basis of a priori principles, such as the causal law and the law of action and reaction (see Butts 1961, p. 165). Empirical laws, besides a priori principles, might be involved in the explanation.

[25] A discussion of Kant's views on hypotheses which is rather different from the reconstruction provided here can be found in Butts (1962).

7.3 Experiments and the Laws of Nature

Let us assume that we have a hypothesis which satisfies the three criteria and is confirmed by experiments. What use shall we make of it? We saw in the introduction that, according to Kant, experiments help us discover the laws of nature. Laws of nature are expressed by necessary statements. Kant regards this as an analytic truth: to say that something is a law is to say that it is necessary.[26] Accordingly, in order to know that a statement expresses a law of nature, we must know that it is necessarily true. However, experiments cannot confer certainty to a hypothesis, because experience cannot establish that a statement is necessarily true.[27] Therefore, just by means of experiments, we cannot prove that a hypothesis is a law of nature (Kant 1783, p. 294).

Does this mean that experiments will never enable us to achieve the purpose they are meant to serve, that is, discovering the laws of nature? Kant does not draw such a pessimistic conclusion. He holds that testing hypotheses by means of experiments is necessary, but not sufficient, to discover the laws of nature. Experimental confirmation of a hypothesis is only the first step toward the discovery of a law of nature.

To illustrate this point, let us consider Copernicus' heliocentric theory of planets. Copernicus' "first thought" was a preliminary judgement: "the observer revolve [s]," whereas "the stars [...] [are] at rest" (Kant 1781/1787, Bxvi). "Copernicus assumed" this thought "only as a hypothesis" (Bxxii n.) This hypothesis proved to be superior to the Ptolemaic hypothesis and to Tycho Brahe's hypothesis. Kant holds that, unlike the Tychonian hypothesis, the Copernican hypothesis does not require supplementary hypotheses (Kant 1966e, p. 559; 1800a, pp. 85–86). Unlike the Ptolemaic hypothesis, the Copernican hypothesis explains all the phenomena. This confers a high degree of probability to the Copernican hypothesis (Kant 1966f, p. 887), but it does not make it certain. What made it certain is the fact that it was subsumed under a body of laws:

> The central laws of the motion of the heavenly bodies established with certainty what *Copernicus* assumed at the beginning only as a hypothesis, and at the same time they proved the invisible force (of *Newtonian* attraction) that binds the universe [...] (Kant 1781/1787, Bxxii n.)

The same applies to the hypotheses that are confirmed by experiments. Experiments make them probable. The integration with a body of a priori laws and principles makes them certain. Thus

> Reason, in order to be taught by nature, must approach nature with its principles in one hand, *according to which alone the agreement among appearances can count as laws*, and, in the other hand, the experiments thought out in accordance with these principles [...] (Kant 1781/1787, Bxiii, italics added).

---

[26] Kant associates the term "law" to necessity (e.g. 1781/1787, A126) and universality (e.g. 1783, p. 310). Kant (1781/1787, A126) qualifies laws as objective rules and other passages (e.g. 1783, p. 301) relate objectivity to necessity and universality.

[27] See Sect. 7.2.3.

What principles must we employ to transform appearances into laws and how can we carry out this process? To answer this question, we must enter the complex territory of Kant's theory of the laws of nature. Two caveats are in place here. First, my outline will be highly selective. I will reconstruct Kant's theory only to the extent which is necessary to understand how experiments enable us to discover new laws of nature. Second, my comments apply more neatly to Kant's view of physical laws than to his view of biological laws. I privilege physical laws because most of Kant's references to specific experiments relate to the field of physics, and not to biology.[28]

There are laws of nature at three levels. At the first and highest level, we encounter the principles of pure reason. We can call them transcendental principles. They are eight highly general statements which describe features of our perception and features of natural objects and phenomena. For instance, a transcendental principle which describes features of our perception is: every sensation has a variable degree of intensity (Kant 1781/1787, B207). A principle which describes features of natural phenomena is the causal principle: "[a]ll alterations occur in accordance with the law of the connection of cause and effect" (B232). Among the transcendental principles, those which apply to natural objects and phenomena extend to physical phenomena and material objects, as well as to psychological phenomena (Kant 1783, p. 295). For example, the causal principle does not only apply to the alterations of material objects. It also applies to mental states. Every change in our mental states is determined by a cause.

According to Kant, the transcendental principles can be proven to be true a priori, at least given Kant's peculiar notion of a priori (Kitcher 2000, p. 17). This means that, although the proofs of the transcendental principles depend on premises concerning our cognitive capacities, they do not rely on any premise concerning *particular* experiences or features of objects that we can only know through experience. By contrast, a principle will be a posteriori in Kant's sense if its proof relies on assumptions concerning particular experiences or features of objects that we can only know through experience.

Kant prefers to call the transcendental principles principles, rather than laws, in order to stress that they lie at the basis of the system of our knowledge. To the extent that they concern natural objects, the transcendental principles are the most basic laws of nature (Kant 1783, p. 319). "[T]hey are not themselves grounded in higher and more general cognitions" (Kant 1781/1787, A148/B188).

At the second level, there are the metaphysical principles of natural science. Among these are Newton's laws of inertia and of action and reaction. Unlike the transcendental principles, the metaphysical principles do not apply to psychological phenomena, but only to material bodies. Like the transcendental principles, the metaphysical principles are known a priori.

[28] Kant also mentions several chemical experiments, but he denies that the empirical generalizations of chemistry can achieve the status of laws (1786, p. 468).

This does not mean that experience or experiments are irrelevant to the metaphysical principles. First, we must have experience of the external world in order to acquire the concept of material body, without which we cannot even formulate the metaphysical principles. Second, experience might be necessary in order to discover them. Kant could well agree that we cannot discover the law of action and reaction without making appropriate experiences, such as the pendulum experiments that Newton details in the *Principia*.[29] However, those experiments are not sufficient for us to *prove* the metaphysical principles. They cannot be, to use Locke's famous expression, "Principles that Matter of Fact justifie" (Locke 1693, p. 248). This is because, being laws, they apply necessarily to every object that exists in space and time. Yet experience, experiments, and matters of fact cannot confer necessity to any judgement. Even if we discover the metaphysical principles empirically, we must prove them a priori, as Kant attempts to do in the *Metaphysical Foundations of Natural Science*. If no a priori proof were available, we would have to deny that they are laws in the proper sense of the term.

At the third level, we find the empirical laws of nature. They apply to specific material objects and physical phenomena. We discover the empirical laws of nature a posteriori, on the basis of observations and experiments (Kant 1781/1787, B263). We can prove them only by making reference to observations or experiments. Therefore, the empirical laws of nature, unlike the transcendental and metaphysical principles, are established only a posteriori.

Empirical laws of nature raise the same difficulty that we encountered earlier with reference to experiments. On the one hand, empirical laws "can only be known [. . .] empirically" and therefore, at least from the point of view of our understanding, "are contingent" (Kant 1790, p. 184, see pp. 179–80). On the other hand, empirical laws, being laws, must contain an element of necessity. Our understanding must think of them "as laws (i.e., as necessary)" (1790, 184). How can we confer necessity to laws that we can know "only empirically"? Since necessity cannot be established a posteriori, it must be established a priori. In Kant's view, we have reason to regard empirical rules as necessary laws if we can relate them to the transcendental and metaphysical principles (Friedman 1992, p. 174). In order to confer necessity to an empirical rule, we must show that it follows from the a priori principles, in addition to empirical premises.[30] Let us consider three examples to see how this process unfolds.

---

[29] E.g. Bk. 3, Prop. 6.

[30] This combination of [*a*] empirical and [*b*] a priori premises is reflected in Kant's statement that "the empirical laws can only [. . .] be found [*a*] by means of experience, and indeed [*b*] in accord with its original laws [a priori principles], in accordance with which experience itself first becomes possible" (1781/1787, A216/B263). Empirical laws can be inferred from the transcendental principles, but not "completely" (B165). "Experience must be added in order to come to know particular laws *at all*" (B165).

1. We have determined certain empirical regularities concerning the melting of wax and we want to formulate an empirical law of nature. We could then prove that the statements describing those regularities follow from a priori principles, such as the causal law and the law of action and reaction, together with empirical statements describing properties of wax.
2. We have formulated statements describing how a billiard ball moves when it is hit by other balls. We could then derive those statements from the causal principle, metaphysical principles, and information on the weight and shape of billiard balls.
3. Faced with Kepler's laws of planetary motion, we could prove that they follow from metaphysical principles such as the law of inertia and the law of action and reaction (Friedman 1992, pp. 175–80).

In each of these cases, we start from empirical regularities which we have established a posteriori. Then we attempt to subsume them under a priori principles. If we succeed in this attempt, we will have reason to regard the statements describing those regularities as "something radically new" (Friedman 1992, p. 178), that is, as descriptions of necessary features of reality.[31]

The process which leads from a hypothesis to an empirical law of nature is represented in Fig. 7.1. The point of departure for this process is a hypothesis which is derived by induction. To begin with, we should assess whether the hypothesis satisfies the three requirements discussed in Sect. 7.2.4. If it does not, we must abandon it. Otherwise, we must approach it holding, in Kant's words, experiments in one hand and principles in the other (Kant 1781/1787, Bxiii). First, we must perform experiments or carry out observations to confirm or disprove the hypothesis. Then, if the hypothesis is confirmed, we must attempt to relate it to the principles which are the basis of natural science. If we succeed in doing so, then the hypothesis could be an empirical law of nature.

I stated that it *could* be a law of nature, and not that it will certainly be a law of nature, because other operations may be involved in our search for natural laws. For instance, we may have to choose between two incompatible hypotheses, both of which are confirmed by the observational evidence and can be integrated with a priori principles. Alternatively, we may have to submit our experiments, hypotheses, and explanations to the scientific community. We may be entitled to claim that a hypothesis is a law of nature only when some degree of intersubjective agreement is reached. This would conform to Kant's repeated claim that people's agreement, and especially the agreement of the learned community, with our opinions provides a reason to hold them true (Kant 1781/1787, A820–21/B848–49; 1783, p. 298).

[31] Typically, empirical laws outline necessary features of specific natural kinds: see Kreines (2008).

Fig. 7.1 Process from hypotheses to empirical laws

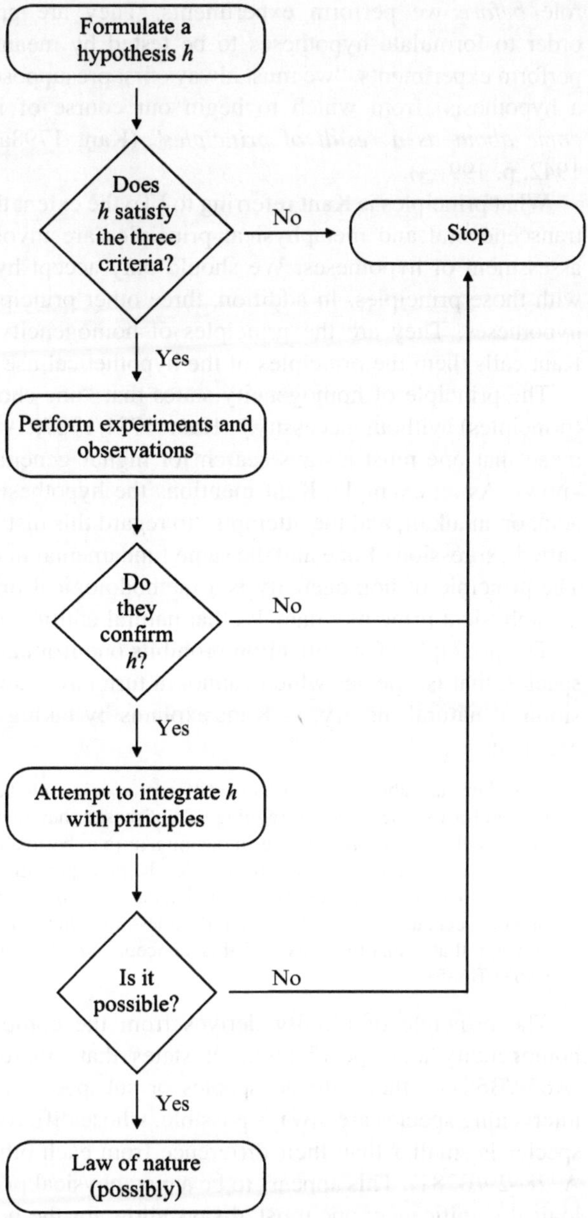

7.4 Experiments and Heuristic Principles

In the previous section we have seen that, for Kant, principles play a role *after* we perform experiments. They enable us to convert experimentally confirmed hypotheses into laws of nature. However, there are also principles which play a role *before* we perform experiments. They are principles that we follow in order to formulate hypotheses to be tested by means of experiments. When we perform experiments, "we must always first presuppose something here (begin with a hypothesis) from which to begin our course of investigation, *and this must come about as a result of principles*" (Kant 1798a, p. 223, italics added; see 1942, p. $199_{1\text{–}3}$).

What principles is Kant referring to? To the extent that we are aware of them, the transcendental and metaphysical principles are involved in the formulation and assessment of hypotheses. We should only accept hypotheses that are consistent with those principles. In addition, three other principles guide the formulation of hypotheses. They are the principles of homogeneity, specification, and affinity. Kant calls them the principles of the hypothetical use of reason.

The principle of homogeneity states that "one should not multiply beginnings (principles) without necessity" (Kant 1781/1787, A652/B680). Kant takes it to mean that one must always search for higher genera for all the species that one knows. As an example, Kant mentions the hypothesis that every salt is either an acid, or an alkali, and the attempts "to regard this distinction as merely a variety or varied expression of one and the same fundamental material" (A652–53/B680–81). The principle of homogeneity is a methodological principle which presupposes a metaphysical principle: namely, that natural entities belong to common kinds.

The principle of specification prohibits one from assuming that there are lowest species, that is, species which cannot in turn have sub-species. This is a presupposition of natural inquiry, as Kant explains by taking different types of soil as an example:

> That there are absorbent earths of different species (chalky earths and muriatic earths) needed for its discovery a foregoing rule of reason that made it a task for the understanding to seek for varieties, by presupposing nature to be so abundant that it presumes them. For we have an understanding only under the presupposition of varieties in nature, just as we have one only under the condition that nature's objects have in themselves a sameness of kind, because it is just the manifoldness of what can be grasped together under a concept that constitutes use of this concept and the business of the understanding. (A657/B685)

The principle of affinity derives from the combination of the principles of homogeneity and specification. It states that "there is a continuum of forms" (A659/B687): "there are no species or subspecies that are proximate [...], but intervening species are always possible, whose difference from the first and second species is smaller than their difference from each other" (A659–60/B687–88; see A228–29/B281). This appears to be a metaphysical principle, but it has a methodological significance: one must always allow for the possibility that there are intermediate species between the species that one already knows. In this instance, as was the case for the two previous principles, we have a methodological prescription (always look for intermediate species), which depends on a metaphysical presupposition (there is a continuum of forms).

The principles of the hypothetical use of reason, like the metaphysical principles of natural science, are not inductive generalizations that we form on the basis of experience. We do not derive them from the discovery that known genera have lower species, that lower species belong to higher genera, and that there are intermediate species between any two known species. In Kant's view, we would not find higher genera, lower species, and intermediate species in the first place, unless we previously assumed the principles of the hypothetical use of reason as guides for the formulation of hypotheses (A660/B688). For instance, we perform experiments to test the hypothesis that several phenomena obey the same law because we assume that there is a higher genus for every given set of phenomena.[32]

We do not have to be fully conscious of this assumption in order to formulate hypotheses and to test them with experiments. We often exercise our mental powers according to rules of which we are unaware. "The exercise of our powers [. . .] takes place according to certain rules that we follow, *unconscious* of them as first, until we gradually arrive at cognition of them through experiments and lengthy use of our powers" (Kant 1800a, p. 11; see 1966e, p. 502; 1966f, p. 790). This applies, for instance, to the rules of grammar, which we discover long time after we started following them. The same applies to the principles of the hypothetical use of reason.

Although we often follow these principles unconsciously, Kant does not think that we always follow them (Kant 1786, p. 472). They "do not say what happens, i.e., in accordance with which rule our powers of cognition actually perform their role and how things are judged, but rather how they ought to be judged" (Kant 1790, p. 182). They are rules or maxims that must guide the formulation of hypotheses (Kant 1781/1787, A666/B694). Like hypotheses, the principles of the hypothetical use of reason are "heuristic fictions" (A661/B689) which precede and guide experimental activity.

To be sure, the principles of the hypothetical use of reason are not the sole presuppositions of experimental activity. Kant argues for the existence of other regulative principles that direct empirical research. They are: the assumption that we cannot have sensory perceptions of any absolute or insurmountable temporal or spatial limits (A508–9/B536–37); the assumption that nature is organized as if it were designed by an intelligent being (A826/B854); and the assumption that living beings are constituted as if they obeyed final causes (Kant 1790, p. 387).[33] Moreover, the principles of the hypothetical use of reason are not *only* presuppositions of experimental activity. They also underlie the systematic organization of cognitions. We employ those principles when we formulate a hypothesis which explains

[32] Kant argues that, in order to discover natural laws, we must assume that nature is ordered into genera and species, in conformity with the principles of the hypothetical use of reason (1790, p. 185). To show this, Kant explains how the principle of affinity was at work in the reasonings that lead to Newton's discovery of universal gravitation (Kant 1781/1787, A662–63/B690–91).

[33] See Guyer (1990).

several phenomena, even if we are unable to perform experiments or observations to test it. If the hypothesis provides the best explanation of the phenomena, we should accept it despite the absence of experimental or observational confirmation. However, if experiments confirm the hypothesis, we have "a powerful reason to take as well grounded the unity that is hypothetically thought-out" (Kant 1781/1787, A661/B689).

7.5 Conclusion

According to Bacon, Boyle, and Hooke, experiments serve mainly to collect data in view of the future construction of natural philosophical theories. Experiments were not typically assigned the function of testing theories and hypotheses. On the contrary, the experimenter's prior natural philosophical beliefs and persuasions were looked upon with suspicion, as potential sources of prejudices which could contaminate his experimental activity. On the whole, data collection and experimentation were seen as theory-free activities.

By contrast, according to Kant, experiments cannot serve to build a base of data which are independent from theories. This is because experiments are always conceived of and carried out in the light of our assumptions, expectations, and heuristic principles. These assumptions and principles depend in part on the nature of the human mind and in part on the experimenter's convictions and purposes. They give rise to preliminary judgements and hypotheses which guide us in the design and performance of experiments.

Kant's emphasis on the importance of preliminary judgements and hypotheses for experiments goes hand in hand with his denial of the Baconian view that prejudices always play a negative function. According to Kant, prejudices (or more precisely, preliminary judgements) are indispensable for many human activities, including experimentation. However, it is necessary to test and assess them in order to either reject them as false, or else to transform them from mere opinions to certain truths.

On the whole, compared with Bacon, Boyle, and Hooke, Kant has elaborate views on the one hand, on how our theoretical and pre-theoretical assumptions bear on experimental practice, and on the other hand, on how the results of experimental activity can be integrated within the body of our theories in order to advance our knowledge of nature. However, Kant seems to have overstated the dependence of experiments on theories.

First, *pace* Kant, some experiments are performed in absence of a clear theoretical framework. Their aim is not testing hypotheses, but exploring new areas of inquiry or circumscribing new phenomena (Steinle 1997). Kant must have read the discussion of these *exploratory* experiments in a work that he knew well, Johann Heinrich Lambert's *New Organon* (1764, vol. 1, p. 355). However, Kant never discusses exploratory experiments.

Second, as a consequence of his disregard for exploratory experiments, Kant overlooks the benefits of the creative interplay of experiments designed to test theories and hypotheses with experiments which have a life of their own. Bacon, Boyle, and Hooke also overlooked the benefits of that interplay, but for the opposite reason: namely, because they focused too much on experiments having a life of their own.

Third, some experiments test hypotheses which contrast with our currently accepted theories. They lead us to replace them with new theories, incompatible with the previous ones. Kant does not provide any account of how these *revolutionary* experiments lead to theory change or revision.[34] On the contrary, he requires that our hypotheses are coherent with the body of our previous knowledge[35] and that experimental results are integrated within a given system, based on the a priori foundations of natural science.[36]

In response to this criticism, Kant could emphasize that exploratory and revolutionary experiments are never wholly independent from our theoretical and pre-theoretical beliefs and hypotheses. However, he should have acknowledged that they enjoy a certain degree of freedom from our theoretical assumptions. This makes them more than handmaids to theory, hypothesis-testing procedures, or preliminaries to the addition of new laws of nature to a static, ever-growing body of natural philosophical knowledge, firmly resting on unshakable Newtonian foundations.

As is well known, developments in science have shaken those foundations and made space for more dynamic conceptions of scientific progress. These conceptions make it easier than it would have been for Kant to accommodate the interaction between theory-testing experiments and exploratory experiments and to explain the roles of experiments for theory change and revision. Arguably, the evolution of Kant's philosophical views points in this direction. The completeness and coherence of the system of natural science becomes more and more a regulative ideal when Kant moves from the Newtonian focus of the *Metaphysical Foundations of Natural Science* to the thorny status of teleological biological explanations in the *Critique of the Power of Judgement* and the reconsideration of chemistry in the *Opus Postumum*. Yet despite these developments, Kant never retracted the claim that he uncovered and enumerated the definitive a priori foundations of natural science.

At any rate, in order to acknowledge the roles of experiments for exploring new territories and establishing new theories, it was necessary to highlight the existence of fecund relations between experiments and theories. Highlighting these relations is a significant contribution of Kant's philosophy of experiment, especially if compared with the views of his British predecessors.[37]

---

[34] On the forms of scientific progress that are compatible with Kant's views, see Malzkorn (2000).

[35] See Sect. 7.2.4.

[36] See Sect. 7.3.

[37] In acknowledging these relations, Kant could rely on the reflections of his German predecessors, from Wolff to Lambert. Reconstructing them and the extent to which they anticipate Kant's reflections is a task that I hope to take up on another occasion. For valuable comments on previous versions of this paper, I would like to thank Peter Anstey, Juan Manuel Gomez, Alan Musgrave, and Kirsten Walsh. The paper benefited from very helpful discussions of audiences at Budapest, Padua, and Sydney.

References

Adickes, E. 1924. *Kant als Naturforscher*, Vols. 2. Berlin: de Gruyter.

Anstey, P.R. 2005. Experimental versus speculative natural philosophy. In *The science of nature in the seventeenth century: Patterns of change in early modern natural philosophy*, ed. P.R. Anstey and J.A. Schuster, 215–242. Dordrecht: Springer.

Anstey, P.R. Unpublished. The Bacon-Boyle-Hooke view of experiment. Unpublished manuscript.

Bacon, F. 1620. Novum organon. In *The Oxford Francis Bacon*, Vol. 11, ed. G. Rees and M. Wakely (2004), 48–448. Oxford: Oxford University Press.

Birch, T. 1756–57. *The history of the Royal Society of London for improving of natural knowledge, from its first rise*, 4 Vols. London: Millar.

Boyle, R. 1662. Defence against Linus. In *The works of Robert Boyle*, ed. M. Hunter and E.B. Davis (1999), Vol. 3, 3–107. London: Pickering and Chatto.

Boyle, R. 1666. The text of Robert Boyle's 'Designe about Natural History'. In *Robert Boyle occasional papers*, no. 3, ed. M. Hunter and P.R. Anstey (2008). Retrieved 22 November 2011, from http://www.bbk.ac.uk/boyle/researchers/works/Occasional_Papers/occasional_paper_3.pdf.

Butts, R.E. 1961. Hypothesis and explanation in Kant's philosophy of science. *Archiv für Geschichte der Philosophie* 43: 153–170.

Butts, R.E. 1962. Kant on hypotheses in the 'Doctrine of Method' and in the *Logik*. *Archiv für Geschichte der Philosophie* 44: 185–203.

Capozzi, M. 2001. *Kant e la logica*, Vol. 1. Naples: Bibliopolis.

Conrad, E. 1994. *Kants Logikvorlesungen als neuer Schlüssel zur Architektonik der Kritik der reinen Vernunft: Die Ausarbeitung der Gliederungsentwürfe in den Logikvorlesungen als Auseinandersetzung mit der Tradition*. Stuttgart: Frommann-Holzboog.

Crusius, C.A. 1747. *Weg zur Gewißheit und Zuverlässigkeit der menschlichen Erkenntniss*. Leipzig: Gleditsch.

D'Alembert, J. 1751. Discours Préliminaire. In *Encyclopédie, ou Dictionnaire raisonné des sciences, des arts et des métiers*, Vol. 1, ed. D. Diderot and J. d'Alembert (1751–77), i–xlv. Paris: Briasson.

Daries, J.G. 1776. *Weg zur Wahrheit*. Frankfurt a.O: Strauß.

Erxleben, J.C.P. 1772. *Anfangsgründe der Naturlehre*. Dieterich: Göttingen.

Friedman, M. 1992. Causal laws and the foundations of natural science. In *The Cambridge companion to Kant*, ed. P. Guyer, 161–199. Cambridge: Cambridge University Press.

Gaukroger, S. 2006. *The emergence of a scientific culture: Science and the shaping of modernity, 1210–1685*. Oxford: Oxford University Press.

Gloy, K. 1996. Kants Philosophie und das Experiment. In *Kant in der Diskussion der Moderne*, ed. G. Schönrich and Y. Kato, 64–91. Frankfurt a.M: Suhrkamp.

Gloy, K. 2009. Die Bedeutung des Experiments bei Kant für die neuzeutliche Naturwissenschaft. In *Kants Philosophie der Natur: Ihre Entwicklung im Opus postumum und ihre Wirkung*, ed. E.-O. Onnasch, 189–201. Berlin: de Gruyter.

Guyer, P. 1990. Reason and reflective judgment: Kant on the significance of systematicity. *Noûs* 24: 17–43.

Hennings, J.C. 1774. *Kritisch historisches Lehrbuch der theoretischen Philosophie*. Leipzig: Weygand.

Hooke, R. 1665. *Micrographia: Some physiological descriptions of minute bodies made by magnifying glasses with observations and inquiries thereupon*. London: Jo. Martyn and Ja. Allestry.

Hooke, R. 1705. A general scheme, or idea of the present state of natural philosophy. In *The posthumous works of Robert Hooke*, ed. R. Waller, 1–70. London: Sam. Smith and Benj Walford.

Hume, D. 1740. An abstract of a book lately published; entitled, a treatise of human nature. In *A treatise of human nature*, ed. D.F. Norton and M.J. Norton (2000). Oxford: Oxford University Press.

Kant, I. 1747. Gedanken von der wahren Schätzung der lebendigen Kräfte. In *Kant* 1900–, vol. 1, 1–181.

Kant, I. 1763. Der einzig mögliche Beweisgrund zu einer Demonstration des Daseins Gottes. In *Kant* 1900–, vol. 2, 63–163.

Kant, I. 1781/1787. *Kritik der reinen Vernunft*, ed. J. Timmermann (1996). Hamburg: Meiner.

Kant, I. 1783. Prolegomena zu einer jeden künftigen Metaphysik, die als Wissenschaft wird auftreten können. In *Kant* 1900–, vol. 4, 253–383.

Kant, I. 1786. Metaphysische Anfangsgründe der Naturwissenschaft. In *Kant* 1900–, vol. 4, 465–565.

Kant, I. 1790. Kritik der Urteilskraft. In *Kant* 1900–, vol. 5, 165–485.

Kant, I. 1793. Welches sind die wirklichen Fortschritte, die die Metaphysik seit Leibnizens und Wolf's Zeiten in Deutschland gemacht hat? In *Kant* 1900–, vol. 20, 253–351.

Kant, I. 1798a. Anthropologie in pragmatischer Hinsicht. In *Kant* 1900–, vol. 7, 117–333.

Kant, I. 1798b. Der Streit der Fakultäten. In *Kant* 1900–, vol. 7, 1–115.

Kant, I. 1800a. Immanuel Kants Logik: Ein Handbuch zu Vorlesungen. In *Kant* 1900–, vol. 9, 1–150.

Kant, I. 1800b. Immanuel Kants physische Geographie. In *Kant* 1900–, vol. 9, 151–436.

Kant, I. 1900–. *Kant's gesammelte Schriften*, ed. Königlich Preußische (Deutsche) Akademie der Wissenschaften. Berlin: Reimer (later: de Gruyter).

Kant, I. 1922. Briefwechsel. In *Kant* 1900–, Vols. 10–13.

Kant, I. 1936–1938. Opus postumum. In *Kant* 1900–, Vols. 21–22.

Kant, I. 1942. Erste Einleitung in die Kritik der Urteilskraft. In *Kant* 1900–, vol. 20, 193–251.

Kant, I. 1961. Vorlesung Philosophische Enziklopädie. In *Kant* 1900–, vol. 29, 3–45.

Kant, I. 1966a. Logik Blomberg. In *Kant* 1900–, vol. 24, 16–301.

Kant, I. 1966b. Logik Busolt. In *Kant* 1900–, vol. 24, 608–686.

Kant, I. 1966c. Logik Dohna-Wundlacken. In *Kant* 1900–, vol. 24, 687–784.

Kant, I. 1966d. Logik Philippi. In *Kant* 1900–, vol. 24, 303–496.

Kant, I. 1966e. Logik Pölitz. In *Kant* 1900–, vol. 24, 502–602.

Kant, I. 1966f. Wiener Logik. In *Kant* 1900–, vol. 24, 790–940.

Kant, I. 1980. Danziger Physik. In *Kant* 1900–, vol. 29, 97–169.

Kant, I. 1998a. Logik Bauch. In *Logik-Vorlesung: Unveröffentlichte Nachschriften*, vol. 1, ed. T. Pinder. Hamburg: Meiner.

Kant, I. 1998b. Logik Hechsel. In *Logik-Vorlesung: Unveröffentlichte Nachschriften*, vol. 2, ed. T. Pinder, 269–499. Hamburg: Meiner.

Kitcher, P. 2000. *Kant's transcendental psychology*. New York: Oxford University Press.

Kreines, J. 2008. Kant on the laws of nature: Laws, necessitation, and the limitation of our knowledge. *European Journal of Philosophy* 17: 527–58.

La Rocca, C. 2003. Giudizi provvisori: Sulla logica euristica del processo conoscitivo. In C. La Rocca, *Soggetto e mondo: Studi su Kant*, 79–119. Venice: Marsilio.

Lambert, J.H. 1764. *Neues Organon oder Gedanken über die Erforschung und Bezeichnung des Wahren und dessen Unterscheidung vom Irrthum und Schein*, 2 Vols. Leipzig: Wendler.

Laudan, L. 1981. The sources of modern methodology: Two models of change. In L. Laudan, *Science and hypothesis: Historical essays on scientific methodology*. Dordrecht: Reidel.

Locke, J. 1693. *Some thoughts concerning education*, ed. J.W. Yolton and J.S. Yolton (1989). Oxford: Oxford University Press.

Locke, J. 1976–89. *The correspondence of John Locke*, ed. S. De Beer. Oxford: Clarendon.

Malzkorn, W. 2000. Kant and the dynamics of science. *Philosophia naturalis* 37: 77–95.

Newton, I. 1687. *The principia: Mathematical principles of natural philosophy*. ed. I.B. Cohen and A.M. Whitman (1999). Berkeley: University of California Press.

Newton, I. 1714/15. An Account of the Book Entituled Commercium Epistolicum Collinii & aliorum, De Analysi promota. *Philosophical Transactions of the Royal Society of London* 342: 173–224.
Parker, S. 1666. *Free and impartial censure of the Platonick philosophie: Being a letter written to his much honoured friend Mr. N.B.* Oxford: W. Hall.
Power, H. 1664. *Experimental philosophy*. London: John Martin and James Allestry.
Reid, T. 2001. *Essays on the intellectual powers of man: A critical edition*, ed. K. Haakonssen and D.R. Brookes. Edinburgh: Edinburgh University Press.
Sato, T. 2008. Chemical affinity in Kant's practical philosophy. In *Recht und Frieden in der Philosophie Kants: Akten des X. Internationalen Kant-Kongresses*, Vol. 3, ed. V. Rohden et al., 359–68. Berlin: de Gruyter.
Seigfried, H. 1989. Transcendental experiments. In *Proceedings of the sixth international Kant congress*, Vol. 2, ed. G. Funke and T.M. Seebohm, 341–50. Washington, DC: Center for Advanced Research in Phenomenology and University Press of America.
Sprat, T. 1667. *The history of the royal society of London for the improving of natural knowledge*. London: J. Martyn.
Steinle, F. 1997. Entering new fields: Exploratory uses of experimentation. *Philosophy of Science* 64 (Suppl): S65–S74.
Turnbull, G. 1740. *The principles of moral philosophy: An enquiry into the wise and good government of the moral world*. London: J. Noon.
Walch, J. G. 1775. *Philosophisches lexicon*, 4th ed., trans and ed. J. C. Hennings. Leipzig: Gleditsch.
Wartenberg, T.E. 1992. Reason and the practice of science. In *The Cambridge companion to Kant*, ed. P. Guyer, 228–48. Cambridge: Cambridge University Press.

Chapter 8
Did Newton Feign the Corpuscular Hypothesis?*

Kirsten Walsh

8.1 Introduction

Newton's famous pronouncement, *Hypotheses non fingo* ('I do not feign hypotheses'[1]), first appeared in 1713, but his anti-hypothetical stance was present much earlier than that. In February 1672, Newton published his first paper, in which he revealed his new theory of light and colours. In this paper, Newton introduced his methodological views for the first time, to explain why he was insisting that his doctrine of colours is a *theory*, not a *hypothesis*:

> For what I shall tell concerning [colours] is not an Hypothesis but most rigid consequence... evinced by ye mediation of experiments concluding directly & without any suspicion of doubt (Newton 1959, pp. 96–97).

Despite these clear anti-hypothetical themes, Newton appeared to hold a *corpuscular hypothesis*.[2] What are we to make of this? It is surprising indeed to

---

* For valuable comments on previous versions of this paper I would like to thank Peter Anstey, Alberto Vanzo and Juan Gomez. The paper also benefited from helpful discussions at the AAPNZ conference and several Otago Postgraduate Workshops.

[1] '*Hypotheses non fingo*' first appeared in 1713 in the 'General Scholium' in the 2nd edition of Newton's *Principia*:

> I have not as yet been able to discover the reason for these properties of gravity from phenomena, and I do not feign hypotheses [*hypotheses non fingo*]. For whatever is not deduced from the phenomena must be called a hypothesis; and hypotheses, whether metaphysical or physical, or based on occult qualities, or mechanical, have no place in experimental philosophy. In this philosophy particular propositions are inferred from the phenomena, and afterwards rendered general by induction (Newton 1687, p. 943 – underlining added).

[2] i.e. The hypothesis that light is a body or an atom.

K. Walsh (✉)
Department of Philosophy, University of Otago, Dunedin, New Zealand
e-mail: walki303@student.otago.ac.nz

J. Maclaurin (ed.), *Rationis Defensor: Essays in Honour of Colin Cheyne*,
Studies in History and Philosophy of Science 28, DOI 10.1007/978-94-007-3983-3_8,

find a hypothesis tucked away in the very paper in which Newton is expressing his anti-hypothetical views for the first time. Is Newton guilty of feigning a hypothesis? Or have we missed something important in Newton's methodological stance?

Newton's methodological views emerged within the context of the *new experimental philosophy* that was being promoted by the Royal Society in the 1660s and 1670s. In this paper I take a fresh look at Newton's first optical papers in light of this emerging experimental tradition. I argue that Newton's *method of hypotheses* should be regarded both as consistent with, and yet a development on, this experimental tradition. I argue that Newton's corpuscular hypothesis is consistent with this method.

8.2 Experimental Philosophy and the Royal Society[3]

The Royal Society officially formed in 1660 as a "College for the Promoting of Physico-Mathematical Experimental Learning" (Bunch and Hellemans 2004, p. 198). Its members were particularly concerned with adopting the best methods for acquiring knowledge about the natural world. To this end, they adopted the new *experimental philosophy*.

Experimental philosophy was usually defended in opposition to *speculative philosophy*. Where experimental philosophy emphasised empirical evidence, speculative philosophy neglected it; where experimental philosophy was concerned with the careful collection and ordering of observations, speculative philosophy was concerned with constructing metaphysical systems. According to the speculative philosophy, natural phenomena could be explained without recourse to systematic observation and experiment. In the sense that observations were made, and explanations appealed to factual matters, the speculative philosophy was empirical. However, such observations were employed to demonstrate or develop the theory; rather than to test or support it.

One key issue for debate was the use and misuse of hypotheses. Present-day philosophers of science distinguish between 'theory' and 'hypothesis' in several different ways. For example, in some cases a 'theory' is the general thesis and 'hypotheses' are subordinate particular theses. In other cases a 'hypothesis' is a general thesis stated prior to there being any empirical evidence; when it has enough empirical support, it becomes a 'theory'. In still other cases the terms 'theory' and 'hypothesis' are used interchangeably to refer both to general theses and subordinate theses. Similarly, in the early modern period, 'hypothesis' referred to generalisations, metaphysical principles, or even explanatory systems. It also began to be used as a synonym for 'conjecture' or 'speculation'. To characterise it broadly, speculative philosophers were said to indulge in hypotheses and conjectures in order to construct speculative metaphysical systems.

In contrast, experimental philosophy emphasised the importance of empirical evidence and the role of the senses in acquiring knowledge. Experimental

[3] In this section I follow Anstey (2004, 2005).

philosophers made several notable claims. Firstly, they claimed that knowledge of nature is derived from facts about nature; not from hypotheses and speculative systems based on metaphysical principles. Secondly, they claimed that those facts about nature are obtained via observation and experiment (i.e. from sensory observation, reports of observations made by others, or by using instruments to manipulate natural objects). Thirdly, they claimed that observation and experiment are the means to an end. Experimental philosophy emphasised observation and experiment only insofar as they produced facts. Observation and experiment were not ends in themselves.[4]

Firm belief in the merits of this experimental philosophy can be seen in passages such as this one from Hooke's *Micrographa,* addressed to the Royal Society:

> The Rules YOU have prescrib'd YOUR selves in YOUR Philosophical Progress do seem the best that have ever yet been practis'd. And particularly that of avoiding *Dogmatizing*, and the *espousal* of any *Hypothesis* not sufficiently grounded and confirm'd by *Experiments*. This way seems the most excellent, and may preserve both *Philosophy* and *Natural History* from its former *Corruptions* (Hooke 1966).

In Britain, the experimental philosophy was adopted and developed in a Baconian form, according to the methods described by Francis Bacon. Thus, in addition to their general commitment to the experimental philosophy, and opposition to the speculative philosophy, there were several more particular features that they agreed upon. In particular, they adopted the Baconian method of natural history as the dominant experimental method. This method involved gathering vast collections of facts (from a wide variety of sources) about natural objects and qualities and using them to construct theories. Thus, natural histories served as the foundation of theorising in natural philosophy. Queries played an important role in the Baconian method. While queries (or questions) already played a role in the scholastic tradition, the Baconian method made the assembling of queries a specific step in the acquisition and development of natural philosophical knowledge. Anstey argues that in the Baconian tradition, the query served to specify the scope of the inquiry and to direct the inquiry towards experimental results.

Newton was certainly aware of the Baconian method of natural history. We can see this from his Trinity notebook from his undergraduate days in the early 1660s (Newton 1983). Newton took careful notes from the *Philosophical Transactions* – the journal of the Royal Society. As I have noted, the Royal Society widely promoted the experimental philosophy, and many issues of the *Philosophical Transactions* contained Baconian natural histories. Newton also took careful notes on Boyle's work, which involved the construction of natural histories. This all suggests that Newton was familiar with the Baconian method of natural history by the time he wrote his first optical paper. Newton didn't construct natural histories in his early work (in fact, later in life he openly opposed the method of natural history). However, other aspects of Newton's early work suggest a Baconian experimental influence. For example, he was concerned with experiments, he made use of queries, and he adopted the familiar

[4] This identifies an important distinction between early modern experimental philosophy and the New Experimentalism promoted by e.g. Mayo (1996).

distinction between theory and hypothesis. In Sect. 8.4, I discuss this distinction, but first, let's take a look at Newton's first optical paper.

8.3 Newton's First Optical Paper

Newton wrote his first optical paper in February 1672. It was read at the meeting of the Royal Society and published in the *Philosophical Transactions* shortly afterwards. The style of the paper is quite anecdotal.[5] Newton begins by describing the experiments he performed in 1666, in which he used glass prisms to examine "the celebrated *Phænomena* of *Colours*" (Cohen and Schofield 1958, p. 47). He tells us that what began as "a very pleasing divertisement, to view the vivid and intense colours produced thereby" produced some unexpected results (Cohen and Schofield 1958, p. 48). The coloured image produced by the passage of white light through the glass prism was shaped like an elongated circle, and yet, "according to the received laws of Refraction" (Snell's Law[6]), the image should have appeared circular (Cohen and Schofield 1958, p. 48). Newton sets out to explain this discrepancy between accepted theory and experimental result.

First, Newton tells us that, on the assumption that the laws of refraction are correct, he wondered if the shape of the image may be explained by the experimental setup or apparatus. He conducted an experiment to find out if the length of the image varies with the thickness of the glass or the size of the aperture. He found that it doesn't. Then, he conducted experiments to see if the unevenness in the glass, or some other contingent irregularity, causes the image to lengthen. He concluded that the elongated image is not caused by any contingent irregularity.

After ruling out experimental faults, and still assuming the laws of refraction, Newton says he wondered if any of his other assumptions were incorrect. Firstly, Newton considered his assumption that all the parts of the beam of light have the same angle of incidence (more or less). He noticed that the rays coming from the sun hit the prism at differing angles of incidence, depending on which part of the disc they come from, and he wondered if this difference is great enough to account for the elongation of the image. He says he found that even the greatest possible difference is too small to account for the experimental result. Secondly, Newton says he considered his assumption that light travels in a straight line. He wondered if the prism was causing the rays of light to curve, "and according to their more or less curvity tend to divers parts of the wall" (Cohen and Schofield 1958, p. 50). But he concludes that the rays are straight rather than curved. Thirdly, Newton says he considered his assumption that all the parts of the beam of light have the same

[5] Although it is almost certainly a 'rational reconstruction' of the events that took place during the *anni mirabilis* (1665–1666). For discussion on this point, see e.g. Whiteside (1966).

[6] Snell's Law states that the ratio of the *sines* of the angles of incidence and refraction is a constant that depends on the media through which the light passes.

refrangibility.[7] To test this assumption, Newton performed his *Experimentum Crucis*.[8] From this experiment, he says he was able to establish his first conclusion: that light is a "*Heterogeneous mixture of differently refrangible Rays*" (Cohen and Schofield 1958, p. 51 – Newton's italics).

From this conclusion, Newton was able to expand his theory of light and colours. He lays it down in 13 propositions. The main points are:

(a) Light is a *heterogeneous mixture of differently refrangible rays* (from his *Experimentum Crucis*);
(b) There is a *one-to-one correspondence* between refrangibility and spectral colour (Proposition 2);
(c) Colour is an *original property* of rays of light (Proposition 1);
(d) *Original colours* are constant, unchangeable and homogeneous (Proposition 3);
(e) New (*compound*) colours are created by combining original colours (Propositions 4, 5 and 6);
(f) White light is a mixture of all original colours *in equal amounts* (Propositions 7, 8 and 9); and finally
(g) All coloured bodies obtain their colour by the process of refraction and reflection, in accordance with (a) – (f) (Propositions 10, 11, 12 and 13).

After laying down his doctrine, Newton concludes his paper with a discussion of two more experiments. The first one is supposed to support (f) and the second one is supposed to support (d). In his final paragraph, he says:

> This, I conceive, is enough for an Introduction to Experiments of this kind; which if any of the *R. Society* shall be so curious as to prosecute, I should be very glad to be informed with what success: That, if any thing seem to be defective, or to thwart this relation, I may have an opportunity of giving further direction about it, or of acknowledging my errors, if I have committed any (Newton 1959, p. 102).

Thus, Newton's main goal in this first paper appears to be to show that he has overwhelming experimental evidence in favour of his new theory. He is not trying to show merely that his theory fits the experimental data. Rather, he is trying to show that his theory is grounded in experiment; that this theory is the only way to interpret the evidence.

8.4 Newton's Method of Hypotheses

Newton doesn't use the term 'theory' at all in this first paper. The only use of the term 'theory' comes from Oldenburg's commentary in the *Philosophical Transactions*, when he refers to Newton's 'theory of light and colours' – Newton

[7] The degree to which a beam of light can be refracted.

[8] A single experiment that is supposed both to refute the old theory and confirm the new theory.

calls it a ‘doctrine’. Moreover, he only uses the word ‘hypothesis’ twice in this first paper. The first time he uses it to refer to Snell’s Law:

> But because this computation was founded on the Hypothesis of the proportionality of the *sines* of Incidence, and Refraction, which though by my own & others Experience I could not imagine to be so erroneous... (Newton 1959, p. 93).

In this case, Newton is using the term ‘hypothesis’ to describe a proposition that is only approximately true. The second time, he contrasts hypothesis with rigid consequence:

> For what I shall tell concerning them is not an Hypothesis but most rigid consequence, not conjectured by barely inferring ’tis thus because not otherwise or because it satisfies all Phænomena (the Philosophers universall Topick,) but evinced by ye mediation of experiments concluding directly & wthout any suspicion of doubt (Newton 1959, p. 96).

In his reply to Hooke (in June 1672) Newton uses ‘hypothesis’ in a third way: he says that hypotheses are not necessary in order to understand his theory, “I do not think it needful to explicate my Doctrine by any Hypothesis at all” (Newton 1959, p. 177). In this case, hypotheses are not simply theories that lack certainty or empirical support. For Newton, theories and hypotheses have different roles and different jurisdictions. Theories explain physical properties; hypotheses give causal explanations and explain the nature of the phenomenon.

Thus, in Newton’s early optical papers, a ‘hypothesis’ is:

H1. Something that is, at best, only approximately true;
H2. A conjecture or speculation – something not based on empirical evidence; or
H3. A causal explanation – something concerning the nature of the phenomenon, rather than its physical properties.

In contrast, a theory:

T1. Is certainly true, because it is supported by (or *deduced* from) experiment;
T2. Is experimental – that is, it has *testable* consequences; and
T3. Concerns the *physical properties* of the phenomenon, rather than its nature.

The criteria for ‘theory’ are very strict, and it is hard to imagine that anything could meet them! But Newton thinks his doctrine *does* meet them.

Newton considers his goal of absolute certainty to be closely related to mathematics. He says:

> A naturalist would scearce expect to see ye science of those become mathematicall, & yet I dare affirm that there is as much certainty in it as in any other part of Opticks ... evinced by ye mediation of experiments concluding directly & without any suspicion of doubt (Newton 1959, pp. 96–97).[9]

[9] This passage was omitted from the published version of the paper. It can be seen in the version that Newton sent to Oldenburg (Newton 1959, pp. 96–97). Hooke refers to this passage in his reply to Newton (which wasn’t published in the *Philosophical Transaction*). Similarly, when Newton replies to Hooke’s reply, the passage was also removed.

In a letter to Hooke on 11 June 1672, Newton retreats a little from his earlier claims of absolute certainty.[10] He says that he did not mean to suggest that the science of colours was as certain as a mathematical demonstration. Rather, the science of colours (just like the rest of optics),

> depend[s] as well on Physicall Principles as on Mathematicall Demonstrations: And the absolute certainty of a Science cannot exceed the certainty of its Principles (Newton 1959, p. 187).

Therefore, to achieve certainty in science, two things are required:

1. To establish experimentally physical principles that are certainly true; and
2. To establish physical principles that are such that they can be used as premises in mathematical arguments.

Newton thinks it is possible to achieve both of these things, because he has done so. When he describes his experimental results, he does so with precision:

> Its distance from the hole or Prisme was 22 foot; its utmost length 13 ¼ inches; its breadth 2 $^5/_8$ inches; the diameter of the hole ¼ of an inch; the angle, which the Rays, tending towards the middle of the image, made with those lines, in which they would have proceeded without refraction, 44 deg. 56′... (Newton 1959, p. 93).

And when he extrapolates to a general theory, he does so with quasi-geometrical reasoning: deducing his propositions from certain physical properties in accordance with the methods of geometry.

> I drew up a series of such Expts on designe to reduce ye Theory of colours to Propositions & prove each Proposition from one or more of those Expts by the assistance of common notions set down in the form of Definitions & Axioms in imitation of the Method by which Mathematicians are wont to prove their doctrines (Newton 1959, p. 237).

Therefore, he argues, the science of colours is "*Mathematicall* & as certain as any part of Optiques" (Newton 1959, p. 188), and absolute certainty is closely related to experiment.

This process can be seen in Newton's *Experimentum Crucis*,[11] which demonstrates Newton's concern with establishing certain physical properties. Refer to Fig. 8.1, where white light travels from the Sun (S), through the first aperture (F), through the first prism (ABC), where it is refracted for the first time, producing an image on the first board (DE). A small amount of light passes through the second aperture (G), producing an image on the second board (*de*). A small amount of light passes through the third aperture (*g*), through the prism (*abc*), where it is refracted for the second time, producing an image on the screen (MN). Newton manipulated the first prism, twisting and turning it so that different parts of the refracted image could pass through aperture G to the second prism. He took careful note of where each image appeared on the board MN.

[10] The whole section, entitled "*12. That the Science of Colours is most properly a Mathematicall Science*" (Newton 1959, pp. 187–188), is omitted from the published version of the letter.

[11] I thank John Bigelow and Ian Lawson for discussing this experiment with me.

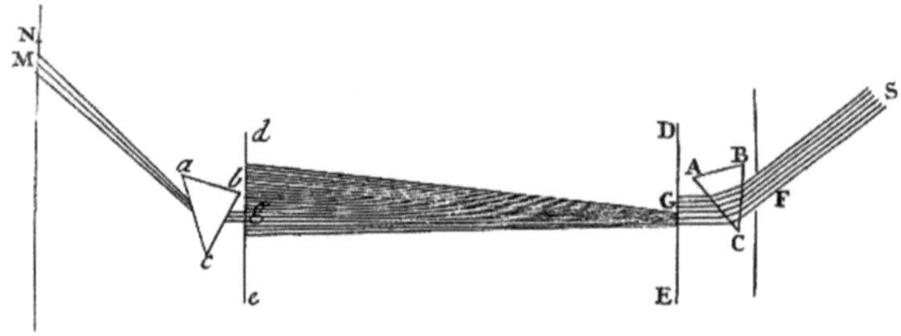

Fig. 8.1 *Experimentum Crucis* (Newton 1730, p. 47)

Newton found that each time a particular ray passes through a prism it refracts to precisely the same degree. For example, light that refracts to 15° at the first prism refracts to 15° at the second prism as well. Newton argues that this shows that refrangibility is an original and constant property of light. Recall that Newton claims to prove that refrangibility is an original property of the light, not an effect of the prism:

> . . .the true cause of the length of that Image was detected to be no other, then that *Light* consists of *Rays differently refrangible*, which, without any respect to a difference in their incidence, were, according to their degrees of refrangibility, transmitted towards divers parts of the wall (Newton 1959, p. 94).

Notice how Newton focuses on *refrangibility*, not colour, in this experiment.[12] He is concerned with geometrical factors such as the length of the image, the position of the image on the wall, the distance from the aperture to the prism and the prism to the wall, and the angles of incidence and refraction. Newton wants to establish physical properties with certainty, so it should come as no surprise that he eschews talk of colour to focus on these other things. It would have been difficult to measure precisely changes in colour,[13] but Newton was able to measure precisely the degrees of refraction by measuring the positions of the image on the wall. Newton recorded observations that are measurable, quantitative and precise, for *this* data could be used to establish physical principles that are certainly true, and could be used in mathematical arguments. Why does Newton think he has achieved absolute certainty? Because he has deduced his propositions from certain physical properties in accordance with the methods of geometry. In his first optical paper,

[12] Newton's *Experimentum Crucis* was heavily criticised by his contemporaries. Hooke, for example, argued that this experiment is not a crucial experiment, because it does not prove that colour is an original property of light (Newton 1959, p. 111). Hooke believed that light becomes coloured as it passes through the prism, and Newton's experiment did not convince him otherwise. This line of criticism is extremely common (e.g. Sabra (1967) and Bechler (1974)), despite the fact that Newton doesn't mention colour in his discussion of this experiment.

[13] In his 2nd paper, of December 1675, Newton makes a statement to this effect, saying "my owne eyes are not very criticall in distinguishing colours" (Newton 1959, p. 376).

Newton only provides a list of propositions, which appear to stand in some sort of deductive relationship to one another. But later, he does this more carefully.

The conclusion of the *Experimentum Crucis*, in turn, provided the first premise of Newton's argument, which amounts roughly to the following argument:

P1. White light is composed of rays of every refrangibility.
P2. There is a one-to-one correspondence between refrangibility and spectral colour.
C. White light is composed of rays of every spectral colour.

And Newton claims that he has established (C) with absolute certainty. The quasi-geometrical 'proof' of his theory of colours is set out in his reply to Huygens (Newton 1959, pp. 292–294).

To summarise, Newton's mathematical method and his experimental method are linked by his notion of absolute certainty. Newton claims his theory of colours is certainly true, because (1) his physical principles are established experimentally and are certainty true, and (2) he can use these physical principles as the basis of his mathematical proof.[14]

Given Newton's strict definitions, one could be forgiven for assuming that Newton is advocating a wholesale rejection of hypotheses. But this is not the case in these early optical papers. Here, he is recommending that hypotheses be used sparingly in accordance with his method of hypotheses. The clearest and most concise description of this method can be found in his reply to Pardies' second letter in July 1672. Newton says:

> The best and safest method of philosophizing seems to be, first to inquire diligently into the properties of things, and establishing those properties by experiments and then to proceed more slowly to hypotheses for the explanation of them (Cohen and Schofield 1958, p. 106).

This tells us, firstly, that experiment must precede hypotheses. One must first establish physical properties, and then seek to explain them. Secondly, this tells us that there are two levels of explanation: physical and metaphysical (or causal).

He goes on to say:

> ...hypotheses should be subservient only in explaining the properties of things, but not assumed in determining them; unless so far as they may furnish experiments (Cohen and Schofield 1958, p. 106).

For Newton, hypotheses could be useful in science, for example, when they suggest further experiments. There is a note of caution here: hypotheses should always be removed from the final version of the theory.

It also becomes clear in his reply to Huygens in April 1673 (Cohen and Schofield 1958, p. 144), and in his second paper in December 1675 (Newton 1959, p. 363), that hypotheses can be useful when they assist in understanding a theory. Sometimes theories are difficult to understand, for they can be quite abstract, describing

[14] That a lengthy and sometimes heated debate followed Newton's original paper, shows that his opponents weren't as convinced by his careful demonstration as he was!

physical properties without any causal explanation. So a hypothesis might be useful to 'illustrate' the theory. Once again, Newton encourages his readers to exercise caution: the hypothesis must not be confused with the theory; rather the two must be kept separate.

In short, Newton is working with a clear distinction between theory and hypothesis, and a fairly explicit method of hypotheses. He prefers theories to hypotheses, but thinks that hypotheses are acceptable, even useful, for two purposes:

1. To illustrate the theory; and
2. To suggest experiments.

The overarching rule is nicely summarised by Newton, when he says: "I shall not mingle conjectures with certainties" (Cohen and Schofield 1958, p. 57).

8.5 Newton's Corpuscular Hypothesis

So far we have seen that there are clear anti-hypothetical themes in Newton's work, supported by his stated aim of absolute certainty. So it is surprising indeed to find that, underneath Newton's self-consciously experimental doctrine, there is a hypothesis about the nature of light.

In this first paper, it's subtle. For example, Newton hints at a corpuscular view of light when he employs the tennis ball analogy to explore the idea that light might curve as it is refracted:

> And for the same reason, if the Rays of light should possibly be globular bodies, and by their oblique passage out of one medium into another acquire a circulating motion, they ought to feel the greater resistance from the ambient Æther, on that side, where the motions conspire, and thence be continually bowed to the other (Cohen and Schofield 1958, p. 50).

But near the end of the paper, he is more explicit, when he suggests that light is substantial:

> For, since Colours are the *qualities* of Light, having its Rays for their intire and immediate subject, how can we think those Rays *qualities* also, unless one quality may be the subject of and sustain another; which in effect is to call it *substance*. We should not know Bodies for substances, were it not for their sensible qualities, and the Principal of those being now found due to something else, we have as good reason to believe that to be a Substance also (Cohen and Schofield 1958, p. 57).

Here, Newton is arguing that, since colour is a property of light, light must be a substance. That is, it must consist of substantial bodies. But even here, Newton exercises caution. He says that although he has established that light is heterogeneous, it is much more difficult,

> to determine more absolutely, what Light is, after what manner refracted, and by what modes or actions it produceth in our minds the Phantasms of Colours (Cohen and Schofield 1958, p. 57).

He says that he is not willing to speculate any further on these matters.

Despite its subtlety, his opponents spotted the hypothesis straight away. So Newton defended his theory, while at the same time leaving no doubt that he believed light to be corpuscular:

> 'Tis true that from my Theory I argue the corporeity of light, but I doe it without any absolute positivenesse, as the word *perhaps* intimates, & make it at most but a very plausible consequence of the Doctrine, & not a fundamentall Supposition, not so much as any part of it, wch was wholly comprehended in the precedent Propositions (Cohen and Schofield 1958, p. 118).

Moreover, "Had I intended any such Hypothesis I should somewhere have explained it" (Cohen and Schofield 1958, p. 118).

In December 1675, almost 4 years after he sent his initial paper to the Royal Society, Newton sent in his second paper. Given his denial of his corpuscular hypothesis, this paper has the unlikely title: "an hypothesis explaining the Properties of Light" (Newton 1959, p. 362). Here, Newton explicates his hypothesis on the nature of light for the first time. For the remainder of this section, I shall consider two striking features of this paper:

(a) Newton's careful detachment from his hypothesis; and
(b) Newton's emphasis on experiment.

From the outset, Newton is careful to remain detached from his hypothesis. He insists that he is only discussing this hypothesis for the purposes of illuminating his theory. He says:

> I have observed the heads of some great virtuoso's to run much upon Hypotheses, as if my discourses wanted an Hypothesis to explain them by, & found that some when I could not make them take my meaning, when I spake of the nature of light & colours abstractedly, have readily apprehended it when I illustrated my Discourse by an Hypothesis... (Newton 1959, p. 363).

Moreover, Newton insists that he is not asserting that this hypothesis is true, nor using it to support his theory. He says, "Were I to assume an Hypothesis it should be this..." (Newton 1959, p. 636), implying that he does not, in fact, assume any hypothesis.

The hypothesis Newton lays down in this paper is composed of the following claims:

1. There is an 'aethereal medium', which is similar to air, but rarer, more penetrating and more strongly elastic (Newton 1959, p. 364).
2. Aether vibrates, carrying sounds, smells and light. While the vibrations differ in size, they are on the whole smaller and swifter (by far) than the vibrations of air (Newton 1959, p. 366).
3. Aether penetrates and passes through the pores of solid substances such as crystal, glass and water. But aether is less dense within the pores than without (Newton 1959, pp. 366–367).
4. Light is neither the aether itself, nor the vibrations, but a substance that is propagated from 'lucid' bodies and travels through the aether (Newton 1959, p. 370).

5. Light warms the aether and the aether refracts the light. Thus, the mutual action of light on aether, and aether on light, explains how light is reflected and refracted (Newton 1959, p. 371).
6. The rays (or bodies) of which light consists differ from one another physically. These physical differences are unchangeable and cause the rays to be different colours. This explains how it happens that colour and refrangibility are unchangeable properties of light (Newton 1959, p. 376).

When he discusses (4), Newton is careful not to push too forcefully for any particular account of light. He says one might suppose light to be "an aggregate of various peripatetic qualities", or "unimaginably small and swift" corpuscles of various sizes, or "any other corporeal emanation or impulse or motion of any other medium diffused through the body of the aether" (Newton 1959, p. 370):

> Onely whatever Light be, I would suppose, it consists of Successive rayes differing from one another in contingent circumstances, as bignes, forme or vigour... And further I would suppose it divers from the vibrations of the aether (Newton 1959, p. 370).

And just in case we were unconvinced by his studied detachment to his own hypothesis, he ends the paper by saying:

> If you now ask how rays are reflected without impinging on the parts of a body, & how those which impinge on its parts may be stopped & stifled, it requires an Hypothesis to explain it by, the description of which is besides my designe. And so the manner how several rays are unequally refrangible & reflexible & originally indued wth severall colours remaines to be explained Hypothetically: But I shall content my self with haveing shewn that *de facto* the rays of light are indued with those properties (Newton 1959, p. 386).

The other striking feature of this second paper is Newton's emphasis on experiment. For example, when Newton makes his first claim, that there is an aethereal medium, he gives an account of a new electrical experiment which seems to support his claim (Newton 1959, pp. 364–365).[15] When Newton makes his third claim, that aether is less dense within the pores of solid objects than without, he discusses the implications for Boyle's tadpole experiments (Newton 1959, p. 367). And finally, when he discusses colour, Newton uses his hypothesis to suggest and interpret experiments. For example:

> Whence, if the Glasses in this posture be looked upon, there ought to appear at *A*, the contact of the Glasses, a black spott, & about that many concentric circles of light & darknesse, the squares of whose semidiameters are to sense in arithmetical progression (Newton 1959, pp. 378–380).

These experiments, known as 'Newton's rings', led to significant advances in Newton's theory of light and colours.[16]

[15] This account caused quite a stir at the meeting of the Royal Society during which the paper was read.

[16] For a detailed discussion of Newton's rings, see Shapiro (1993).

Even when explicitly engaging in the practice of hypothesising, Newton remains the Experimental Philosopher. He follows his method of hypotheses – keeping his corpuscular hypothesis well separate from his theory, using it to illustrate but not support his theory, and to suggest experiments. He does not appear to be 'feigning' the corpuscular hypothesis.

8.6 Conclusion

I have argued firstly, that Newton has a clear method of hypotheses. Secondly, that Newton's methodology is consistent with the experimental philosophy as it was advocated by the Royal Society in the 1660s and 1670s. And thirdly, that in his early optical work, Newton distinguishes between theory and hypothesis. He gives his theory of colours higher epistemic status than his corpuscular hypothesis, only using the hypothesis to illustrate his theory and to suggest and interpret experiments. I conclude that he does not feign the corpuscular hypothesis.

Newton's later methodological views seem to echo these early ones. Every edition of the *Opticks* begins with (Newton 1704, 1718, 1721, 1730, p. 1):

> My Design in this Book is not to explain the Properties of Light by Hypotheses, but to propose and prove them by Reason and Experiments: in order to which I shall premise the following Definitions and Axioms.

References

Anstey, P. 2004. The methodological origins of Newton's queries. *Studies in History and Philosophy of Science* 35:247–269.

Anstey, P. 2005. Experimental versus speculative natural philosophy. In *The science of nature in the seventeenth century*, eds. P. Anstey and J. A. Schuster, 215–242. Netherlands: Springer.

Bechler, Z. 1974. Newton's 1672 optical controversies: A study in the grammar of scientific dissent. In *The interaction between science and philosophy*, ed. Y. Elkana, 115–142. Atlantic Highlands: Humanities Press.

Bunch, B.H., and A. Hellemans. 2004. *The history of science and technology: A browser's guide to the great discoveries, inventions, and the people who made them, from the dawn of time to today*. Boston: Houghton Mifflin.

Cohen, I.B., and R.E. Schofield. 1958. *Isaac Newton's papers and letters on natural philosophy and related documents*. Cambridge: Cambridge University Press.

Hooke, R. 1966. *Micrographia: Or, some physiological descriptions of minute bodies made by magnifying glasses, with observations and inquiries thereupon*. New York: Dover.

Mayo, D. 1996. *Error and the growth of experimental knowledge*. Chicago: University of Chicago Press.

Newton, I. 1687. The principia: Mathematical principles of natural philosophy. In *A new translation*, ed. I.B. Cohen and A.M. Whitman. 1999. Berkeley: University of California Press.

Newton, I. 1704. *Opticks: Or, a treatise of the reflexions, refractions, inflexions and colours of light. Also two treatises of the species and magnitude of curvilinear figures*. London: Printed for Sam Smith and Benj Walford, Printers to the Royal Society, at the Prince's Arms in St. Paul's Church-Yard.

Newton, I. 1718. *Opticks: Or, a treatise of the reflections, refractions, inflections and colours of light. The second edition, with additions.* 2nd ed. London: Printed for W and J Innys, printers to the Royal Society, at the Prince's-Arms in St Paul's Church-Yard.

Newton, I. 1721. *Opticks: Or, A treatise of the reflections, refractions, inflections and colours of light. The third edition, corrected.* 3rd ed. London: Printed for William and John Innys at the West End of St. Paul's.

Newton, I. 1730. *Opticks: Or, A treatise of the reflections, refractions, inflections and colours of light. The fourth edition, corrected.* 4th ed. London: Printed for William Innys at the West End of St. Paul's.

Newton, I. 1959. In *The correspondence of Isaac Newton, vol. 1*, ed. H.W. Turnbull, J. F. Scott, A.R. Hall, and M.B. Hall. Cambridge: Published for the Royal Society at the University Press.

Newton, I. 1983. Part II. Transcription and expansion of questiones quaedam philosophicae. In *Certain philosophical questions: Newton's trinity notebook*, ed. J.E. McGuire and M. Tamny. Cambridge: Cambridge University Press.

Sabra, A.I. 1967. *Theories of light from Descartes to Newton*. London: Oldbourne Book Co Ltd.

Shapiro, A. 1993. *Fits, Passions and Paroxysms*. Cambridge: Cambridge University Press.

Whiteside, D.T. 1966. Newton's marvellous year: 1666 and all that. *Notes and Records of the Royal Society of London* 21: 32–41.

Chapter 9
The Progress of Scotland and the Experimental Method

Juan Gomez

9.1 Introduction

The eighteenth century witnessed the emergence of a great number of learned societies all across Europe. The Royal Society of London (RSL) was the paradigm case of this kind of societies, but by no means the only one. With the RSL in mind as their inspiration, over 200 societies were established between 1700 and 1799,[1] some being 'specialized' societies (i.e. literary, cultural, agricultural, economic, political, poetical, scientific, and antiquary), some discussing a mixture of topics. Despite the great number and variety of learned societies, there have been only a few in depth studies regarding them, of course excluding the vast amount of literature available on the Royal Society. There are very few texts that deal with Scottish societies at length: H. Lewis Ulman edited and published the minutes of the Aberdeen Philosophical Society in 1990, Davis McElroy wrote his PhD thesis on Scottish societies and later published a book (McElroy 1952, 1969), and historian Roger L. Emerson has a number of articles (Emerson 1973, 1979, 1981, 1985, 1988) on the Philosophical Society of Edinburgh, besides his extensive and insightful research on Scotland in the enlightenment period. Other than these academics, the societies are only briefly mentioned in anthologies on Scotland and the enlightenment, or in biographies of the intellectuals of the eighteenth century, usually being little more than an anecdote.

One reason for this lack of research on the learned societies is the scarcity of empirical evidence, since only few of the original minute books, discussions, or publications from several societies have survived. Even so, there is enough material

[1] According to the database of the Scholarly Societies Project of the University of Waterloo, a total of 210 societies are listed as being established between 1700 and 1799. (http://www.lib.uwaterloo.ca/society/overview.html).

J. Gomez (✉)
Department of Philosophy, University of Otago, Dunedin, New Zealand
e-mail: gomju504@student.otago.ac.nz

J. Maclaurin (ed.), *Rationis Defensor: Essays in Honour of Colin Cheyne*, Studies in History and Philosophy of Science 28, DOI 10.1007/978-94-007-3983-3_9,

worth looking into. This lack of a detailed research is shocking, especially because most of the intellectuals in the second half of the eighteenth century were members of at least one of such learned societies. Studying them will not only enhance our understanding of the Scottish enlightenment, but it could also help us shed light on the intellectual context and development of some of its most relevant figures.

In this paper I investigate two of the Scottish learned societies, The Philosophical Society of Edinburgh and the Select Society of Edinburgh, from a specific focus point: I am interested in the role played by the experimental method. This method, originating in Bacon's works and adopted and developed by intellectuals throughout the second half of the seventeenth and the whole of the eighteenth century, was regarded by most of the British philosophers to be the only method that made the acquisition of knowledge possible. I believe that this method played a very important role in the development and success of these learned societies, and an examination of the available material will provide enough evidence to support my claim. I will first introduce a relevant methodological framework for our present purpose: the experimental/speculative distinction. With this terms in mind, we can explore the main inspiration for most learned societies, Bacon's college described in his *New Atlantis*. These two aspects will provide us with a proper framework to examine the rules, question lists, and minute books of the societies to discover how the distinction played a prominent role. This is followed by an examination of some of the discourses and discussions to show that the use of the distinction was not just rhetorical, but expresses a deep commitment to the experimental method as *the* only way to acquire knowledge. Finally, we will take a look at the most important feature the experimental method could offer to the societies: the practicality that would allow them to contribute to the progress of Scotland as a nation.

9.2 The Experimental/Speculative Distinction

For the past 2 years I have been part of a research project at the University of Otago. It is titled "Experimental Philosophy and the Origins of Empiricism", led by Peter Anstey, with Alberto Vanzo, Kirsten Walsh, and myself contributing to the project. We are arguing for new terms of reference to approach philosophy in the early modern period. The distinction between rationalism and empiricism has been the traditional way of talking about philosophy in the eighteenth century. We believe that the distinction between speculative and experimental philosophy can provide us with a better understanding of the early modern period. Besides being the actual terms used by late seventeenth and eighteenth century philosophers (the terms 'rationalism' and 'empiricism' in their modern sense only emerged after Kant), they provide us with a more accurate interpretation of the development of early modern philosophy.

The base for the project can be found in Peter Anstey's 'Experimental vs. speculative Natural Philosophy' (Anstey 2005). I will briefly present how the experimental/speculative distinction plays out, leaning on Anstey's article. There are five main claims Anstey argues for:

1. This distinction is in evidence, in some form or other, from the late 1650s until the early decades of the eighteenth century.
2. This distinction provides the primary methodological framework within which natural philosophy was interpreted and practised in the late seventeenth century
3. This distinction is independent of disciplinary boundaries within and closely allied to natural philosophy
4. This distinction crystallised in the 1690s when opposition to hypotheses in natural philosophical methodology intensified
5. This distinction provides the terms of reference by which we should interpret Newton's strictures on the use of hypotheses in natural philosophy (Anstey 2005, p. 216).

Anstey's article focuses on the distinction within natural philosophy, but our research project goes beyond that. Roughly, when we talk about speculative philosophy we are referring to "the development of explanations of natural phenomena without prior recourse to systematic observation and experiment;" experimental philosophy consisted in "the collection and ordering of observations and experimental reports with a view to the development of explanations of natural phenomena based on these observations and experiments" (Anstey 2005, p. 215). I believe that such distinction also plays a role outside of natural philosophy: as a rhetorical device and as the expression of a methodological commitment to the experimental method. This commitment implied adopting an anti-hypothesis attitude and a strict emphasis on facts and observations as the sources of our explanations.

It also important to mention here that in Britain most intellectuals sided with experimental philosophy and in their works the term 'speculative' even comes charged with a derogatory tone. Not only was experimental philosophy the only valid path to knowledge, but those practicing speculative philosophy were doing nothing more than the mere construction of fables and fantasies that had nothing to do with reality. This is not to say that the distinction was used by the British as no more than a means of justifying the superiority of their intellectual inquiries; on the contrary, their deep methodological commitment to the experimental method was what justified their use of the term 'speculative' as derogatory.

The origins of the distinction (and the commitment to facts and observation) can be found in the works of Francis Bacon, but as Anstey points out, not in the same form it appeared in the last decades of the seventeenth century (Anstey 2005, p. 217). Bacon is relevant for the attitude he embodied, since he "famously opposed idle speculation and promoted the derivation of natural knowledge from experiment" (Anstey 2005, p. 217). This approach to the acquisition of knowledge got impressed in his *New Atlantis* which was the inspiration for the establishment of the Royal Society and subsequently most of the eighteenth century philosophical societies.

9.3 Bacon's New Atlantis and Philosophical Societies

It is evident that the institution of the Royal Society of London was inspired by, dedicated to, and based on Bacon's work, as we can tell from Cowley's ode to the society and Sprat's *History of the Royal Society*:

> Some few exalted Spirits this latter Age has shown
> That Labour'd to assert the Liberty
> (From Guardians, who were now Usurpers grown)
> Of this Old Minor still, Captiv'd Philosophy;
> But 'twas Rebellion called to fight
> For such a long oppressed Right
> Bacon at last, a mighty Man, arose
> Whom a wise King and Nature chose
> Lord Chancellour of both their Laws,
> And boldly undertook the injur'd Pupils caus (Sprat 1959).

Besides these first-hand accounts, the connection between Solomon's College of the *New Atlantis* and the Royal Society has been constantly mentioned in the scholarship.[2] As for the Scottish Societies in question, we know they also found their inspiration in Lord Bacon as well as in the Royal Society. It is therefore of our interest here to examine the nature of Bacon's Solomon College in order to shed light on the character of the Scottish learned societies we are investigating.

Lord Bacon's *New Atlantis* is a fable, according to Rawley's preface,

> devised to the end that He [Bacon] might exhibite therein a modell or Description of a Colledge instituted for the Interpreting of Nature and the Producing of Great Marveilous Works for the Benefit of Men: under the name of Salomon's house, or the Colledge of the Six Dayes' Works (Bacon 1900, p. ix).

Rawley also mentions that the work is incomplete, since Bacon had in mind the description of a model for the best state as well, but preferred to complete the description of the college first. It is a story about a ship crew that sets sail from Peru towards China, by the 'South Sea,' but gets lost on the way and stumbles upon the island of Bensalem. This country is unknown to the Europeans of the time, but the stranded sailors find a prosperous and civil nation that receives them and treats them as best as it is possible to imagine. After some days in the island, the stranded sailors start settling in, and discovering the day to day life of the island. They hear about Solomon's College, described as the "very Eye of this Kingdom," and our narrator eventually obtains a hearing with one of the fathers of the house, who describes the college and its workings.

In what follows I want to show, by relying on Bacon's text, how the learned societies can be viewed as the instantiation of Solomon's College. Most of the description given by the delegate from the college is occupied by a detailed listing of all the different houses, gardens, pools, caves, mountains and buildings they use

[2] For example see Hunter (1989), Lynch (2001).

for their experiments. Though very interesting, in this paper I am more interested in the specific end of the college and the description of the functions of the fellows. The purpose of the college is summed up in the following sentence:

> The End of Our Foundation is the Knowledge of Causes, and Secret Motions of Things; And the Enlarging of the bounds of the Humane Empire, to the effecting of all Things possible (Bacon 1900, pp. 34–35).

From this statement we can identify two aims of the college, the first being the actual science, and the second one the improvement of mankind (based on the achievements of science of course). The Edinburgh societies were committed to carrying out both this goals, but it was the second one that allowed them to contribute greatly to the progress of Scotland.

Bacon not only establishes the goals such an academy should aim for, but also gives us some guidelines on how to achieve them. The college assigned a number of functions to their fellows to be carried out for this purpose. They are of special interest because they summarize nicely what the Scottish societies established as the tasks they had to carry out. There are nine main functions described in the text (Bacon 1900, pp. 44–45):

1. Travel to foreign countries for the exchange of knowledge.
2. The written collection of experiments. (Scientific)
3. The written collection of experiments. (Non-scientific, e.g. liberal sciences)
4. Trying out new experiments.
5. Categorization of experiments. (Titles and tables)
6. Practical output of experiments.
7. Direct new experiments based on the former. (More penetrating into nature)
8. Execution of experiments.
9. Raise the discoveries of the experiments into Axioms and Aphorisms.

The remaining paragraphs of the description of the college are dedicated to the ordinances and rites, which are not of our present interest. The brief description we have given of the goals and functions of the college will help us sketch a picture of the Edinburgh societies that reflects Bacon's proposals.

The description given above of the end and functions of the college holds the essence of the experimental method. It was this approach to the study of nature that the British philosophers believed would give us knowledge and bring about the progress of humanity. This experimental method of the new philosophy was the driving force behind the Scottish societies. Solomon's college, more than a general model for the societies, embodied the advantages and benefits of following an experimental approach to the study of nature and the acquisition of knowledge in general, as opposed to a purely speculative approach that was detached from facts and observation. I will rely on the rules, minute books and some of the available essays to show that (a) The experimental/speculative distinction was employed in the societies, and (b) The societies were fully committed to the experimental method of the new philosophy in order to contribute to the progress of Scotland as a nation.

9.4 The Evidence

Although there are few written records from the time these Edinburgh societies functioned, we still have access to some very relevant texts and manuscripts, and hopefully we'll be able to identify more in due time. Regarding the Edinburgh Philosophical Society we have the eight volumes of collected essays published by the society between 1742 and 1771, and five volumes containing discourses solely from the absorbed Medical Society (1731–1737). Other than these collections of essays, we can rely on some mentions of the society in the correspondence and biographies of some of the members. The case with the Select Society is even worse, since they had no publications, though their minute book is available. The work done by Roger Emerson has been remarkable and has served as a perfect guide for this research. Still, from the primary sources we can find enough passages to shed light on our understanding of them.

Since there are no surviving minute books, records, membership lists, etc. belonging to the Philosophical Society of Edinburgh, we have to rely on the prefaces to their five published volumes of essays for information on the way the society was run and any kind of methodological statements. Besides these prefaces, the *Proposals for the Regulation of a Society for Improving Arts and Sciences and particularly Natural Knowledge* was published in the General index to the first 34 of volumes of the Transactions of the royal Society of Edinburgh. The *Proposals* outline the administrative regulations for the society, as well as certain rules regarding topics and guidelines for discussion. One rule in particular shows the strictly Baconian and experimental approach they had in mind:

> Authority is to be held of no weight in their reasonings. The shew of Learning, and Quotation of Authors sparingly used in their Papers. Things to be minded not words. Arguments to be chiefly drawn from proper Experiments and clear Consequences deduced from them or from evident Propositions. Metaphysical Subtilties not be insisted on (Emerson 1979, p. 165).

This rule highlights the attitude characteristic of experimental philosophy, rejecting any sort of speculation detached from observation ("metaphysical subtilties," emphasis on words and authority) and calling for a due attention to facts and experiments. It seems that this attitude was carried on from the *Proposals* to the established society as we can tell from the prefaces of their published volumes. The Philosophical Society of Edinburgh emerged in 1737 when Colin MacLaurin proposed to Alexander Monro, secretary and founding member of the Medical Society of Edinburgh, to enlarge the project of the society and include natural philosophy instead of just medicine. The Medical Society had published from 1731 to 1737 a number of volumes of essays, which were reprinted by the newly formed Philosophical Society. The second edition of the first volume contains a number of essays that are highly charged with the rhetoric of experimental philosophy, constantly referring to the emphasis on facts and observation. The volumes are also dedicated to Sir Hans Sloane and the RSL, which he presided. In the preface

Monro mentions that they follow the great example of the Royal Society, with the disclaimer that they would focus only on medical topics. Monro also includes in the preface a list of regulations for the acceptance of papers. Among them we find statements that show their adoption of the attitude of experimental philosophy and their method, highlighting the importance of facts and observations and rejecting any kind of a priori speculation:

> 2. (Regulations for papers) Whatever Relation may be found between the Changes in our Atmosphere and the epidemick Diseases, all seem to agree, that there are certain Circumstances and Symptoms which distinguish the Return of similar Constitutions and point out the most probable Method of Success in treating Diseases while such Constitutions prevail. The only way of discovering certainly these Circumstances and Symptoms, is a long continued Series of Observations, which we hope our Work will supply.
>
> 2. (Regulations for correspondants) The Descriptions and Virtues of simple Drugs, are to be clearly and succinctly told, without enlarging on arguments *a priori*, which are to liable to lead into Error (Philosophical Society of Edinburgh 1737–1738, pp. 17 and 19).

The society published the final volume of Medical essays in 1752, and the first volume of their *Essays and observations, physical and literary* was published in 1754. It was followed by two other volumes, one in 1756 and the last one in 1771, along with a second edition of the two first volumes. The preface to the first volume briefly summarizes the story of the society, mentioning that it follows the same goals which other established societies (RSL) have pursued:

> The object of this society is the same with that of the other academies, which have been established in other parts of Europe, the promoting of natural philosophy, and of literature, by communicating to the public such dissertations as shall be transmitted to them, either by their own members or by others. 'Tis allowed, that these two branches of learning, especially the former, are more promoted by the observation of facts than by the most ingenious reasonings and disputations (Philosophical Society of Edinburgh 1754, p. v).

It is worth noting that the writers of the preface (Monro and Hume[3]), tell us that the method should also be used outside natural philosophy, mentioning literature in particular. This supports the claim that their society was fully committed to the experimental method, regardless of the area of inquiry. Their praise for experimental philosophy is confirmed after the above passage, where there is a reference most likely to Bacon and Newton and their contribution to natural philosophy:

> There arise, from time to time, bold and happy geniuses, who introduce method and simplicity into particular branches of science; and reducing the scattered experiments to more general theorems, abridge the science of nature. Hints of this kind, we hope, may be able to pass thro' our hands; and at worst, our collections will be a species of magazine, in which facts and observations, the sole means of true induction, will be deposited for the purposes of philosophy (Philosophical Society of Edinburgh 1754, p. vi).

[3] Although the authorship of the preface is usually ascribed only to Hume, I am working on a paper that shows that this is not the case.

Although these passages show their full commitment to the experimental method, the society did feel that some topics had to be banned certain from being discussed at the society:

> The sciences of theology, morals, and politics, the society are resolved intirely to exclude from their plan. However difficult the inferences in these sciences, the facts, on which they are founded, are extremely obvious; and we could not hope, by our collections, to be, in this respect, of any service to the public (Philosophical Society of Edinburgh 1754, p. vii).

Even though they still describe them as 'sciences' and founded on facts, they exclude them because they want to avoid topics that will lead to personal disputes, and not because they could not be treated like the other sciences. This particular exclusion is most likely one of the reasons why a number of the members of the Philosophical Society of Edinburgh decided to establish the Select Society of Edinburgh, which debated questions in morals, politics, economy, arts and literature. There is nothing in the rules annexed to the minute book, or any methodological statement that can tell us explicitly that they also followed the experimental method. There is a rule on the topics to be debated that only excludes topics on "revealed religion, or which may give occasion to vent any Principles of Jacobitism," presumably for the same reason their parent society excluded political and religious topics. Roger Emerson mentions in his paper on the society that they followed what he calls 'empirical methodological norms,' praised the work Locke and Newton, and that experience was their guide in moral science. But there is nothing in the minute book or question book that points this out. Emerson's conjecture is probably based mostly on the character and other works of the famous members of the society, like David Hume and Adam Smith.

What we can say of the Select Society from the available manuscripts is that they were fully committed to Lord Bacon's purpose where the society was supposed to contribute for the progress of the nation. Besides the minute book, the society kept a question book which contained the topics proposed by the members to be discussed at the meetings. Among them we find many that were strictly related with the actual state of the country and ways to contribute to its progress. Many members of the society were Edinburgh businessmen involved with trades and manufactures and were interested in debating questions directly related to the improvement of their enterprises, and in consequence the economic and industrial progress of Edinburgh. The question book shows that the members proposed and debated a number of questions related to trades, taxes and manufactures. The following are just a small sample of them: "Whether Bounties on the exportation of Corn be advantageous to Trade and manufactures as well as to agriculture?", "Whether the Numbers of Banks now in Scotland be useful to the trade of their Country? And whether paper Credit be advantageous to a nation?", "Whether the Bounty should be continued on the Exportation of low priced linens made in Scotland?", "Do the Laws in Scotland relating to Coalers and Salters promote the Interest of this Country?" (Minutes of the Select Society of Edinburgh, MSS 170).

Besides the vast number of questions recorded that have something to do with the economic and social progress of Scotland, the society created two off-shoot societies specifically designed to promote the progress of Scotland: the Edinburgh Society for the encouragement of arts, sciences, manufactures and agriculture, and The Society for promoting the reading and speaking of the English language in Scotland. The former gave premiums in different categories, with the following purpose in mind:

> To encourage genius, to reward industry, to cultivate the arts of peace, are objects deserving the attention of public-spirited persons.
>
> That the inhabitants of Scotland may become diligent in labour, and excellent in arts, is the concern of all who indeed love their country (Edinburgh Society for the Encouragement of Arts, Sciences, Manufactures and Agriculture, p. 3).

They gave prizes for the best essay in taste, best essay in vegetation, best discovery in science, best printed book, best printed cotton cloth, best beer, best carpet, best crop of potatoes, of tulips, and so on (An account of the Edinburgh Select Society 1755, pp. 126–130). The members of the society believed that this was the encouragement needed to promote the Scottish industry. But the efforts of the society to contribute to the progress of Scotland did not stop at the awarding of prizes. There is one situation in particular that shows the practical outcome of the debates of the society. The society debated a question regarding the roads of Scotland, as we know from the list of questions given in the *Scots Magazine* for March 1757, namely, "What is the best method of getting public highways made, and repaired: whether by a turn-pike, as in many places in G. Britain? By county or parish work? By a tax? Or by what other method?" (Questions treated in the Edinburgh Society 1757, p. 164). The same periodical published in 1759 part of a document that related the outcome of the debate, which resulted in a set of proposals and a plan that eventually resulted in the improvement of the Scottish roads.

The other off-shoot society, for the English language, was created due to the poor prose of the Scottish university students and the desire of government officials to be more eloquent. For this purpose the society subsidized a series of lectures on the English language given by Thomas Sheridan. I quote the *Scots Magazine* for 1761 to show the success of the lectures:

> They were attended by more than 300 gentlemen, the most eminent in this country for their rank and abilities; who expressed no less satisfaction with the ingenuity and justness of his [Sheridan's] sentiments, than with the elegant and interesting manner in which he deliver them (Lectures on the English tongue by Mr. Sheridan 1761, p. 390).

The rules and prefaces give us a picture of the beliefs of the founders of the societies regarding the purpose of their institutions and the method to carry it out. Even though it is clear that they followed the experimental method, the available discourses and essays provide even more evidence of the commitment to the experimental method the members of such societies had adopted. Most of the essays included in the three collected volumes express in some way the rejection of mere speculation and the promotion of the experimental method. I will only refer to some of the more explicit passages that illustrate this.

The first two essays of the first volume are on the laws of motion. The first one is by Henry Homes, Lord Kames, and the second one by John Stewart as a reply to

Kames' essay. Even though they disagreed, we can see that the discussion is always held within the bounds of experimental philosophy, both of them relying on experiment and observation for their arguments. Lord Kames' essay starts with the following statement where he shows his apathy towards speculative philosophy:

> Nothing has more perplexed philosophy, than an unlucky propensity, which makes us grasp at principles, without due regard to facts and experiments...This bent of the mind is productive of manifold errors. Prepossessed once by a favourite principle, we are no longer open to conviction. Every phenomenon must be accommodated to that principle, and every opposite fact, however obstinate, must go for nothing.
>
> Even in Natural Philosophy, theory was introduced before experiment, and every philosopher urged his own notions, without regard to truth or reality. This produced a mass of undigested and contradictory theory; which at length could not fail to bring on the discovery, that the whole was a little better than a fancy and chimera.
>
> But tho' our only sure guides to truth are fact and experiments, it is however expedient to keep the end in view. Facts and Experiments are useless lumber, if we are not to reason about them, nor draw any consequences from them...Theory becomes only a source of error, when we indulge in it too much, or attach ourselves to it beyond what facts and experiments can justify. In short, theory is vain without experiment, and experiments are best understood by applying them to theory (Philosophical Society of Edinburgh 1754, pp. 1–3).

Lord Kames is constantly referring in his essay to facts and observations, and to the rejection of false hypothesis. He concludes his essay by reminding us that his comments are based on facts and observation and not mere conjectures:

> I shall conclude the essay with the following observation, that the powers I have ascribed to matter, are in nothing similar to occult qualities. The error of those who dealt in the doctrine of occult qualities was, in attributing every different effect to some quality or cause confined to that single effect...This was not advancing a single step in knowledge, but amusing one's self with words in place of things (Philosophical Society of Edinburgh 1754, p. 68).

John Stewart's reply is also written using the rhetoric of the experimental method, always talking about experiments and observation as the only sources of knowledge, but there is no explicit methodological statement like the ones found in Kames' essay. We do find such statements in an essay by Charles Alston on the sexes of the plants. His aim is to give an overview of the available theories regarding the sexes of the plants and reject them for not being based on facts and experiments. After reviewing most of the argument from different botanists, he even rejects arguments from analogy in natural philosophy, which he believes are not reliable:

> Thus I think I have sufficiently answered all the arguments for the sexes of plants, taken either from the structure of flowers, or experiments of any consequence that I could meet with. But since no small stress seems still to be laid on the analogy between plants and animals, as much favouring this doctrine; I must beg leave a little to consider it also... the method of reasoning by analogy, is but too apt to lead us into mistakes...for mere analogy, based on facts, and extended by conjecture, however plausible, can at most, but furnish motives for a reasonable doubt, and further inquiry (Philosophical Society of Edinburgh 1754, p. 270).

We can see here that Alston rejects the argument from analogy since all it does is extend facts by conjectures, showing his full commitment to the experimental method and rejecting even the more plausible theories regarding the sexes of the plants. We find similar expression of this commitment in two essays by a fellow teacher of Alston at Edinburgh, the chemist Andrew Plummer. His first essay in the collection relates some observations on chemical solutions. Plummer introduces the essay by clearly stating his methodology:

> I shall lay down these remarks by way of propositions, and, after each, shall mention the facts or experiments which gave occasion to the remark, or which confirm and illustrate the proposition (Philosophical Society of Edinburgh 1754, p. 284).

Similarly, in his second essay Plummer contrasts speculative and experimental philosophy. The essay is about neutral salts and in it he uses some principles of motion. He concludes the essay with the following remark:

> These principles of motion in matter, are not the vain fictions of men merely speculative in philosophy, but evidently deduced from observations and experiments on a great variety of bodies in many different circumstances (Philosophical Society of Edinburgh 1754, p. 340).

This is perhaps the most explicit expression of the experimental/speculative distinction in the collection of essays, although there are many instances where the praise for the experimental method is present. Thomas Melville reinforces this sentiment in his essay on light and colors. He contrasts Euler's theory with Newton's, preferring the latter for the method and consistence with experience. Here is Melville praising Newton:

> As it is of great consequence in philosophy, to distinguish between facts and hypotheses, however plausible; it ought to be observed, that the various refrangibility, reflexibility, and inflexibility of the several colours, and their alternate dispositions at equal intervals to be reflected and transmitted, which are the whole ground-work of the Newtonian system, are to be considered as certain facts deduced form experiment (Philosophical Society of Edinburgh 1754, p. 50).

Not only does he praise the Newtonian method, but later on when mentioning Descartes he shows his aversion to speculative philosophy:

> From the lazy method of philosophizing in the closet, among books and diagrams, there never arose, never will arise, any discovery of consequence (Philosophical Society of Edinburgh 1754, p. 88).

There are plenty more passages where the allegiance of the authors can be detected, either from their rejection of vain speculation or their constant references to facts, observation, and experiments as the only foundation for their claims. The quotes provided here illustrate nicely the attitude held by the Society and its members. But the adoption of the experimental method by them was not just expressing an attitude and following a methodology. The Scottish societies found great value in the emphasis on practical outcomes contained within the experimental method.

9.5 The Progress of Scotland

So far we have examined the commitment of the society to the experimental method, but not much has been mentioned of their impact on society. This is where Bacon's Solomon's college plays an influential role. If we recall Bacon's description, acquiring knowledge through the experimental method was only one part of the purpose of the college. Knowledge was not enough unless it was used for the progress of mankind. Nowhere was this attitude as important as it was for the Edinburgh societies. Their desire and willingness to follow the functions proposed by Bacon and develop practical uses for their experiments was attempted and achieved with success in some cases. In what follows I will show some examples of how some of the undertakings of the society evolved into practical contributions for the development of Scotland.

One of these undertakings was the proper mapping of the country, specifically of Orkney, the Shetland Islands, and the northern coasts. The main reason for carrying out this task was that there was a lack of accurate maps for large areas of the country, and completing them would benefit trade and commerce by accurately mapping out trade routes.

The society ceased to be active after the death of Colin MacLaurin in 1746, but it was revived around 1750 and started meeting again. With the revival a change of vice-presidents was in order, and this shifted the societies focus to a much more industry-oriented output. Lord Kames and Robert Whytt occupied the positions, the former being the one that drove the societies experiments and discussions to have some impact in the public life of Scotland. The Philosophical Society discussed a number of papers on fisheries, the uses of peat, a proposal for an engine to raise water from coal or lead mines, as well as all sort of questions on improvements to make mines more efficient.

A letter from Alexander Dick shows this drive to put their experiments into practical use:

> I have been wishing for the Aera to see those collections of facts and circumstances (I have been making in relation to our entails, our high roads, the employment of the industrious poor, & the increase of our population, in which my good Lord kames & some others have given me great assistance) brought into real utility. I have no other ambition for the years I have to live but to contribute all in my power to improve Medicine... (Cited in Emerson 1981, p. 154).

Sir Alexander also carried out, with the help of the chemists of the society, an alternative to linen rags which supply was decreasing. They engaged in paper-making from a plant found at the ponds around Edinburgh. This alternative helped not only stop the decrease, but it actually contributed to arise in linen production, as Emerson points out (Emerson 1981, p. 163). Description of operations and diseases helped advancements in medicine, specially the work on fevers by Robert Whytt, William Cullen, and John Pringle. The Monro's also read a number of papers on operations and dissection, which proved to be helpful for the medical students and increase the reputation of the Edinburgh Medical School.

These are just some examples of the impact the activities of the society had on Scotland's public life. They show that the society was concerned with the utility of their discussions, turning facts and observations into practical uses for the progress of the country. This practical consideration was attached to the adoption of the experimental method, and without this methodological commitment the societies would not have had the impact they had on eighteenth-century Scotland.

References

An account of the Edinburgh Select Society. 1755. *Scott's magazine*, March, vol. xvii, pp. 126–130. Retrieved 15 August 2011, from http://books.google.co.nz/books?id=_lwAAAAAYAAJ&source=gbs_navlinks_s.

Anstey, P.R. 2005. Experimental vs. speculative natural philosophy. In *The science of nature in the seventeenth century*, ed. P.R. Anstey and J.A. Schuster, 215–242. Dordrecht: Springer.

Bacon, Francis. 1900. *The New Atlantis*, ed. G.C. Moore Smith. Cambridge: Cambridge University Press.

Edinburgh Society for the Encouragement of Arts, Sciences, manufactures, and Agriculture. 1755. *Rules and orders of the Edinburgh Society for the encouragement of arts, sciences, manufactures, and agriculture*. Eighteenth Century Collections Online. Gale. Retrieved 15 August 2011, from http://find.galegroup.com/ecco/infomark.do?&contentSet=ECCOArticles&type=multipage&tabID=T001&prodId=ECCO&docId=CW124934741&source=gale&userGroupName=otago&version=1.0&docLevel=FASCIMILE.

Emerson, R.L. 1973. The social composition of enlightened Scotland: The select society of Edinburgh, 1754–1764. *Studies on Voltaire and the Eighteenth Century* 114: 291–329.

Emerson, R.L. 1979. The Philosophical Society of Edinburgh, 1737–1747. *The British Journal for the History of Science* 12(2): 154–191.

Emerson, R.L. 1981. The Philosophical Society of Edinburgh, 1748–1768. *The British Journal for the History of Science* 14(2): 133–176.

Emerson, R.L. 1985. The Philosophical Society of Edinburgh, 1768–1783. *The British Journal for the History of Science* 18(3): 255–303.

Emerson, R.L. 1988. The Scottish enlightenment and the end of the Philosophical Society of Edinburgh. *The British Journal for the History of Science* 21(1): 33–66.

Edinburgh Society for the Encouragement of Arts, Sciences, manufactures, and Agriculture. 1755. Rules and orders of the Edinburgh Society for the encouragement of arts, sciences, manufactures, and agriculture. Eighteenth Century Collections Online. Gale. Retrieved 15 August 2011, from http://find.galegroup.com/ecco/infomark.do?&contentSet1/4ECCO Articles&type1/4multipage&tabID1/4T001&prodId1/4ECCO&docId1/4CW124934741&source1/4gale&userGroupName1/4otago&version1/41.0&docLevel1/4FASCIMILE.

Hunter, M. 1989. *Establishing the new science: The experience of the early Royal Society*. Wolfeboro: Boydell press.

Lectures on the English tongue by Mr. Sheridan. 1761. *Scott's magazine*, July, vol. xxiii, p. 390. Retrieved 15 August 2011, from http://books.google.co.nz/books?id=QD0oAAAAYAAJ&dq=scots+magazine+1761&source=gbs_navlinks_s.

Lynch, W.T. 2001. *Solomon's child: Method in the early Royal Society of London*. Stanford: Stanford University Press.

McElroy, D. 1952. *The literary clubs and societies of eighteenth century Scotland, and their influence on the literary productions of the period from 1700 to 1800*. Unpublished PhD thesis, University of Edinburgh.

McElroy, D. 1969. *Scotland's age of improvement: A survey of eighteenth-century literary clubs and societies*. Pullman: Washington State University Press.

Minutes of the Select Society of Edinburgh. *MSS 170*, 1754–1763. National Library of Scotland.

Philosophical Society of Edinburgh. 1737–1738. *Medical essays and observations, revised and published by a society in Edinburgh. ... Volume 1. The second edition corrected.* Eighteenth Century Collections Online. Gale. Retrieved 15 August 2011, from http://find.galegroup.com/ecco/infomark.do?&contentSet=ECCOArticles&type=multipage&tabID=T001&prodId=ECCO&docId=CW108824207&source=gale&userGroupName=otago&version=1.0&docLevel=FASCIMILE.

Philosophical Society of Edinburgh. 1754. *Essays and observations, physical and literary. Read before a society in Edinburgh, and published by them. Volume I.* Eighteenth Century Collections Online. Gale. Retrieved 15 August 2011, from http://find.galegroup.com/ecco/infomark.do?&contentSet=ECCOArticles&type=multipage&tabID=T001&prodId=ECCO&docId=CW107037053&source=gale&userGroupName=otago&version=1.0&docLevel=FASCIMILE.

Questions treated in the Edinburgh Society. 1757. *Scott's magazine*, March, vol. xix, p. 164. Retrieved 15 August 2011, from http://books.google.co.nz/books?id=fF0AAAAAYAAJ&dq=scots+magazine+volume+xix&source=gbs_navlinks_s.

Sprat, T. 1959. *History of the Royal Society*, ed. J.I. Cope and H.W. Jones. London: Routledge & Kegan Paul.

Part III
In Metaphysics

Chapter 10
Propositions: Truth vs. Existence

Heather Dyke

10.1 Introduction

The view that propositions exist as part of the furniture of the world is one that I think we should be suspicious of. There is an underlying tension in the notion of a proposition arising from the fact that some of the roles propositions are normally thought to play, such as being the primary truth-bearers, require them to be essentially representational, while others, such as being the objects of knowledge, require them to be non-representational. I argue that this tension maps onto an important distinction: that between the truth of a proposition and its existence. Finally, I argue that attending carefully to this distinction permits a nominalist view of propositions according to which we can have all the benefits of proposition-talk without the drawbacks of being committed to their robust existence.

10.2 The Standard View of Propositions

What I take to be the standard view of propositions is a collection of theses typically asserted by philosophers, both those who believe in them and those who don't.[1] They comprise a view which is intended to be taken as neutral between competing ontological theories of propositions. The theses can be divided into two groups: those that identify the principal roles that propositions are normally thought to play,

[1] There are, of course, particular theses that are rejected by some friends of propositions, but by and large, they adhere to a majority of these theses.

H. Dyke (✉)
Department of Philosophy, University of Otago, Dunedin, New Zealand
e-mail: heather.dyke@otago.ac.nz

J. Maclaurin (ed.), *Rationis Defensor: Essays in Honour of Colin Cheyne*, Studies in History and Philosophy of Science 28, DOI 10.1007/978-94-007-3983-3_10,

and those that state what features propositions are normally thought to possess. I will outline the two groups of theses in turn.

10.2.1 The Characteristic Roles of Propositions

(1) Propositions are the fundamental, or primary bearers of truth and falsity. They are also the primary bearers of the modal properties of necessity, possibility, contingency and impossibility. And they are the primary relata of the various logical relations, such as entailment and contradiction.
(2) Propositions are the objects of knowledge, belief and other intentional attitudes such as desire, fear, hope etc.
(3) Propositions are the objects of acts of assertion. When I utter a declarative sentence such as 'The window is broken', my utterance asserts a proposition, namely, the proposition *that the window is broken*.
(4) Propositions are what declarative sentences express or mean. Alternatively, propositions are expressed by declarative sentences, and they *are* the meanings of those declarative sentences. The sentence 'The window is broken' expresses the proposition *that the window is broken*, and that proposition is the meaning of that sentence.
(5) Propositions are representational entities. They represent the world, or things in the world, as being a certain way. The proposition *that the window is broken* represents the window being broken.

10.2.2 The Characteristic Features of Propositions

(6) Propositions are mind-independent. They are not produced by minds, nor are they mental states. They would have existed even if minds had never evolved. There are many propositions that have never been thought by anyone, and never will be, but they exist nonetheless.
(7) Propositions are language-independent. They are not linguistic entities, but extra-linguistic entities. They can be expressed by linguistic entities, such as utterances and written sentence-tokens, but they would have existed even if language users had never existed.
(8) Propositions are abstract entities. A broken window is a concrete entity, but it is not the same thing as the proposition *that the window is broken*.
(9) Propositions exist necessarily. The proposition *that there is a tooth fairy* is false in this world and in many other worlds too, but there are worlds in which it is true. So it must exist in every world in order to possess there the truth-value that it does.
(10) Propositions exist eternally. The proposition *that hydrogen is less dense than helium* was true before anyone discovered, thought or uttered it, and will

remain true long after language users die out. Thus, the proposition itself exists at all times.

(11) Propositions have their truth conditions essentially. The proposition *that the window is broken* is true if and only if the window is broken, and it must have that particular truth condition to be the proposition that it is. Sentences, on the other hand, have their truth conditions only contingently. If we used the word 'window' to mean teapot, then the sentence 'The window is broken' would be true if and only if the teapot is broken.

(12) Propositions have their truth-values eternally. Sentences that appear to change their truth-value over time, such as 'It is now raining', in fact express different propositions at different times of utterance, for example, when uttered at *t* it expresses the proposition *that it is raining at t*, and those time-indexed propositions have their truth-values eternally.

(13) Propositions are publicly accessible. Two or more people can stand in the 'believing' relation to one and the same proposition.

I don't mean to suggest that every realist about propositions is committed to every one of these theses. However, I do think that, by and large, realists about propositions believe a substantial majority of them. Next I want to highlight some tensions in the standard view as encapsulated by the foregoing 13 theses.

10.2.3 Some Tensions in the Standard View[2]

According to (1) propositions are the fundamental bearers of truth and falsity, and according to (4) they are the meanings of declarative sentences. But prima facie, meanings are not the sorts of things that can have truth-values. Perhaps it would make more sense to say that propositions *have* meanings. That way they could possess both truth-values and meanings. But then we are left without an explanation of what meanings are. And the thought that meanings are propositions was attractive precisely because it gives an objective and publicly accessible account of what meanings are.

Thesis (4) also comes into tension with thesis (2) that propositions are the objects of intentional attitudes such as belief, and with thesis (3) that propositions are the objects of acts of assertion. When I believe that the window is broken, prima facie, the object of my belief is not a meaning. In general, we don't believe meanings, still less do we desire, hope or fear them. Neither do we assert, deny or prove meanings (Thomson 1969, p. 739).

Thesis (2) also comes into tension with thesis (5) that propositions are representational entities. When I believe that the window is broken, I don't stand in the

[2] Wrigley (2006) alludes to the objection to realism about propositions that the roles they play are "potentially incompatible" (p. 158) but does not expand on the nature of this incompatibility.

'believes' relation to some sort of representation of the window's being broken. Rather, I believe that the window itself (the familiar concrete entity) is actually broken. In other words, my belief is about the world, not about some representational intermediary.

Something like this thought makes Russell's early (1903) view of propositions as structured entities seem attractive. According to that view the proposition that the window is broken consists of the window itself, the concrete entity made of glass possessing the property of being broken. Propositions so conceived seem like very concrete, worldly entities. One might even go so far as to think that on this view there is no representational intermediary between the belief and the fact[3] in the world that the belief is about; that this view has done away with propositions altogether. Of course, as is well known, this view runs into difficulty when faced with the question of how to deal with false beliefs. Suppose the window is not broken, but I believe that it is. The object of my belief cannot be the window possessing the property of being broken, because the window does not possess that property. It may be tempting to identify propositions with facts in the world so long as one is considering only true propositions, but false propositions cannot be identified with facts, because the facts with which we would like to identify them do not exist.

It may be considerations of false propositions that encourage the thought that propositions are abstract, or thesis (8). Since there are no facts corresponding to false propositions, we cannot identify a false proposition with the fact that it represents. It may be natural then to think that we should move to the realm of abstracta, and there we will find our false propositions. But, the thought continues, false propositions are not different in kind from true ones, so if the false ones are abstract, the true ones must be abstract too. Now that we are in the position of thinking that the objects of our beliefs are abstract, rather than the concrete entities that they are about, it makes sense to think of them as representing the world. Then we can say that all propositions represent the world, some of them truly, and some of them falsely. But now we have come right back to our point of departure: the tension between theses (2) and (5). When I believe that the window is broken, I believe that the window itself is broken, whether I believe truly or falsely. I do not believe some representation or picture of the window's being broken. So thesis (8) exacerbates the tension between (2) and (5).

There is a further tension between theses (5), that propositions are representational, and (8) that they are abstract, which is that it is difficult to see how anything that is abstract can be representational at all (see, for example, Jubien 2001, p. 54). In order for a to represent b, a must at the very least be numerically distinct from b, and must stand in some relation that can be construed as representing to b. And this 'can be construed as' is essential to the nature of representation.

[3] The term 'fact' is, unfortunately, a theoretically loaded one. My use of it is intended to be as theoretically neutral as possible. By 'fact' I simply mean some extra-linguistic, non-truth-bearing entity or collection of entities.

Representation is a relation that an intentional being confers on a representing entity and what it represents. So it is implausible to think that an abstract, non-spatio-temporal entity could do that all by itself. What is needed for successful representation is beings like us, conscious beings who can construe a proposition as representing the world being a certain way.

But if its representational nature is conferred on it from an external source, then we have a further tension with thesis (11), that propositions have their truth conditions essentially. The proposition *that the window is broken* is true if and only if the window is broken. That it has this truth condition is essential for it to be the proposition that it is. If it had some other truth condition, then it would be a different proposition. But then it seems to follow that it is essential to that proposition that it represent the window's being broken which, if this concern about the representational capacities of abstract entities is justified, is not possible if propositions are abstract. So we have a three-way tension between theses (5), (8) and (11).

A final tension I want to mention is that between theses (1) that propositions are the primary truth bearers, and (2) that they are the objects of knowledge, belief, and other intentional attitudes. Suppose I believe the proposition *that the window is broken*. I then come to know that that proposition is true. What is involved in my transition from belief to knowledge? I must come to know the fact that it represents. But since I know the fact, the proposition is now redundant (see, for example, Ryle 1929, p. 107). So most propositional attitudes, (for example, belief, desire, hope etc.) require a proposition as intermediary between the mental state and the fact it is about, but for at least one, viz. knowledge, that propositional intermediary is redundant.

Since we can have false beliefs, and we can desire and hope for things that won't turn out to be realized, it is easy to suppose that beliefs, desires, hopes and so forth, represent the world, and they can represent truly or falsely. But when we turn to consider knowledge, and the possibility of knowing falsely is eliminated, it no longer seems reasonable to suppose that the object of knowledge is something that represents the world being a certain way. Now it seems more natural to suppose that the object of my knowledge is something in the world itself, and there is no longer any need for an intermediate proposition.

A common theme to these tensions is that propositions are, on the one hand, representational and the primary truth-bearers, and on the other, objective, mind-independent, language-independent, and the objects of intentional attitudes. In order to occupy both of these sets of roles they are required to be both representational and non-representational; a fundamental conflict. Propositions are taken to be intermediaries between language and thought on the one hand, and the external world on the other. But the foregoing suggests that this is not a position that can happily be occupied by any single entity, or perhaps by anything at all. There is pressure from the 'language and thought' side of the divide, requiring propositions to be representational, and there is pressure from the 'external world' side of the divide requiring them to be objective, mind-independent and language-independent, and this is at the very least in tension with their being representational.

10.3 The Tension Manifest in Some Extant Theories

The fundamental conflict in the notion of a proposition that I have identified manifests itself in a number of different realist theories of propositions as the source of problems that they face. In this section I present some examples of this.

A common criticism of Russell's (1903) view of propositions is that it collapses true propositions into the facts that make them true. The proposition *that the sea is blue* has the sea (the vast, watery entity itself) and the property of blueness as constituents; the proposition *that the Burj Khalifa is the tallest building in the world* has the Burj Khalifa (the building itself) and the property of being the tallest building in the world as constituents. And while this makes it easier to see how such propositions can be the objects of knowledge (when I know that the Burj Khalifa is the tallest building in the world I know that the Burj Khalifa, the building itself, has that very property), it makes it very much harder to see how there can be false propositions, and how there can be false beliefs. There are no facts for false propositions to collapse into. If I mistakenly believe that Taipei 101 is the tallest building in the world, that proposition couldn't have Taipei 101 being the tallest building in the world as its constituents, because it no longer possesses that property.

This objection illustrates the tension I identified that arises out of propositions occupying the role of intermediaries between language and thought on the one hand, and the extra-linguistic world on the other. There is pressure exerted on propositions from both directions. Russell's view submits to the pressure from the extra-linguistic side of the divide, identifying true propositions with the facts that they are about, thus making more sense of the idea that the objects of our knowledge are facts in the world itself, but making less sense of the notion of how a proposition can be representational. How can the Burj Khalifa itself represent that the Burj Khalifa is the tallest building in the world? In order for something to represent something else it must, as I argued above, be numerically distinct from it. It also effectively rules out the possibility of false propositions, and raises questions about the objects of our intentional attitudes other than knowledge, where there is the possibility that we may believe, judge, fear or hope something that isn't the case.

Other philosophers who have collapsed the distinction between true propositions and the facts that are ordinarily thought to make those propositions true are McDowell (1994) and Hornsby (1997). According to McDowell "There is no ontological gap between the sort of thing one can mean, or generally the sort of thing one can think, and the sort of thing that can be the case. When one thinks truly, what one thinks *is* what is the case" (1994, p. 27). Hornsby, in her identity theory of truth, claims that each true proposition is identical with its truthmaker. By yielding to the pressure to identify propositions with the external world, this approach involves abandoning the representational role of propositions. If true propositions are identical with their truthmakers, how can they represent those truthmakers? And if they are not representational, how can they be the bearers of truth and falsity?

Recently, Jeff King (2007) has offered a variant of this kind of view that attempts to address this problem. King identifies propositions with facts that have as constituents the objects, properties and relations that those propositions are about. But the fact that is the proposition is a different fact from the fact that is its truthmaker (if it has one). The sentence 'Ruby smiles' is made up of the words 'Ruby' and 'smiles' standing in a certain syntactic relation to each other. The proposition expressed by this sentence is made up of the semantic values of those words (Ruby and the property of smiling) standing in semantic relations to words that stand in a syntactic relation to each other. So the constituents of the proposition are Ruby, the property of smiling, and a complex of semantic and syntactic relations holding them together. That proposition is true if and only if Ruby possesses the property of smiling, so the proposition is not identical with its truthmaker, and if Ruby is not smiling the proposition would still exist, but be false.

King's view therefore manages to resist the pressure to collapse true propositions into the facts in the world that make them true. Furthermore, he claims, they manage to be representational in a straightforward sense, as the proposition has a structure that directly mirrors the structure of the sentence that expresses it. However, by resisting the pressure to collapse the distinction between propositions and the extra-linguistic world, this view cannot help but significantly close the gap between propositions and the language we use to express them. As he readily acknowledges, King's view makes propositions utterly dependent on language, such that a proposition doesn't exist until a sentence expressing it is produced. His view thus flouts theses (7), (9) and (10) of the standard view. He presents detailed defenses of the implications of his view that propositions do not exist necessarily or eternally, and is happy to accept that they are more language-dependent than proponents of propositions are normally prepared to concede.

However, by narrowing the gap between propositions and the language we use to express them, his account of the objects of our intentional attitudes becomes less palatable. According to King, what I believe when I believe that Ruby smiles is a structured entity made up of Ruby, the property of smiling, and a complex semantic/syntactical relation holding them together. But this is almost as implausible as the suggestion that what I believe when I believe that Ruby smiles is a mathematical function or a set of possible worlds.[4] And even if that objection can be overcome, it still creates an unwanted distance between what it takes to be the object of my belief, and what I intuitively want to say is the object of my belief, namely, Ruby herself smiling. We return again to the fundamental tension between propositions

[4] I allude here to an objection to theories that reduce propositions to sets of possible worlds or mathematical functions, according to which they cannot perform many of the roles of propositions because they are entities of the wrong sort.

being representational on the one hand, and objective, extra-linguistic objects of knowledge and other intentional attitudes on the other.

Another problem for King's view also arises out of this fundamental tension. King identifies propositions with facts, albeit not the facts that are their truthmakers, and thus makes a concession to the pressure to have propositions be objective features of the external world. But how is it that a fact can be a truth-bearer? Facts are not normally the sort of thing that we think of as being true or false; on the contrary, they obtain or are the case, and they make truth-bearers true. And even if there is an explanation as to how a fact can possess a truth-value, how is it that some facts possess truth-values, but not all facts? If the answer is that it is only those facts that are propositions that possess truth-values, then we have a circular, and vacuous explanation.

An alternative view of propositions holds that they are unstructured, sui generis, primitive, Platonic entities (for example, Bealer 1998). This kind of view is immediately subject to the problem I discussed in Sect. 10.2.3 of explaining how an inert, abstract, non-spatio-temporal entity can be representational. Views of this kind are motivated by the requirement on propositions that they be objective, mind-independent and language-independent, but by acquiescing to these requirements in the way that they do, they sacrifice the capacity of propositions to represent. In contrast to the Russellian view that also succumbs to the pressure to make propositions objective, mind-independent and language-independent, but ultimately does so by identifying them with the concrete facts that they are about, this view takes the alternative route, rendering them abstract, non-spatio-temporal and inert, thus exacerbating their inability to represent.

10.4 The Truth and the Existence of Propositions

I have argued that the standard view of propositions is subject to a number of tensions, chief among them being the tension between a proposition's being essentially representational, and its being an objective, extra-linguistic entity that can function as the object of knowledge and other intentional attitudes. It is a proposition's role as truth-bearer that requires it to be representational. It is its objective existence, independent of minds and language, together with its role of being the object of intentional attitudes, that exerts pressure on it to be non-representational. This tension maps onto an important distinction in the notion of a proposition: that between the truth and the existence of propositions. Unlike for entities of any other kind, the question of a proposition's existence is closely bound up with the notion of truth. It is important to keep these notions apart, for reasons that I will now explain.

When considering whether a proposition is true there are two interpretations of what we may be doing. We may be asking whether there exists a truth-bearer that possesses the truth-value *true*, or we may be asking about how the world (or a part of the world) is arranged. One of the motivations for postulating propositions in the first place is the idea that many truths are true independently of our recognizing them to be true. For example, the truth that hydrogen is less dense than helium was true before any scientist discovered that it was true, and indeed, would have been true even if observers had never discovered it, or even evolved. This motivation compels philosophers to insist that the truth that hydrogen is less dense than helium must be utterly independent of its being recognized, observed, thought or uttered. The inference is then made that, since this truth is independent of any human thought or utterance that expresses it, there must be a truth-bearer whose existence is independent of any human thought or utterance that expresses it; a truth-bearer that existed before humans came into existence, and would have existed even if humans had never come into existence. That truth-bearer is the proposition *that hydrogen is less dense than helium*, a mind-independent, language-independent, eternal, necessary, abstract entity. But this last inference, from something's *being true* independently of any human thought or utterance, to there *existing* independently of any human thought or utterance, a truth-bearer, is fallacious.

The mistake is to think that, just because it was true before humans existed that hydrogen is less dense than helium, there must have been a truth-bearer, an abstract proposition, that existed then that possessed the truth-value *true*.[5] But this is to slide between assertions concerning the *existence* of a proposition and assertions concerning its *truth*. All that is required for it to have been true that hydrogen is less dense than helium in the past, (before humans evolved, let's say) is for hydrogen to have been less dense than helium then.

Let's look more closely at this example. Suppose we want to say that the proposition *that hydrogen is less dense than helium* was true before humans evolved. What is required for that to be the case? Well, for the proposition to be true, its truth condition must be satisfied. The proposition *that hydrogen is less dense than helium* is true if and only if hydrogen is less dense than helium. So, in order for that proposition to be true at that time, all that must be the case is that hydrogen be less dense than helium at that time. There is no further linguistic requirement that some truth-bearer exist and possess the truth-value *true*.

To think that the existence of the truth-bearer at a past time is required to account for a truth such as this is to conflate the following two notions:

(1) 'P' is true independently of us;
(2) The world is such that P.

The world's being the way it is, is not dependent on there being truth-bearers expressing that it is the way it is. But, unfortunately, we can express the second notion by means of the first, which then seems to license the inference to there existing independently of us truth-bearers that possess the truth-value *true*. This conflation is exacerbated by the fact that we often use the word 'true' and its

---

[5] I put forward a similar argument to this in Dyke (2002) and (2008).

cognates in such a way that they are ambiguous between these two senses. We sometimes say, for example, ‘it is true that the earth is spherical’, and this can mean either that the truth-bearer (whatever it is) that expresses that the earth is spherical is true, or it can mean that the world is such that the earth is spherical.

A related point also encourages the thought that there must have been truth-bearers at times, and at worlds, at which there are no observers. It is the thought that if we deny that any propositions were true before humans evolved, we are thereby denying that anything was the case, or that the world was a determinate way. But this is not the case. To draw this inference is to conflate the following two notions:

(3) There were no truths before humans evolved;
(4) The world was not a determinate way before humans evolved.

It is to think that if (3) is true, then (4) must be true too. But (3) contains an instance of the ambiguous term ‘truth’. If ‘truths’ in (3) is taken to mean facts, then (3) and (4) are equivalent. But understood in this way, (3) does not follow from the claim that there were no true propositions before humans evolved. If, on the other hand, ‘truths’ in (3) is taken to mean truth-bearers, then (3) does not entail (4). It does not follow from the claim that no truth-bearers exist at a time or a world, that nothing is the case at that time or that world.

The existence of a proposition is a question of ontology. Is the world such that it contains entities that play roughly the roles and possess roughly the features outlined in the standard view? The truth of a proposition, on the other hand, is a matter of whether or not its truth condition is fulfilled; whether or not the world, or a part of it, is arranged in a determinate way (for further discussion of this point see Iacona (2003, p. 338)).

We cannot answer the ontological question about whether propositions exist merely by pointing to the fact that there are many truths about the world. If, by ‘truths’ here, we mean truth-bearers or propositions, then what we are saying is vacuously true. If, on the other hand, by ‘truths’ we mean facts in the world, or ways the world determinately is, then this has no implications whatsoever for the ontological question of whether propositions exist.

10.5 Whither Propositions?

I began by outlining the standard view of propositions. This was made up of 13 theses, some of them stating roles that propositions are normally thought to play, others stating features that propositions are normally thought to have. Not all of these theses are accepted by all proponents of propositions. But any proponent of propositions will adhere to a substantial majority of them. What I have tried to show is that there are many tensions underlying the notion of a proposition as an ontological entity of any sort, which quickly rise to the surface as soon as an attempt is made to construct a coherent account of propositions.

One conclusion that we might draw from this is that it is unlikely that there is one kind of entity that can play all the roles and possess all the features of propositions as outlined in the standard view. If we were to draw that conclusion there would be two alternatives. One could be a pluralist about propositions, and argue that different entities play each or some of the different roles that propositions are normally thought to play. Or one could be an eliminativist, and hold that there is nothing that plays all the roles that propositions are normally thought to play. I have not explored the pluralist option, but if it were to be pursued it would be, in my view, advisable to offer an account that ties the representational and non-representational roles to entities of different kinds so as to avoid the tensions I have discussed here.

It is often said that if one wants to be an eliminativist about some entity, one must provide a reductive analysis of all our talk of the entity in question in terms of something else, or somehow dispense with or eliminate our talk of the entity in question (see, for example, Grayling (1982, p. 24)). So, if one wanted to deny the existence of propositions one would have to be able to analyze all our talk of propositions in terms of some other kind of entity; utterances, or sentences perhaps. I have argued (Dyke 2008) that this is an unreasonable requirement that does not have to be met. Just because an expression proves indispensable for certain purposes, it doesn't follow that it has an ontological counterpart. Instead, truths involving that indispensable expression may have truthmakers involving entities of a different sort.

Talk of propositions certainly seems to be indispensable in many fields: logic, linguistics and cognitive psychology, to name just three. It is extremely unlikely that we will be able to completely excise proposition-talk from our language. But it doesn't follow that there are mind-independent, language-independent, abstract, necessary, eternal, non-spatio-temporal entities that are the primary truth-bearers, the objects of our intentional attitudes and acts of assertion, the meanings of our declarative sentences, and that represent the world as being a certain way. The truths of proposition-discourse can be true even if there are no such entities.

In Sect. 10.4 I drew attention to what I think is the most serious ambiguity at the heart of the notion of a proposition; the one that underlies many of the tensions, discussed earlier, and the one that is very likely responsible for propagating the notion of a proposition in the first place. It is that the term 'truth' can mean either a true proposition (or other truth-bearer) or a fact. Our talk of truths can either be talk of true representations of the world, or it can be talk of the world itself. Thus, we introduce entities that are supposed to be intermediaries between our representations and the world. But different contexts of use exert pressure on these entities in both directions. Sometimes they are required to be more like the world than a representation, for example, when they play the role of objects of knowledge. Other times they are required to be more like a representation than the world, for example, when they play the role of primary truth-bearer.

This, in turn, gives rise to the ambiguity between the truth of a proposition and its existence. Talk of truths as true propositions can be interpreted as either about the ontological status of propositions, or about their truth. Questions about the truth of propositions are, in effect, questions about whether their truth conditions are satisfied, which are questions about the way the world is. So an investigation into which propositions are true is really an investigation into the world.

This ambiguity has been noticed before. (See, for an early example, Robinson (1931, p. 77).) The right response to it, in my view, is to give up on the idea of propositions as ontological entities in their own right. But this need not require us to excise from our language all reference to propositions. As Mackie (1973) pointed out, "The words 'statement' and 'proposition' are just terms that enable us to speak generally about what is said, what is believed, what is assertible or believable, and so on" ((Mackie 1973, p. 21), see also Thomson (1969, p. 738)). In other words, the term 'proposition' can be seen as a mere façon de parler, a term of art, that facilitates investigation in the many fields in which proposition-talk has proved indispensable. But the indispensability of proposition-talk need have no implications for the question of whether propositions constitute part of the furniture of the world.

References

Bealer, G. 1998. Propositions. *Mind* 107: 1–32.
Dyke, H. 2002. Tokens, dates and tenseless truth conditions. *Synthese* 131(3): 329–351.
Dyke, H. 2008. *Metaphysics and the representational fallacy*. New York: Routledge.
Grayling, A.C. 1982. *An introduction to philosophical logic*. London: Duckworth.
Hornsby, J. 1997. Truth: The identity theory. *Proceedings of the Aristotelian Society* 97: 1–24.
Iacona, A. 2003. Are there propositions? *Erkenntnis* 58: 325–351.
Jubien, M. 2001. Propositions and the objects of thought. *Philosophical Studies* 104: 47–62.
King, J.C. 2007. *The nature and structure of content*. Oxford: Oxford University Press.
Mackie, J.L. 1973. *Truth, probability, and paradox*. Oxford: Oxford University Press.
McDowell, J. 1994. *Mind and world*. Cambridge, MA: Harvard University Press.
Robinson, R. 1931. Mr. Ryle on propositions. *Mind* 40: 73–78.
Russell, B. 1903. *Principles of mathematics*, 2nd ed. New York: W.W. Norton and Company, Inc.
Ryle, G. 1929. Are there propositions? *Proceedings of the Aristotelian Society* 30: 91–126.
Thomson, J.F. 1969. Truth-bearers and the trouble about propositions. *Journal of Philosophy* 66(21): 737–747.
Wrigley, A. 2006. Abstracting propositions. *Synthese* 151: 157–176.

Chapter 11
Against Advanced Modalizing*

Josh Parsons

11.1 The Modal Status of Modal Realism

Suppose that modal realism is necessarily true. By modal realism, I mean the thesis that there is a plurality of worlds, that those worlds are isolated spacetimes just as real as our own, that the worlds satisfy some principle of recombination, that some worlds have as parts blue swans, talking donkeys, etc. Roughly, modal realism is the content of David Lewis's book *On the plurality of worlds*.

It is a logical consequence of modal realism that there are blue swans. After all, if there is some world that has a blue swan as a part, then there are blue swans. Abbreviating the thesis that there are blue swans "*Swans*", and the thesis of modal realism "*PW*", we may say *Swans* is a logical consequence of *PW*. Since that is the case, we should also say the material conditional "if *PW* then *Swans*" is necessarily true. But, then, since *PW* is necessarily true, *Swans* is necessarily true.

But then it follows that necessarily there are blue swans! And that is absurd. Surely the modal realist's view should be that blue swans are contingent existents – they exist in some worlds and not in others.

Let's look at the argument in more detail. My point is that the following set of propositions is inconsistent (here I use □ to abbreviate "it is necessary that"):

1.1 *PW* logically entails *Swans*
1.2 (the N rule) from "*p* logically entails *q*" you may infer $\Box\,(p \supset q)$

* Special thanks to Dan Marshall, discussions with whom prompted me to write this paper. A number of the ideas of this paper are derived from that discussion, and it is now hard to recall who came up with what. Thanks also to Ben Caplan, John Divers, Dana Goswick, Hud Hudson, Ed Mares, my colleagues and students at the University of Otago, and the members of the UC Davis philosophy department modality reading group.

J. Parsons (✉)
Corpus Christi College, Oxford University, Oxford, UK
e-mail: josh.parsons@ccc.ox.ac.uk

J. Maclaurin (ed.), *Rationis Defensor: Essays in Honour of Colin Cheyne*, Studies in History and Philosophy of Science 28, DOI 10.1007/978-94-007-3983-3_11,

1.3 (the K rule) from $\Box p$, $\Box (p \supset q)$ you may infer $\Box q$
1.4 $\Box$ *PW*
1.5 $\neg\Box$ *Swans*

There is a problem here. Modal realists have shown signs of believing all five of these. But the five are inconsistent. It is open to the modal realist to reject members of this inconsistent set – and which ones they choose will make a difference to what their theory says.[1]

Before I turn to surveying ways of rejecting premises, let me say a bit about what signs there are that modal realists endorse them.

11.1.1 That PW Logically Entails Swans

Modal realism is not just a theory about the truth conditions of modal vocabulary. It is also a theory about what possible worlds there are, and what they are like. Roughly put, the modal realist wants to say that "*every* way a world could possibly be, that way some world *is*" (Lewis 1986, pp. 2, 86). That is a rough way of putting it, because when combined with a possible worlds analysis of modality, it will turn out to be trivial. What the modal realist means by saying this kind of thing, though, is that insofar as we think there might have been blue swans, we should think of "there is a world containing blue swans" as part of the theory of modal realism. Since "there is a world containing blue swans" entails *Swans*, modal realism entails *Swans*.

The other way in which modal realists say things as part of their theory that commit them to theses like *Swans* is via principles of recombination (Lewis 1986, pp. 87–92; Divers 2002, pp. 100–103). These are intended to, in a non-trivial, non-modal way, specify at least a lower limit to the number and nature of worlds. They are difficult to state exactly, but the intention is that they should entail the existence of all the worlds needed to make sense of the commonest assertions of possibility. Take a swan and dress it in the feathers of a peacock. Principles of recombination entail that there is such a creature somewhere among the possible worlds.

You might complain that, insofar as (1.1) is true, it's unfair to describe *PW* as "modal realism". *PW* is a comprehensive theory of modality; perhaps "modal realism" would be better used to mean just one plank of that theory, the view that possible worlds are isolated spacetimes just as real as our own – a thesis about the nature of possible worlds is, rather than a thesis about what's possible. In that sense "modal realism" would be compatible with the thesis that there are no blue swans

[1] This type of problem for modal realism was first pointed out to me by Dan Marshall, and is discussed in his unpublished paper "A puzzle for Genuine Modal Realism". Though Marshall's argument, when I saw it, was different in structure from my own (for example, it makes a particular analysis of modality a premise, and depends on the idea that analyses give rise to necessary biconditionals) I believe that he is pointing to another aspect of the problem I am describing here.

(not even in any other possible world). I have no argument against modal realism, in that sense.[2] My thesis is there is a contradiction in the large bundle of theses that philosophers associate with modal realism (whether it is fair to call this bundle "modal realism" or not – it is surely fair to call it "*PW*"), and that Divers' Advanced Modalising does not succeed in removing that contradiction.

11.1.2 The Validity of the N Rule and the K Rule

These are inference rules that may be fairly regarded as among the most central to ordinary patterns of modal inference. They are, for example, valid in all "normal" modal logics. Though they are not explicitly part of modal realism (which is, after all, a metaphysical, rather than a logical theory) I think it would be bad if modal realism entailed that they must be rejected.

The charge of violating "laws of modal logic" is one that has often been raised against modal realism; never, I think, convincingly. Most of these charges have exploited odd corners of modal logic that are connected with its more metaphysical aspects. Thus, for example, there are objections having to do with violations of the S4, S5, B, or E axioms of modal logic; objections having to do with the nature of quantification into modal contexts, such as the validity or invalidity of the Barcan Formula and its converse; objections having to do with non-standard and controversial extensions to modal logic, such as the logic of actuality.[3] All of these should be "spoils to the victor" – hostages to a correct metaphysical account of modality.

Another way of making this last point is to say, again, that modal realism is not a theory about modal logic. Many of the aspects of modal logic in question are artefacts of the logical machinery which do not have anything to correspond to them in ordinary modal language. For example, how one feels about the intuitive plausibility of principles concerning interactions between quantifiers and modal

[2] Indeed, as an anonymous referee pointed out to me, the argument of this paper could be used to defend an alternative comprehensive theory of modality in a modal realist spirit. It could for example be used to defend the "modal noneism" discussed in Sect. 11.4, or Divers' (2004, 2006) agnosticism about possible worlds, both of which are versions of "modal realism" in the sense under consideration in this paragraph.

[3] For a survey of objections to modal realism based on the question of whether it sustains certain theorems of standard quantified modal logic (QML) such as S4, S5, B, and Barcan formulas, see Divers (2002, pp. 142–144). For an objection to modal realism based on the logic of actuality, see Fara and Williamson (2005). Fara and Williamson would no doubt insist that the logical features of actuality that they complain modal realism and counterpart theory cannot sustain are as basic and uncontroversial as my examples of the N, K, T, and distribution rules. What is controversial about the logic of actuality though, is whether "actually" should be thought of as a modal operator, rather than a mechanism for disambiguating the relative scopes of modal operators and predicates. Though Fara and Williamson point to ordinary language usages that cannot be represented in QML without such an operator, there are other ways of extending QML – for example by adding first order plural (FPO) quantifiers – that would suffice to represent these usages.

operators can vary depending on how the modal operators are "read" in ordinary language. For example, if you read the atomic formula $\Box\ Fx$ as "x couldn't have failed to be F", you'll look with less favour on the Barcan formula, $(\forall x)(\Box\ Fx) \rightarrow \Box(\forall x)(Fx)$ and with more favour on its converse than if you read that formula as "it is necessarily the case that x is F". The modal realist does not need modal logic – they can paraphrase ordinary language directly into possible world talk, and avoid debates about the validity of the Barcan Formula and its friends. As Lewis says: "What is the correct counterpart theoretic interpretation of the modal formulas of… quantified modal logic? – Who cares? We can make them mean whatever we like… We needn't be faithful to the meanings we learned at mother's knee – because we didn't." (Lewis 1986, p. 12).

In short, "wolf!" has been cried many times over modal realism's adherence to modal inference patterns. My appeal to the N and K rules is different. First, I have in mind not principles of modal logic, but inference patterns in ordinary language. From "*p* entails *q*" one may validly infer "It is necessarily true that either *q* or not *p*"; from "*p* is necessarily true" and "It is necessarily true that either *q* or not *p*", one may validly infer "*q* is necessarily true". Second, these are not controversial matters of iterated modality; they correspond to rules valid in every normal modal logic. Third, these are not controversial matters of quantification in or out, they are strictly *de dicto*. Fourth, these are not controversial matters of extensions to the well understood core of modality, of talk of actuality, or of conditionals. Everything I am saying in this paper applies only to modal realism as a theory of Quine's "first grade of modal involvement" – a theory of necessity and possibility as properties of sentences. Here, the logic is simple indeed.

11.1.3 That PW Is Necessary

It seems obvious to many people that *PW* should be, if true, necessary. It does not seem so obvious to me. My favoured solution to the problem is to deny this premise. But this is not without costs.

Much of the attachment to the idea that *PW* must be necessary if true seems to be connected with some idea that metaphysical theories must be metaphysically necessary. But why should this be? Perhaps the idea is that just as physics discovers laws that are physically necessary, so metaphysics should discover laws that are metaphysically necessary. This is a poor analogy. "Physically necessary" just means "compatible with the laws of physics". But "metaphysically necessary" does not just mean "compatible with the laws of metaphysics". Perhaps the idea is that because *PW* is a theory about different "strengths" of truth, it should assign itself the strongest kind of truth available. This is an attractive idea, but it is a good question whether this is actually a consequence of the modal realist's analysis of modality. I address this question in the next section, and conclude that it is not. I doubt that there is any good reason to hold that *PW* is necessary if true; but many metaphysicians of modality I've spoken to seem to be motivated by these considerations. For them, denying this premise would be a cost.

Another reason for thinking that *PW* is, if true, necessary is connected with epistemic objections to the theory. Faced with the challenge to show that we should use pragmatic criteria to decide ontological questions, Lewis gave an analogy with the ontology of mathematics. Faced then with the objection that sets and numbers are abstract, but possible worlds are concrete, he replied that the distinction between abstracta and concreta is obscure, but that, in contrast, there is a clear point of analogy between possible worlds and mathematical entities – both exist necessarily. "Modal and mathematical knowledge together fall on the right side of the line. Our *contingent* knowledge that there are donkeys at *our* world requires causal acquaintance with the donkeys… Our *necessary* knowledge that there are donkeys at *some* worlds… does not" (Lewis 1986, p. 112). I read this as saying that consequences of *PW* itself – that there are donkeys at some world – are necessary.

I think that this was a moment of weakness on Lewis's part. Nowhere else does he make such definitive statements on the modal status of his theory of modality. However, without this move, his reply to the epistemic objection to modal realism is somewhat blunted. I leave it to the reader to decide how severe a problem for modal realism that is.

11.1.4 That It Is Not Necessarily True That There Are Blue Swans

I take it that modal common sense is firmly behind this premise.

However, modal realists are not afraid of denying common sense. Many modal realists (or their fellow travelers) will think that, in some subtle way, this is the premise to deny. That might be one way of understanding John Divers' doctrine of advanced modalizing, which I discuss in a later section. My objections to that are too complex to preview here.

Instead, I will give a reason for thinking that modal realists should accept this premise. Recall the principles of recombination discussed above. These are part of *PW* and should entail that if there is a world that contains a blue swan, then there is a world that does not contain one. The modal realist's analysis of modality should make that sufficient for "there is a blue swan" being contingent. I say "should" rather than "does" because the exact details of this analysis are part of the problem. In the next section I tackle the difficulties of formulating such an analysis.

11.2 The Analysis of Modality

Another way of posing the problem I am discussing is to ask for a detailed statement of the modal realist's analysis of modality. If we had such a statement, we could use it to analyse the modal sentence "*PW* is necessarily true", and see whether what we get is a truth according to the modal realist.

The task of stating the analysis is complicated in two ways. First, the "modal realist analysis of modality" is a vastly complicated affair – a research programme being worked on by metaphysicians, linguists, and philosophers of language giving possible worlds treatments of the many and complicated modal idioms of ordinary language. Fortunately for us, a very much cut down corner of this programme will suffice. All we need is the modal realist's analysis of the first grade of modal involvement, of necessity and possibility construed as predicates of non-modalised sentences (Quine 1966). From here on, that is what I will mean by an "analysis of modality".

Second, it is already well known that the most obvious and simple analyses of modality have serious problems when turned on the subject matter of modal realism itself. Take, for example, this, the *simple analysis of modality*:

S is possibly-true iff for some world *w*, $[S]^w$ (where $[S]^w$ is the result of restricting all quantifiers contained in *S* to parts of *w*).[4]
S is necessarily-true iff for every world *w*, $[S]^w$.

Suppose we apply this analysis to ask about the modal status of the sentence "there are at least two possible worlds". Given the isolation of possible worlds that is part of *PW*, this turns out to be impossible! In fact, on this analysis, it is necessary that that there is precisely one possible world.

One might hope that some, more complicated, analysis would not have this problem. There are two ways to proceed here. You could start making up analyses by trial and error. A far more rewarding procedure, however, is to lay down some criteria for a successful analysis, and use logical means to determine whether those criteria are jointly satisfiable. One would like at least these three criteria:

2.1 That the analysis be compatible with the simple and literal truth of *PW* and its consequences.
2.2 That the analysis be compatible with validity of basic patterns of ordinary modal inference (such as the N and K rules).
2.3 That the analysis, in conjunction with *PW*, be compatible with modal common sense.

The simple analysis fails to satisfy (2.1) and (2.2) together. Consider a modal realist who was prepared to bite the bullet and hold that *PW* is necessarily-false (as the simple analysis reports) but also simply true. This isn't totally crazy.[5] After all, *PW* isn't about what happens at just one world. It has a modal status that

[4] As I emphasised earlier, all the problems I am describing here can arise with an extremely logically restrained version of modality – one in which the only modal vocabulary consists of predicates of non-modalized sentences. For that reason my modal operators are written "is necessarily-true". The hyphen emphasises that this predicate should not be imagined to be the result of applying a modal operator "necessarily" to the predicate "is true". The operation represented by square brackets and a superscript term, as in "$[S]^w$", is a syntactic transformation on sentences which has the effect of restricting quantifiers. This notation is adapted from Lewis (1968).

[5] Noonan (1994) and Hudson (1997) both recommend a view like this to the modal realist.

transcends mere necessity – if a neat way to mark that status is as "impossible but true" then so be it. Nor is this position contradictory. It only becomes contradictory if the modal realist who holds it accepts that validity of the T rule: from *p* you may infer "possibly *p*". And that's the problem – like the N and K rules, T is an obviously valid pattern of ordinary modal inference.

There is an analysis of modality that satisfies all three of the above criteria. For lack of a better name, I'll call it the *T-preserving analysis*. It also preserves N and K:

S is possibly-true iff *S*, or for some world *w*, $[S]^w$.
S is necessarily-true iff *S*, and for every world *w*, $[S]^w$.

As far as the analysis of possibility and necessity goes, this is equivalent to taking the whole pluriverse to be an extra, large world of many island universes. A very similar approach would to be combine the simple analysis with revisions to *PW* that allowed the pluriverse (or perhaps any fusion of worlds) to count as a world, thereby abandoning the isolation of worlds. That, however, would lead to a problem which my proposed analysis of modality does not. To a modal realist, "the actual world" means "the world we are in". But on the revised version of *PW* where the pluriverse is a world, we are in two worlds, the actual Lewis-world, and the pluriverse. On the revised version of *PW* where every fusion of worlds is a world, we are in infinitely many worlds. Which is the actual one? In some of these worlds there are blue swans. Are they actual?[6]

The T-preserving analysis does not have this problem because it does not revise the internals of modal realism (by counting different things as worlds from what modal realism would count as a world) but only the treatment of "possibly-true" and "necessarily-true". I don't think that this analysis has much to intuitively recommend it. Its advantage is not that it makes particular sense or coheres better with the underlying metaphysics than the simple analysis. It's just that it's the only way to get *PW* to be true compatibly with those three criteria.

It has the consequence (in conjunction with *PW* itself, of course) that *PW* is possibly-true. But when the quantifiers contained in a statement of *PW* are restricted to a single world, the result of that restriction will be false. So the denial of *PW* is also possibly-true on this analysis. *PW* is contingent.

Not everyone will like that. Is there a way to get an analysis of modality that would render *PW* necessarily-true – to satisfy criterion (2.4)?

2.4 That the analysis be compatible with the necessary truth of *PW*.

No analysis can satisfy (2.4) while also satisfying (2.2) and (2.3). That is what my argument of the previous section showed. For any analysis that made *PW* out to be necessarily-true would either violate (2.2) by omitting to validate the N and K rules, or violate (2.3) by making "there are blue swans" necessarily true. Contingent truth is the best that the modal realist can consistently hope for.

---

[6] On this worlds-within-worlds variant of modal realism, and especially on the problem about actuality, see Yagisawa (1992), Bricker (2001), Divers (2002, pp. 103–105), and Parsons (2007).

11.3 Against Advanced Modalizing

The problems mentioned in the previous section, of turning the modal realist's analysis of modality on modal realism itself, are not new. It has been widely supposed, however, that some complicated revision to the analysis of modality will save the day, delivering the result that *PW* is necessarily true. I think I have already shown, in general terms, that that is not possible. However, such a general argument might not seem very compelling when held up against a detailed account of a revision to the analysis that claims to do this. A good way for me to show the force of my argument is to apply it to a worked out sophisticated analysis of modality that is intended to be compatible with the necessary truth of *PW*. Such an analysis is John Divers' (1999, 2002, pp. 47–50) theory of *advanced modalizing*.

According to Divers, the correct analysis of de dicto modality comes in two cases. In the *ordinary* case, we should use the simple analysis already mentioned:

S is possibly-true iff for some world *w*, $[S]^w$.
S is necessarily-true iff for each world *w*, $[S]^w$.

In the *extra-ordinary*, or *advanced*, case – where we apply modal language to the subject matter of modal realism itself – we should use a *redundancy analysis*:

S is possibly-true iff *S*
S is necessarily-true iff *S*

The idea is that when we ask "is it necessary that there be blue swans?", we are engaging in ordinary modalizing, and it gets the former analysis; when we ask "is it necessary that there be a plurality of worlds", that is advanced modalizing, and it gets the latter.

Talk of "cases" here papers over an important distinction. There are two ways we might understand Divers' proposal. On the first of these, the *ambiguity* view, Divers is saying that there are really two senses of possibility. One of these senses is correctly analysed by the ordinary analysis, and the other of them is correctly analysed by the redundancy analysis.

The second way of understanding Divers is the *disjunctive* view. According to it, the correct analysis of first grade modality is neither of those given above, but something like this, the *disjunctive analysis of modality*:

S is possibly-true iff either (a) *S* contains some vocabulary relating to worlds, sets, properties, propositions, or other predicates of modal realist metaphysics[7] and *S* or (b) *S* contains no such vocabulary, and there is some world *w*, such that $[S]^w$.
S is necessarily-true iff either (a) *S* contains some vocabulary relating to worlds, sets, properties, propositions, or other predicates of modal realist metaphysics and *S* or (b) *S* contains no such vocabulary, and for each world *w*, $[S]^w$.

[7] Sets, properties, and propositions are included on this list because, according to Lewis, these are objects that do not exist wholly in any one possible world. As Divers points out, the same kind of problems we are discussing here arise for "It is possible that there be natural properties" – this sentence turns out to be false on an ordinary analysis of "it is possible that", and true on an advanced one.

The difference between these two might be illustrated by comparison with other examples of ambiguity and disjunctiveness. The word "bank" is *ambiguous* – it has two independent senses, one having to do with rivers, the other with finance. Competent users of the word "bank" always intend one or the other. The word "jade" by contrast is implicitly *disjunctive*. This is the harder notion to nail down, but the idea is that, in some sense, the correct analysis of "jade" goes by cases. A thing is jade iff it is jadeite *or* nephrite. A competent user of "jade" need not know this, and even if she does, need not intend to mean one or the other. The ambiguity view of advanced modalizing holds that "is necessarily-true" is like "is a bank"; the disjunctive view holds that it is more like "is jade". It's not clear to me from Divers' writing which of these he meant. Unfortunately neither will work.

One drawback of the ambiguity view is that it is vulnerable to empirical refutation. It is obvious to anyone competent with both senses of "bank" that the word is ambiguous. By contrast, competent users of "is necessarily-true" do not hear the phrase as ambiguous (or at least, not ambiguous between an ordinary and advanced sense), even if they are disposed to believe that *PW* is necessary if true.[8] It took the

[8] There are a number of well-known linguistic tests for ambiguity, including the *conjunction reduction* test. If "John works at a bank" and "Jane works at a bank" are both true, and "bank" in both cases applies to financial institutions, then competent speakers will find it appropriate to infer "John and Jane both work at banks". But if the sentences are true because John works at a financial institution and Jane works at the side of a river, then competent speakers find it *inappropriate* to say "John and Jane both work at banks".

We might apply this test to "It is necessarily true both that all bachelors are unmarried and that there are at least two possible worlds". If the ambiguity view were right, speakers competent with "necessarily true" who believe *PW* would find this type of thing inappropriate, as it equivocates on "necessarily true". Though this is an empirical matter, when I put myself into a modal realist mood, I find it hard to hear this as equivocatory.

Recall that the reason Lewis wanted *PW* to be necessary was so that it would enjoy the same quasi-apriori status as mathematics. Lewis has luckily provided us with a natural experiment in conjunction reduction. He says "Modal and mathematical knowledge *together fall on the right side of the line* [i.e. are *both necessarily true*]" (Lewis 1986, p. 112, my emphasis). This suggests that he found no equivocation between the sense in which he is saying that modal realism is necessary, and the sense in which more familiar necessary truths are necessary.

This is, sadly, not decisive as an objection to Divers, as it is part of his view that the sense in which mathematics is necessary is the extra-ordinary sense. That buys him other problems however, and brings us back to the question of whether the folk find "is necessarily true" ambiguous. Philosophers who've never heard of modal realism, and even ordinary people who've reflected on the matter, tend to regard mathematics as necessary. If it is only necessary in an extra-ordinary sense, then we should expect them to resist "It is necessarily true both that all bachelors are unmarried, and that two plus two is four".

Perhaps Divers' view is that the folk (and common philosophical opinion) are mistaken about the sense in which mathematics is necessary – people believe mathematics to be necessary in the ordinary sense, which it is not, but mathematics *is* in fact necessary in the extra-ordinary sense. It is as if they think that Jane works at the financial institution when in fact she works by the river. But if that were so, then Lewis's argument that *PW* is quasi-apriori would fail, for he could not expect his interlocutor to agree that *PW* is necessary in the same sense that mathematics is.

discovery of the failings of the ordinary analysis, and Divers' own treatment of advanced modalizing to discover the difference between these senses. That was not a discovery of an ambiguity in the meaning of "is necessarily-true" – at best, it was a discovery of the need for one. Philosophers cannot wish linguistic phenomena into existence.

What philosophers (like anyone) can do is create ambiguity in their own idiolects, by introducing new words that are spelled and pronounced like old ones. A fan of advanced modalizing could make "is necessarily-true" ambiguous for themselves and their conversational partners by introducing a new and hitherto unknown meaning for that term. But this would not solve any problems connected with the old meaning. For we would still use the simple analysis to analyse the ordinary sense of the modal idioms. As we have seen, the simple analysis is incompatible with criteria (2.1) and (2.2), because on the simple analysis "there are at least two possible worlds" comes out necessarily-false. That is, if *PW* is true, then ordinary modalizing invalidates the T rule – "there are at least two possible worlds" is both true and (ordinarily) necessarily false! To say that there is some other, technical, extraordinary, advanced, sense of possibility that does not have this problem does not remove it from the ordinary sense. Only rejecting *PW*, the ordinary analysis, or the T rule (as applied to ordinary modality) could do that.

On the disjunctive view, we would reject the simple analysis, putting in its place a disjunctive analysis with two cases, as given above. The trouble with this is that the view that modality is disjunctive is bound to do funny things to the canons of modal inference.

Consider this sentence:

3.1 There are at least two possible worlds and a blue swan.

Since this sentence contains the requisite vocabulary, it gets to be necessary under the extraordinary disjunct of the disjunctive analysis. So:

3.2 It is necessary that there are at least two possible worlds and a blue swan.

Now, if the most familiar and basic modal inference patterns were valid, we should be able to infer from here:

3.3 It is necessary that there is a blue swan.

Here I am using the rule of distribution of necessity over conjunction – if "p and q" is necessarily-true, then p is necessarily-true and q is necessarily-true. Like the N and K this is very ordinary indeed, and valid in all normal modal logics.[9] But of course, (3.3) is false according to the disjunctive analysis. The disjunctive analysis invalidates the distribution rule.

[9] In fact the distribution rule is derivable from the N and K rules. $(p \wedge q) \supset p$ is a classical tautology, so by N, $\Box((p \wedge q) \supset p)$. Suppose $\Box(p \wedge q)$. Then, since $\Box((p \wedge q) \supset p)$, by K, $\Box p$. So $\Box(p \wedge q) \vdash \Box p$.

It might be thought that a more complicated disjunctive analysis could block the move from (3.1) to (3.2). But, as I showed in Sect. 11.2, no analysis of modality can satisfy all of the criteria (2.1 through 2.4) that Divers seems to be demanding. The failure of the disjunctive analysis to satisfy (2.2) is simply a special case of this.

A striking feature of the argument I just gave is that the canons of modal inference are used to infer from an "advanced" modality into "ordinary" modality, and this feels like an equivocation. But to describe the argument in that way is to fall back into the ambiguity view. It's ambiguities that give rise to equivocation. The disjunctive view is precisely that there is one univocal, but disjunctive, analysis of metaphysical modality. What I have shown is that that univocal analysis does not plausibly validate the canons of modal inference.

We should not get too hung up on the canons of modal inference. Perhaps Divers could reject the move from (3.2) to (3.3). He can even say why it's an exception to the distribution rule – it is an inference from a sentence containing some extraordinary vocabulary to a sentence containing only ordinary vocabulary. But the same move could be used to save the simple analysis, as I described in Sect. 11.2. Let us simply say that "There is a plurality of worlds" is both true and necessarily-false, and say that the T rule is inapplicable in this case because of the sentence's extraordinary vocabulary. Advanced modalizing, if I have understood what it is, no better off than the simple analysis.[10]

11.4 Against Advanced Quantification

There is another kind of alleged ambiguity that might help to get *PW* to be necessary. This would not be an ambiguity in the modal idioms, but in quantificational ones. When the modal realist says "there are at least two possible worlds" their quantifiers, it may be said, are extraordinary – they range over all things, in all possible worlds, and over trans-world individuals, things not in one world, but scattered over many. When common sense, however, says "there are no blue swans, but there might have been", its quantifiers are ordinary – they range only over the contents of the actual world (or, in modal contexts, the contents of some other world).

This ambiguity would allow us to save the simple analysis of modality. That analysis speaks of "restricting quantifiers". But, on the current proposal, this is

[10] Perhaps a believer in advanced modalizing might hold that the T rule is more important to save than the distribution rule. But the disjunctive view is committed to denying the T rule as well. Consider the sentence "There are two blue swans which are spatio-temporally isolated from one another". This sentence, according to *PW*, is true. This does not contain any talk of worlds, sets, properties or propositions, so when assessing its modal status, only the ordinary disjunct of the disjunctive analysis will be relevant. So the sentence will be both true and necessarily-false, because no one world contains two isolated swans. That is a counterexample to the T rule.

ambiguous. Which quantifiers does it mean? Let it mean the ordinary quantifier. When an extraordinary quantifier appears in a modal context, it will be untouched by the analysis. Sentences like "there are at least two possible worlds" will also be ambiguous. Read in the extraordinary way, this will be true and necessarily so. Read in the ordinary way, it will be false, and necessarily so. The N, K, T, and distribution rules are all respected, provided no equivocation between the two quantifiers takes place.

The problem of Sect. 11.1 also appears to be solved. Both *PW* and *Swans* are ambiguous between extraordinary and ordinary versions. On either disambiguation, (1.1) is true. As already pointed out, the standard inference rules are valid provided no equivocation takes place, so (1.2) and (1.3) are true. (1.4) is intended to be a claim made by the modal realist, using the extraordinary quantifier only. In that sense it is true. (1.5) is intended to be a claim of modal common sense. Since common sense knows only the ordinary quantifier, (1.5) is true. The puzzle, it seems, relies on an equivocation.

Perhaps it is something more like this that Divers means with his talk of advanced modalizing. He sometimes says things that suggest it: "[modal realism] treats our ordinary claims about donkeys, swans, and planets, etc. as implicitly world restricted claims" (Divers 2002, p. 48) "[The modal realist] appeals to the extraordinary interpretation of modal claims whenever she intends or interprets the associated non-modal content as content that is not world-restricted." (Divers 2002, p. 50) Perhaps what we should take from this is that it is not the non-logical vocabulary of a sentence that determines whether it gets the ordinary or extraordinary branch of the disjunctive analysis, but the style of quantification. As I pointed out above, this would obviate the need for anything beyond the simple analysis. All we need to do is understand the simple analysis so that it only touches "ordinary" quantifiers, leaving extraordinary ones alone. If this were Divers' view, he would do better to describe it as "advanced quantification" than "advanced modalizing".

The trouble with this proposal can be seen from Divers' quote about "ordinary claims" above, if we understand it in the way I am now considering. Modal realism treats these ordinary claims as restricted in a sense – but it does not claim, as would be plainly false, that the common sense view that (e.g.) "there are no blue swans" is *intended* to be restricted to anything less than all there is.

On the view I am currently considering, there are two *semantically different* quantifiers between which the ordinary language "there is"/"there are" are ambiguous. One of these is the ordinary quantifier which common sense uses to truly say "there are no blue swans." The other quantifier, supposedly, is only used in discussion of the ontology of modality – for example by the modal realist to say "there are at least two possible worlds". Since the ordinary quantifier is intended to be an unrestricted quantifier – a quantifier that ranges over all there is – the only way to understand the extraordinary quantifier is as intended to range over more than all there is – over all there is and some of what there is not.

Let me put this point another way. Modal realists are not Meinongians. They do not hold that they have discovered a new, possibilist quantifier, unused by actualist common sense, with which they can state a theory not statable with ordinary actualist quantification. Meinongians claim to agree with common sense about what exists, but they add that what exists is not all that there is. The modal realist does not say this. The modal realist is using the unrestricted quantifier that common sense uses to say "there are no blue swans", and she is disagreeing "to an extreme extent" (Lewis 1986, p. 133) with common sense about what there *is*. She does not, therefore, treat common sense claims about swans, donkeys, etc. are implicitly restricted. (Not at least if that restriction is *semantic* – part of *what* is *meant* by the quantifier).

There is another sense in which "there are no blue swans" might be said to be restricted to the actual world. Modal realists may say that when common sense says "there are no blue swans" it is "getting at" something true – but something that would be better said with a quantifier restricted to the actual world. This is not to say that, when the folk say "there are no blue swans", their quantifier is implicitly restricted, so that what they say is true. Rather, what they say is literally false, according to the modal realist. But they may succeed in conveying information that might be truly expressed by "there are no blue swans in the actual world", even though that is not what they said (nor even what they intended to say). This is *pragmatic*, rather than semantic, quantifier restriction. But this phenomenon does not establish an *ambiguity* between two different senses of the quantifier, which is what would be needed for the proposal currently under consideration.

Extraordinary quantification then, is suitable only for non-Quineans about ontological commitment. We might call the view that has been described in this section *modal noneism*. According to modal noneism, possible worlds are Meinongian nonexistent objects. They are just the way that the modal realist imagines them to be, except that (with the exception of the actual world) they do not exist.[11] The modal noneist enjoys all of the fruits of possible worlds semantics without (they would say) believing in merely possible worlds. For their view is that merely possible worlds do not exist. Modal noneism avoids all of the problems described in this paper. But it is not modal realism.

11.5 Metalanguage and Object Language

Trouble results from the view that *PW* is necessarily-true. Maybe the right reaction to this is not to say that *PW* is contingent, but to say that the whole question of its modal status is an ill-formed question. Questions about the modal status of a sentence can

[11] The word "noneism" comes from Richard Sylvan. Noneism, in general, is the view that "None of space, time or location – nor for that matter other important universals such as numbers or attributes – exist. . . but these items are not therefore nothing, they are each something" (Routley 1979, p. 1). Noneism is intended to be a view comparable to, and rivaling, realism and nominalism about various subject matters. Modal noneism is the application of this programme to possible worlds.

only be posed in a modal language. But, it might be said, the modal realist's analysis of modality has the form of truth conditions in a non-modal metalanguage for sentences of a modal language. In the extensional language of modal realism, the question of the modal status of true sentences like "there are at least two possible worlds" cannot arise. In the modal language being analysed, "there are at least two possible worlds" means something false – it claims that there are island universes.

Lewis (1968) gives this type of analysis. There he gives translations of the sentences of quantified modal logic (QML) into sentences of first order logic (FOL) that the modal realist would accept. But QML is a different language from FOL. In QML, the question of whether *PW* is necessarily-true cannot be raised because it does not have the expressive resources to state modal realism (in particular, it lacks a quantifier that can range over more than the contents of one world at a time). In FOL, the question cannot be raised because it does not have modal operators.

But Lewis was not claiming in this paper that QML is the logical form of ordinary modal language, still less that the quantifiers of QML are the quantifiers of ordinary language and of common sense. That would be to fall back into the "advanced quantification" view. Rather, he was making the rather more technical point that counterpart theory lacks none of the expressive power of QML.

More generally, though I am attracted by the claim that questions of the modal status of modal realism are somehow ill-formed questions, I do not believe that it can be well cashed by this object language/meta-language distinction. First, modal realism is *not* stated in some non-modal metalanguage. It is stated in English, and in other natural languages. Those are the languages in which philosophy papers stating modal realism are written, and they are modal languages.

Second, this approach would have the vices of advanced quantification. Suppose that the modal realist did intend for their theory to be statable only in some technical, non-modal subset of English. The idea would be that in that metalanguage "there are at least two possible worlds" would be true, while in ordinary English that sentence would make a false claim about actually existing island universes. But this could only be so if the quantifiers of the metalanguage had a larger domain than the unrestricted quantifiers of ordinary English. That is to say, the quantifiers of the metalanguage would have to range over more than what exists, with merely possible worlds among the things that do not exist. And as I said earlier, that would not be modal realism, but modal noneism.

11.6 Modal Eliminativism

My conclusion is that modal realists should regard their theory of modality as contingent. But this should not be regarded as a deep or profound fact. It is simply a matter of book-keeping and consistency – an odd, if unobjectionable, consequence of the only modal realist analysis of modality that is compatible with the truth of modal realism itself and with the canons of basic modal inference.

I am tempted by a stronger view. Perhaps the modal realist should stay silent on questions of the modal status of *PW*. They should state their theory of possible worlds solely in non-modalized language – which they can do – and offer no systematic translation of all of modal English into talk of possible worlds. They might offer ad hoc paraphrases – perhaps systematic translations of *parts* of modal discourse – but offer no paraphrase of others. Perhaps "*PW* is necessarily-true" is one of the sentences that would not get a paraphrase. This kind of theorist would say: "I cannot analyse modality, but I can paraphrase all of it *that has any clear content*. I can make no sense of questions about the modal status of theories of modality. Most people would agree that senseful questions about modal status can be posed in the non-modal terms of possible worlds. If you can make sense of these questions, paraphrase them into possible worlds terms yourself, and then I will answer them."

This theorist is not a modal realist. Unlike the modal realist, she abstains from using modal discourse entirely. Unlike the modal realist, she cannot claim to have an *analysis* of modality. What she has instead is a programme of paraphrasing away. Her view is that modal discourse lets us pose questions that do not really make sense, and we would be better off not to use it for that reason. She is, in short, a Quinean eliminativist about modality, albeit one who has embraced the modal realist's theory of possible worlds. Modal realists who consider the question of whether modal realist is necessary or contingent to be somehow frivolous or ill-formed should ask themselves if they are really realists rather than eliminativists of this kind.

References

Bricker, P. 2001. Island universes and the analysis of modality. In *Reality and Humean supervenience*, ed. G. Preyer and F. Seibelt, 27–55. Lanham: Rowman & Littlefield.

Divers, J. 1999. A genuine realist theory of advanced modalizing. *Mind* 108: 217–240.

Divers, J. 2002. *Possible worlds*. London: Routledge.

Divers, J. 2004. Agnosticism about other worlds: A new antirealist programme in modality. *Philosophy and Phenomenological Research* 49: 659–684.

Divers, J. 2006. Possible-worlds semantics without possible worlds: The agnostic approach. *Mind* 115: 187–225.

Fara, M., and T. Williamson. 2005. Counterparts and actuality. *Mind* 114: 1–30.

Hudson, H. 1997. Brute facts. *Australasian Journal of Philosophy* 75(1): 77–82.

Lewis, D. 1968. Counterpart theory and quantified modal logic. In *Philosophical papers*, vol. 1, 26–39. Oxford: Oxford University Press.

Lewis, D. 1986. *On the plurality of worlds*. Oxford: Blackwell.

Marshall, D. (unpublished). A puzzle about genuine modal realism.

Noonan, H. 1994. In defence of the letter of fictionalism. *Analysis* 54: 133–139.

Parsons, J. 2007. Is everything a world? *Philosophical Studies* 134: 165–181.

Quine, W.V. 1966. Three grades of modal involvement. In *The ways of paradox*. New York: Random House.

Routley, R. 1979. *Exploring Meinong's jungle and beyond*. Canberra: Australian National University.

Yagisawa, T. 1992. Possible worlds as shifting domains. *Erkenntnis* 36: 83–101.

Chapter 12
Spread Worlds, Plenitude and Modal Realism: A Problem for David Lewis

Charles R. Pigden and Rebecca E.B. Entwisle

12.1 Introduction: Modal Realism and Meta-philosophy

David Lewis was famous (among other things) for the meta-philosophical claim that knock-down refutations are rare to non-existent in philosophy. You can argue that the costs of a theory exceed the intellectual benefits but it is difficult, if not impossible, to *prove* that it is false. When it comes to choosing between philosophical theories, it is generally a matter of 'the price is right', and when it comes to refuting a theory the best you can generally do is to drive up the costs whilst diminishing the benefits. There may come a time when the costs are so high and the benefits are so low that the theory can't attract any takers, but this is usually due to a cumulative process of philosophical debate, not to a sudden drop in net value brought about by a single decisive counterargument. When philosophers win, it's nearly always a victory on points and almost never a knock-out. At the same time, Lewis subscribed to a philosophical thesis so outrageous as to cry out for a knock-down refutation, namely that in order to make sense of modal truths, we must postulate an infinity of possible universes, all just as real as the one that we inhabit, but cut off from us in space and time. (This has the corollary that in order to make sense of causality we must postulate an infinity of items that have no causal connection to anything that happens in this universe.) Thus the temptation is to go for the double-whammy, the knock-down refutation of modal realism which is

C.R. Pigden (✉)
Department of Philosophy, University of Otago, Dunedin, New Zealand
e-mail: charles.pigden@otago.ac.nz

R.E.B. Entwisle
Department of Philosophy, University of Otago, Dunedin, New Zealand

Princeton University, Princeton, NJ, USA
e-mail: entwisle@alumni.princeton.edu

J. Maclaurin (ed.), *Rationis Defensor: Essays in Honour of Colin Cheyne*, Studies in History and Philosophy of Science 28, DOI 10.1007/978-94-007-3983-3_12,

also a counterexample to the meta-philosophical claim. This was our aim when we first drafted this paper, but we can't honestly claim the double victory that we originally hoped for. We may have given Lewis's modal realism a drubbing, but we don't pretend to have decked it. Even if we have succeeded in driving up the costs of Lewis's modal realism, it still retains enough benefits to attract some (rational) takers. Thus whatever the fate of Lewis's *philosophical* thesis, the meta-philosophical claim remains intact.

We first wrote this article nearly 20 years ago when Lewis himself was very much alive, but for one reason or another, we never got around to publishing it. Rather than rewriting it to take account of the sad fact of his death we have decided to leave it in the present tense. We dedicate this paper not only to our friend and colleague, Colin Cheyne (who does not think we should believe in entities which cannot causally affect us), but to the memory of David Lewis, a great philosopher and a great teacher.

12.2 Modal Realism: Mad Dogs, Milksops and Primitives

In his metaphysical summa of 1986, *The Plurality of Worlds* (henceforward POW), David Lewis famously defends a doctrine he calls 'modal realism'. The idea is that possible worlds are real, indeed *really* real. They are concrete entities just like our own world, but cut off from us in space and time. To every logically possible set-up there corresponds such a world. Lewis calls this the Principle of Plenitude. In fact, Lewis *identifies* his worlds with such set-ups – a move which makes it difficult to specify the Principle of Plenitude in a non-circular way (POW, pp. 86–92). Lycan (1988) calls this doctrine 'Mad dog modal realism', 'Rape-and-loot modal realism' and even 'Nuclear holocaust modal realism'. This is in contrast to moderate, milksop or, as Lewis describes it, *ersatz* modal realism. Ersatz realists believe in possible worlds all right. It is just that their possible worlds are less ontologically outrageous. Usually, they construct them out of actual entities of some kind: points, sets, sentences or what have you. This is 'paradise on the cheap', since we get all the theoretical advantages of possible worlds without the ontological drawbacks.

Lewis thinks he can do without modal primitives and that the milksop defenders of ersatz worlds cannot. This he holds to be the drawcard of an otherwise unpalatable theory. His possible worlds are admittedly implausible things (especially in such abundance) and the chief reason for believing in them, rather than in ersatz possibilia of some kind, is that they enable us to eliminate modal primitives.

Although sometimes denounced as a Meinongian, indeed a relentless Meinongian (Lycan 1979), Lewis is, in fact, a rather heterodox Quinean. His aim is to explicate the modal concepts in terms of first order logic plus an ontology of possibilia. He retains a Quinean *ideology,* that is, the conviction that the resources of first-order logic, quantifiers, predicates, truth-functional connectives and the possible assistance of truth and satisfaction, are adequate to describe reality.[1] But unlike Quine (who prefers to believe there are no such things[2]) Lewis recognizes modal facts. To accommodate these facts within a Quinean ideology,[3] to *explain* the modalities in terms of quantification theory, Lewis has to invent *more* reality. The price of his ideological restraint is ontological inflation. He must posit an infinity of objects to quantify over. But it is important to realize that despite his ontological exuberance, what Lewis is putting forward is a *reductive* theory. It is just that the reductions are conceptual rather than ontological. The aim is to take the modality out of modality, to reduce modal to non-modal discourse, even if the universe of discourse has to be inordinately expanded to do so. (It is for this reason that Plantinga (1987) denies Lewis his title to modal realism. For Plantinga, a modal realist must be a realist about *modality* and not just *possibilia*.) Moreover, in remaining true to the ideological constraints, Lewis has to ditch other aspects of the Quinean legacy. For Quine it is science that tells us What-There-Is. For Lewis that is true only of *this* world. Science tells us about that tiny portion of reality that is spatiotemporally related to *us*. As for the rest, a logically disciplined imagination is a better guide (POW, pp. 113–115).

The natural response to Lewis is a *tu quoque*. Milksop realists and plain modal sceptics have tried to pin Lewis down with modal primitives. They have tried to argue that he too implicitly relies on primitive notions of necessity and possibility. In which case his theory is no better than that of his milksop opponents, or indeed than the No-Theory Theory of Modality which simply assumes modal operators as primitives and leaves it at that. Indeed it is considerably worse off, since their ontologies are more believable. (See Lycan 1988.) Our criticism is just such a riposte.

We shall argue that (on *one* reading) Lewis's Principle of Plenitude licenses us to assume maverick possible worlds which spread through logical space gobbling up all the rest. Because they exclude alternatives, these worlds undermine Plenitude itself, and worse still, result in contradictions, since different spread worlds are incompatible with one another. Plainly the Principle of Plenitude must be amended to exclude these excluders. But, we maintain, this cannot be done without bringing

---

[1] Lewis's ideological allegiance to Quine is most nakedly displayed in the opening paragraphs of his (1968) 'Counterpart Theory and Quantified Modal Logic'. See also his POW (1986, pp. 1–20), and especially (1986, p. 4). However, in the 1990s there was some degree of relaxation in Lewis's ideological austerity. He came to accept the plural quantification theory of George Boolos. See Lewis (1991) *Parts of Classes*, especially (1991, pp. 62–71). But by adding the apparatus of plural quantification to his ideology, Lewis ceases to be a heterodox Quinean and becomes something close to an apostate. (Boolos is very explicitly an opponent of Quine. See his (1985, pp. 331–334).) We are inclined to think that this undermines the philosophical rationale for modal realism. After all, if we are allowed the ideological indulgence of plural quantification given the rather minimal ontological savings it brings in, why not allow us the further indulgence of modal primitives given that the ontological savings would be massive?

[2] For an excellent survey of Quine's views on modality and some sharp criticisms along the lines suggested see Haack (1978, ch. 10).

[3] See Quine (1961, p. 131).

in modal primitives. And once we admit modal primitives, bang goes the rationale for Lewis's modal realism. We then consider various replies, the chief of which is due to Lewis himself. He insists that Plenitude does *not* license the spread worlds and hence that no amendment is needed. *A fortiori* no amendment is needed which requires him to make use of modal primitives. We answer that in that case his theory is disturbingly unspecific about which things are possible. Its intellectual value is therefore much diminished. Perhaps the price is no longer right.

12.3 The Principle of Plenitude: Some Preliminaries

What is the Principle of Plenitude? It appears in two versions, a *naive* version (absolutely every way a world could possibly be is the way that some world is), and an *official* version which relies on the Principle of Recombination. The official version is required since the naive version is trivialized once Lewis identifies worlds with ways worlds could possibly be. Both the official and the naive versions of Plenitude look like attempts to improve on another principle which we shall refer to as ur-Plenitude. It goes like this: to every consistent set of sentences there corresponds (at least) one possible world. We conceive these sets of sentences as finite and as stated in some language such as English. (We are not, for instance, talking about infinite sets of sentences composed of letter shaped sets of points and inscribed in deep space.) For a set of sentences to be consistent it is not enough for it not to *include* any contradictions. It must not *entail* any contradictions either. Hence ur-Plenitude presupposes some concept of consequence. But logical consequence won't do by itself. For the set of sentences {Oscar is round; Oscar is square.} is *logically* consistent, since no contradiction can be derived from the pair with the aid of logic alone. But it does not represent a genuine possibility. Round squares (let alone round squares called 'Oscar') cannot be consistently supposed. The way to get around this is to specify that a set of sentences is consistent (in this sense) iff it does not entail any contradictions *once the relevant analytic truths or meaning-postulates are added in*. Since it is analytic that squares are not round, we can derive the consequence that Oscar is round and not round from the offending sentence set. Since this is a contradiction, the set does not qualify as consistent under the amended definition.

We suspect that Lewis would find ur-Plenitude sympathetic. Other things being equal (perhaps an important qualification!), he would like to derive ur-Plenitude as a consequence of his theory. Indeed, ur-Plenitude (or something like it) is probably among the intuitions that his theory is tailored to fit. (See POW (p. 154), where the role of such intuitions is discussed.) But ur-Plenitude cannot figure among the constitutive principles of his system. This is partly because it is vague and partly because it presupposes, or seems to presuppose, modal notions. And these must be banished from the foundations of a reductive theory. Ur-Plenitude is *vague* because it relies on the notion of analyticity and the borders of the analytic are notoriously unclear. And it *presupposes modal notions* because *analyticity*, like *consequence*, is difficult to define without recourse to necessity. In the case of consequence the difficulties can perhaps be overcome. Consequence can be defined in model-theoretic terms if we help ourselves to set theory. And the concept of a contradiction poses no particular problems since it can be defined without recourse to modalities. Thus it is analyticity that poses the real problem. This concept is notoriously difficult to elucidate without bringing in necessity. And even a heterodox Quinean like Lewis would hesitate to employ it as a primitive.[4]

Now, as van Inwagen has pointed out, Plenitude in its naive form turns out to be trivial once we identify worlds with ways that worlds could be. If worlds *are* logical possibilities then by definition there would be a world corresponding to each possibility even if there were only 17 worlds or none at all. (POW, pp. 86–87.) This means (among other things) that we cannot derive ur-Plenitude from naive Plenitude. For if there were just 17 worlds, naive Plenitude would be true and ur-Plenitude false. After all, there are more consistent sets of sentences than 17 (and more than 17 sets of such sentences that are *both* incompatible with actuality *and* with each other).

To rescue his theory from triviality, Lewis is compelled to develop a more sophisticated version of the Principle of Plenitude. He does this with the aid of *worlds, duplicates* and a Principle of Recombination. Miller (1989 p. 477), extracts the following definition of a world from Lewis.

(1) Individuals are worldmates if they are spatio-temporally related.
(2) A world is a mereological sum of worldmates.[5]

Now we know *what* worlds are (or at least, what they are supposed to be), the next question is: *which* worlds are there? Part of Lewis's answer is to list some of the weird and wonderful possibilities he thinks are realized somewhere in logical space. This gives the reader some idea of the range of variations permitted by Plenitude. But his more careful answer relies on duplicates and the Principle of Recombination. Duplicates are physically identical copies of things. They must also not be worldmates of the original thing. The Principle of Recombination says that duplicates of anything can coexist, or fail to coexist, with duplicates of anything else (POW, pp. 88–89). So if there is one world with a talking donkey, and another world with a tap-dancing owl, then there is a third world in which a talking donkey and a tap-dancing owl are playing Canasta. Again, if there is one world with a talking head attached to a human body then there is another world in which the head is just as voluble but detached. (It must be rather alarming for the inhabitants of that world, especially if there is only one detached talking head!) It is evidently Lewis's belief that Recombination will underwrite the enormous variety of worlds

[4] Lewis does not discuss ur-Plenitude explicitly. But his arguments against linguistic ersatzism, suggest that he would reject it as a fundamental principle for the reasons we have alleged. (The same goes for the principle of Consistent Describability discussed below.)

[5] This definition did not take a lot of extracting. On page 71 Lewis states that 'things are worldmates iff they are spatiotemporally related', and on page 69 that 'a world is the mereological sum of all the possible individuals that are parts of it, and so are worldmates of one another'.

suggested (though not implied) by the more naive version of Plenitude. By the same token, Recombination should guarantee the truth of ur-Plenitude. Indeed it *ought* to do so, if, as we suspect, Plenitude is an attempt to generalize the idea behind ur-Plenitude, by transcending the limitations imposed by language.

But there is a problem with Recombination. As Denis Robinson put it in conversation, it is something of a sausage machine. Once you have a set of worlds, you can generate new ones by recombining duplicates of their components. Or rather, since the 'new' worlds are supposed to be there already, given a set of worlds, you are entitled to *assume* the existence of other worlds composed of duplicates of their components. (Recombination, despite some loose talk, is an *epistemic* not a *metaphysical* principle, a license to posit, not a license to construct.) But which worlds and which entities are you entitled to posit in the first place? An obvious answer would be worlds composed of duplicates of actual entities.[6] This is quite a generous license. It allows us to assume anything that can be made up out of whatever the fundamental particles of this universe turn out to be. (At present, we understand the smart money is on quarks, gluons and leptons.) And we take it that an enormous variety of non-actual entities could be constructed out of these (or to be precise, their duplicates). But the license is not generous enough. For Lewis is inclined to believe in the possibility of *alien* entities, things possessed of properties not exemplified by anything in this world and not composed (or not wholly composed) of entities exemplifying this-worldly properties. Although an atheist with respect to the actual, he believes that God is possible even though a God could not be manufactured out of duplicates of actually existing things. (God is alleged to be simple and not composed of parts of any kind, let alone parts which are duplicates of actual items.)

But if we can assume alien individuals before applying Recombination, what are the constraints on postulating worlds and individuals to duplicate? Part of the answer is supplied in Lewis's section on 'Isolation' (POW, pp. 69–81). Only spatial or temporal beings can qualify, although Lewis is fairly relaxed about non-standard times, spaces and space-times. But another answer is suggested in Lewis's reply to Forrest. 'If, as I suppose, a being does not have to satisfy some inconsistent description to be a god, then I take the number of Gods to be at least 2^{∞}.' (Lewis 1983, p. xi n4.) If a god does not have to satisfy an inconsistent description in order to be, then somewhere or other He, She or It exists. (This does not mean that in logical space someone can always hear your prayers, let alone your screams, for the Gods of other worlds are deaf to us. Lewis's polytheism is not of a very comforting kind!) The same, we presume, goes for other aliens, from dragons and devils down to talking donkeys and Democritean atoms. If a being does not have to satisfy an inconsistent description to exist, then somewhere in logical space there is such a being. If you've got Consistent Describability then

[6] David Armstrong proposes something similar in *A Combinatorial Theory of Possibility* (1989). However, Armstrong's combinatorial worlds are ways actual objects could be, not ways that their duplicates are.

you have got what it takes to be. What might be called the principle of Consistent Describability entails but is not entailed by ur-Plenitude. Ur-Plenitude says that there is at least one world corresponding to every consistent description. The Consistent Describability principle says there is at least one world corresponding to every *possible* consistent description. Plainly the second entails the first since *actual* consistent descriptions are a mere subset of the possible. But there is a problem with Consistent Describability. Lewis no doubt believes that there are *more* things than *we* can consistently describe (where the 'we' is restricted to the inhabitants of actuality), since some alien entities possess properties which we have not even imagined. What we have not imagined we have passed over in silence, neglecting to invent the appropriate predicates. And in the circumstances, this act of neglect will be difficult to remedy. (There are more things in logical space, Horatio, than we will *ever* dream of in our philosophies. See POW (p. 159).) But Consistent Describability cannot be *added* to Plenitude as an extra clause telling us which alien worlds and entities it is right to posit. The reasons have been touched on already when we discussed ur-Plenitude. If Consistent Describability were properly spelt out it would almost certainly employ a modal primitive. This is not to say that Consistent Describability should not come out *true* according to Lewis's system. Indeed, it might be employed as a sort of a *gloss* to elucidate Plenitude and to give people an idea of which alien entities can be safely assumed. But as *primitive* principles or as *axioms* of Lewis's system both ur-Plenitude and Consistent Describability are out of the running. The closest Lewis can come is the insistence that there are no contradictory worlds, no worlds of which sentences of the form 'P and not-P' are true. For according to Lewis, you cannot tell the truth about *anything* – not magic mountains far away, nor individuals in another space-time – by contradicting yourself. (See POW (p. 7n) and Lewis (1983 p. 21).)

What is the upshot? We have discussed the following principles:

1) ur-Plenitude: to every consistent set of sentences there corresponds at least one world; and
2) Consistent Describability: to every *possible* consistent set of sentences there corresponds at least one world.

Both are intuitively plausible. But neither can be admitted as axioms within Lewis's reductive theory. Yet without them Lewis's principle of Plenitude in its official version seems curiously incomplete. We are allowed to assume *some* alien entities before applying the principle of Recombination, but apart from the proviso that we should not assume worlds or entities of which contradictions are, or would be, true, we don't have much guidance as to which entities are available for duplication. However, we have argued that (pending strong counterarguments) 2) and 1), *ought* to come out true, according to Lewis's theory, and some of Lewis's remarks suggest an explicit commitment to 2). Plenitude should be interpreted in such a way as to underwrite the existence of worlds corresponding to consistent descriptions. There should be at least as many worlds as we can consistently suppose.

12.4 David Lewis Meets Martin Luther

But Plenitude so interpreted is false. There are some worlds which are *consistent* but *impossible*, or less tendentiously, some sets of sentences which are logically consistent, indeed consistent in the stricter sense we have been trying to elucidate, but which *must not* be allowed to describe existent possible worlds. For if these worlds are admitted as genuine, then they induce modal collapse. That is, they reduce the plenitude of worlds to one.

One such world is the *Luther*world. This is the world in which Luther's theology is true (See Luther (1525/1957) *Of the Bondage of the Will*). In that world the following propositions hold:

(i) God exists.
(ii) God foreknows and wills all things, not contingently, but necessarily and immutably (Luther 1957, pp. 83–4).

What (ii) amounts to is: for any proposition p, if it is true then God knows it necessarily. In symbols: $(p)(p \supset \Box GKp)$.

In the Lutherworld all true propositions are necessary. So if Luther's God is possible (in the modal realist sense), then God exists and everything God foreknows will necessarily come to pass. Here is how it works. It is a truism indeed a necessary truth about knowledge that for any (person) x, if x knows that p, then p. In symbols:

$$\Box(x)(xKp \supset p)$$

From which it follows that necessarily if God knows that p, then p, i.e.:

$$\Box(GKp \supset p).$$

The following is a theorem of all modal logics from K on up (indeed, it is often used as an axiom): $\Box(p \supset q) \supset (\Box p \supset \Box q)$. By substituting in we get: $\Box(GKp \supset p) \supset (\Box GKp \supset \Box p)$. Now consider any truth p. From thesis (i) of Luther's theology, it follows that necessarily God knows that p, i.e. $\Box GKp$. Then from a truism about knowledge plus an uncontroversial modal axiom it follows by 'resistless logic' (Luther's phrase) that $\Box p$, that is that p is necessary. Since all truths are known to God and in Luther's world necessarily known, it follows that all truths are necessary, indeed *logically* necessary, necessary in the strongest sense of the word. So far, you may say, all that has been shown is that *in the Lutherworld*, all true propositions are necessary. That does not prove that they are necessary in *this* possible world. And you would be right but for the S_5 principle (accepted by Lewis) that if possibly necessarily p, then necessarily p (or $\Diamond\Box p \supset \Box p$). What this means is that if a proposition p is necessary in the Lutherworld, it is necessary in *all* possible worlds including this one.

Contingency therefore is abolished and there is only one way the world could be, namely the godly and predetermined way that it is.[7] Given Lewis's analysis of modal concepts this means that there is only one possible world, the actual one, and it is the Lutherworld. (Or alternatively there are *lots* of identical worlds and that *all* are Lutherworlds.) We call worlds like the Lutherworld '*spread worlds*' or '*excluder* worlds' since once they are admitted as possible, they spread through logical space and exclude all alternatives.

12.5 Thick Spread Worlds, Thin Spread Worlds and an Ontological Argument

However, we distinguish between *thick* spread or *highly* exclusive worlds, which eliminate all worlds but one (such as the Lutherworld), and *thin* spread or *moderately* exclusive worlds which merely eliminate worlds which lack certain characteristics. The Lutherworld is a *thick* spread world since it excludes all worlds except the one that God foreknows. Plantinga's world by contrast, the one he assumes for the sake of the ontological argument, is a *thin* spread world, since if it is admitted, it only excludes worlds without God.

In his famous (or perhaps infamous) proof, Plantinga proposes a possible world in which unsurpassable greatness is exemplified (Plantinga 1974, p. 216). Unsurpassable greatness amounts to having maximal excellence in every possible world, and maximal excellence amounts to being God-like. This allows us to reformulate his argument as follows.

(i) Possibly there exists a being which is necessarily God-like. In symbols:

$$\Diamond(\exists x)(\Box Gx).$$

(ii) Necessarily there exists a being which is necessarily God-like, if and only if necessarily there exists a being which is God-like. More perspicuously:

$$\Box((\exists x)(\Box Gx) \equiv \Box(\exists x)(Gx)).$$

The second thesis (ii) is crucial to Plantinga's proof. In fact it is one half of the necessary biconditional, (ii') $\Box((\exists x)(\Box Gx) \supset \Box(\exists x)(Gx))$, that does all the

[7] For more on Martin Luther and his repellent opinions see Pigden (1989, pp. 141–2), and Pigden (1990, §2).

work.[8] The trick behind Plantinga's proof (as many people have realized)[9] is to exploit the S_5 maxim:

$$\Diamond\Box p \supset \Box p.$$

Crudely put, the idea is to move from 'Possibly necessarily God exists', to 'Necessarily God exists' – in which case He is actual. But that is a little *too* crude. Plantinga's initial premise (i) is not that 'Possibly necessarily there exists something which is God-like' (in symbols: $\Diamond\Box(\exists x)(Gx)$) but 'Possibly there exists something which is necessarily God-like' (in symbols: $\Diamond(\exists x)(\Box Gx)$). This premise by itself does not allow him to work the old S_5 magic, since there is a quantifier standing between the '$\Diamond$' and the '$\Box$'. These operators must be brought together if $\Diamond\Box p \supset \Box p$ is to do its stuff. What (ii') does is to allow Plantinga to hop the modal operator over the quantifier, so that the '$\Diamond$' and the '$\Box$' are adjacent. In other words, it allows him to derive

(iii) $$\Diamond\Box(\exists x)(Gx).$$

From then on (in S_5) it is all plain sailing. The reason Plantinga requires an extra premise to do this is because he rejects the Barcan formula $\Box((\exists x)(\Box\varphi x) \supset \Box(\exists x)(\varphi x))$ (as does David Lewis[10]). Indeed, any philosopher who supposes that there might be more or less things than there are, *must* reject this principle. The fact that a world contains something which is necessarily φ (i.e. φ in all the worlds in which it exists) does not entail that necessarily there is something which is φ. For the thing which is φ might not exist or might not have a counterpart in other possible worlds. So Plantinga needs (ii') to vindicate in particular the thesis that he rejects in general: Necessarily if there is something which is necessarily God-like then necessarily there is something which is God-like.

---

[8] What is the other half of the biconditional, (ii") $\Box(\Box(\exists x)(Gx) \supset (\exists x)(\Box Gx))$, supposed to do? Why does Plantinga need this? Perhaps to ensure that it is the same entity that is God-like in every possible world. If different beings possessed the God-like properties in different possible worlds, none of them would be truly God. So Plantinga must establish not only that a God-like being exists in every possible world, but that the God-like being is *God-like* in every possible world. It would not do, for instance, if the 'God' of this world were the 'Satan' of another and vice versa. This is not a problem for Lewis. If he were willing to accept Plantinga's premises (which, of course, he is not) it would follow automatically that the 'God' of this world (or better the God-like being) was 'identical with' the 'Gods' in all the other worlds in the rather attenuated sense that it would be a counterpart of all the others. (At least it would follow on the assumption that you can have at most one 'God' per world.) And that is all the trans-world identity there is. Lewis would agree with Plantinga that being a counterpart of x is not really being identical with x, but would try to fob him off with the assurance that there is nothing more to be had.

[9] See for example Mackie (1982, pp. 55–63).

[10] Lewis (1968, 1983, p. 36).

Now the interesting thing is that Plantinga does not need such a strong set of premises to achieve his result. All he needs to do is to strip the initial operators from (i) and (ii'), yielding

(ia) $$(\exists x)(\Box Gx)$$

and

(ii'a) $$(\exists x)(\Box Gx) \supset \Box(\exists x)(Gx)$$

and to conjoin these behind one possibility operator. This produces the thesis

(i-ii') $$\Diamond((\exists x)(\Box Gx)\&((\exists y)(\Box Gy) \supset \Box(\exists y)(Gy)))$$

(Possibly there exists something which is necessarily God-like, and if there exists something which is necessarily God-like then necessarily there exists something which is God-like.)

This entails (iii), and with (iii) and S_5, we can spread God-like beings throughout logical space. But Lewis's system of possible worlds (like Plantinga's) is designed to underwrite S_5. Lewis therefore must deny (i-ii') since if it were true, there would be no world without a duplicate of God. And this is inconsistent with the Principle of Plenitude which states that the duplicates of anything can coexist or fail to coexist with the duplicates of anything else. This means that we can specify a world, the Plantingaworld, which is, or seems to be, consistent, but which Lewis must prohibit as impossible. It is the world (or set of worlds) described by the following sentences:

(ia) $$(\exists x)(\Box Gx)$$

(There exists something which is necessarily God-like.)
and

(ii'a) $$(\exists x)(\Box Gx) \supset \Box(\exists x)(Gx)$$

(If there exists something which is necessarily God-like then necessarily there exists something which is God-like.)

The Plantingaworld is a *thin* spread or moderately exclusive world. Unlike the Lutherworld which excludes *all* alternatives, the Plantingaworld only excludes the Godless ones. There are less worlds than Lewis supposes but still more than one. So far as Plantinga is concerned, possible worlds can be as they please, so long as they include God (and all that that entails). *Thick* spread worlds, therefore, exclude Plurality; *thin* spread worlds merely exclude Plenitude. Both are inimical to Lewis.

12.6 Is God the Problem?

Although we have chosen a pair of godly examples, it should be obvious that God is not really the logical culprit here. Or if He is there are other alleged possibilia that are equally at fault. For we can specify spread or excluder worlds without bringing in God at all. The Plantingaworld provides a sort of schema which allows us to manufacture excluder worlds at will. We simply substitute for 'G' in (ia) and (ii'a) any predicate we like, and straightaway we have a world (or a world-description) which will spread that predicate throughout logical space (at least it will do so if the world in question is admitted as real). If the 'G' stands for 'G-string' there are no worlds without G-strings, if the 'G' stands for gumnut there are no worlds without gumnuts etc. etc. etc. Nor does Plantinga provide the only set of sentences that can be used to perform this trick. The Lutherworld suggests otherwise, and we are sure that ingenious logicians could invent many more. And there is another point to note. Although excluder worlds seem to be internally consistent in the broad sense we have defined, they need not be compossible, that is, compatible with other excluder worlds. There are worlds or world-descriptions such that if either is admitted as possible, the other is automatically excluded. For instance, we can specify *atheistic* excluder worlds; worlds which once admitted as real, rule out a genuinely Divine Being. Again Plantinga shows us how (1974, pp. 218–20). He is discussing properties rather than worlds, but his argument can easily be adapted to our needs. Consider the world described by the following sentences:

(iv) $$(\exists x)(\Box \sim (\exists y)(Gy))$$

(There exists something such that necessarily there is no God-like being.)

and

(v) $$(\exists x)(\Box \sim (\exists y)(Gy)) \supset \Box((\exists x)(\sim (\exists y)(Gy))$$

(If there exists something such that necessarily there is no God, then necessarily there exists something such that there is no God.)

Call this the Shelleyworld (named for Shelley's pamphlet *The Necessity of Atheism*). This spreads godlessness throughout logical space excluding theistic alternatives. It is thus a thin spread world, but one which is not compossible with either the Plantingaworld or the Lutherworld. Indeed, Lewis himself believes in a world (in his view the actual one) which is at least *mildly* exclusive since it rules out both the Plantingaworld and the Shelleyworld. For Lewis believes that the following sentence is actually true (Lewis 1983, p. ix):

(vi) $$\sim (\exists x)(Gx) \& \Diamond(\exists x)(Gx)$$

(It is not the case that there exists a being which is God-like and possibly there exists a being which is God-like.)

But if God is not actual the Plantinga world is impossible and if God is possible then the Shelleyworld is not. This shows us that excluder worlds can be mutually exclusive. So some ban on the promiscuous positing of such worlds is needed, not just to save Plenitude or even Plurality, but to save us from contradiction. We can also see *why* excluder worlds are dangerous, indeed, why it is that they are so exclusive. It is not because the sentences that describe them deal in the Divine, but because they deal in modalities, especially necessities, that excluder worlds spread through logical space edging out their rivals.

12.7 The Problem with Plenitude

'But what *is* the problem exactly?' someone might demand. 'Excluder worlds are incompatible with Plenitude and with each other. Plenitude is true. So there are no such worlds. End of story.'

Imagine a naive set theorist – call him Georg – who accepts Russell's paradox as an elegant proof that there is no class of classes that are not members of themselves but thinks he can leave it at that. Such a response would miss the point. For the existence of the paradoxical class follows from the abstraction axiom which is central to naive set theory. Russell's paradox proves that the abstraction axiom is false and has to be amended. Georg cannot simply dismiss the Russell class and relapse into his dogmatic slumbers – not if he wants to have a consistent theory that is. For much the same reasons, it is not open to Lewis to shrug off our results as a reductio of the existence of certain worlds. It is true that the existence of spread worlds is incompatible with Plenitude. But Plenitude *itself* seems to license such worlds. And if Plenitude implies the existence of worlds which are incompatible with Plenitude, then Plenitude is false and will have to be amended. So the real question is: does Plenitude really entail the existence of spread worlds?

12.8 Revenge of the Modal Primitive from Logical Space

Suppose that it does. Then Lewis must restrict Plenitude so as to exclude excluder worlds. *And he must do this without recourse to modal primitives.* (Otherwise modal realism loses out to its ersatzist competitors in the philosophical price war.) We think this will be very hard to do. The problem is that excluder worlds are consistent but impossible (or as Lewis would have to say in his official theory, consistent but *nonexistent*). So Lewis must define what it is about excluder worlds which makes them dangerous and must add to the Principle of Plenitude the proviso that there are no sums of worldmates like *that*. The obvious way to do it is to deny that there are any worlds corresponding to world-descriptions which entail non-logical, *de dicto* necessities (i.e. box-prefixed sentences which are not logical truths). But apart from the fact that this restriction appeals to the notions of *necessity* and *logical truth*, and probably *logical consequence* and *analyticity* as well, Lewis *wants* some necessary but non-logical sentences to be true at his worlds, namely those which are analytic and those which express the findings of pure mathematics. (Given the ontological commitments of mathematics mathematical truths do not qualify as analytic i.e. true in virtue of logic plus definitions.) Nor would it do to roundly assert that there are no non-logical, *non-analytic*, *non-mathematical*, de dicto necessities, and that therefore excluder worlds are consistent but impossible (Oops sorry! *nonexistent*). For such a proviso obviously relies on several notions it would be difficult to define without recourse to modal primitives, analyticity being the chief.

(This suggests an important difference between ontological and conceptual reductions. If it is ontological reduction that we have in mind then there is no problem about mentioning the items to be reduced. These can figure within the reducing theory which can simply assert that there are no such things or that they are non-basic and are made up out of other things. It is otherwise with conceptual reduction. Of course, the total theory must include translation schemata whereby statements involving the suspect concepts are paraphrased into a dialect which makes no mention of them. But the reducing theory itself must not employ the concepts to be reduced. And you are *using* a concept even when you are *denying* sentences which themselves employ that concept. In the present case Lewis would not succeed in his conceptual reduction if he were forced to say that there are *no* worlds or individuals which entail certain kinds of necessities. Plenitude is supposed to be part of the reducing theory and so should be statable in a language that does not employ modal notions, even in the context of sentences or subsentences which are negated. Conceptual reduction therefore is a ticklish business, in some respects rather more difficult than ontological reduction. In the official reducing theory you can't even talk about what you are trying to do away with.)

Suppose on the other hand that Plenitude does not entail the existence of excluder worlds. Plenitude is not to be read in such a way as to underwrite the Principles of ur-Plenitude and Consistent Describability. Hence it is not (or not provably) inconsistent. We are allowed to posit alien entities subject to the constraints of consistency and to some other constraints which are not specified. These (we must suppose) do not entail the existence of thick or thin spread worlds. Then Lewis's system would escape contradiction and the need to posit modal primitives – but only at a cost.

12.9 Plenitude and the Spread Worlds

David Lewis discussed this argument with us on a number of occasions both orally and in writing. And several other philosophers have been kind enough to furnish us with comments. In this section we shall attempt to deal with Lewis's response. We shall refer other criticisms as we go along.

As we understand it, Lewis's position is this. The Principle of Plenitude does not entail the existence of spread worlds. But Plenitude is true. Thus the fact that spread worlds are incompatible with Plenitude is simply a proof that there are no such worlds. If there is a Lutherworld 'then there isn't also a world in which God foreknows differently, let alone a Godless one – but there are such worlds, ergo there's no Lutheran world'. We have, in effect, performed a reductio on both ur-Plenitude and the principle of Consistent Describability. It turns out to be false that there is a world corresponding to every consistent set of sentences. Hence it is false that there is a world corresponding to every *possible* consistent set. But

Plenitude does not imply either of these false principles and never has. Indeed, in an early article 'Anselm and Actuality'[11] Lewis discusses a variant of Plenitude which comes close to the one we have tried to saddle him with. He calls this *the Principle of Saturation*, and dismisses it precisely because it leads to contradictions of the kind we have suggested (Lewis 1970, pp. 182–183, 1983, p. 16.[12]). So there is no need to add a 'new wrinkle' to the Principle of Plenitude to exclude the spread worlds. A fortiori, there is no need for a 'new wrinkle' which employs a modal primitive (Lewis, 1989, 1990, Private communication).

Well, does Plenitude entail the existence of spread worlds or doesn't it? There are three (not two!) options.

A. The Principle of Plenitude is clear and the existence of spread worlds does indeed follow. We win, Lewis loses. He must add a new wrinkle, and this will be hard to do without recourse to modal primitives.
B. The Principle of Plenitude is *not* clear. Hence it is not clear whether it entails the spread worlds or not. In which case we have struck Lewis a glancing blow. For it turns out that his theory is (to use his own terminology) a magical one. You have a bit of philosophical machinery (Plenitude) which is supposed to do a job (generate roughly the same range of worlds that we would be naively inclined to postulate). But it is not clear how this bit of machinery is supposed to work or whether it does its job. Furthermore, Plenitude remains at risk. It may be that when it is spruced up and clarified it *either* implies the existence of spread worlds *or* employs a modal primitive.
C. The Principle of Plenitude *is* clear and *does not* entail the spread worlds. We lose and Lewis wins. Even so, the victory may be Pyrrhic. Perhaps Plenitude turns out not to be so attractive once we understand it properly. Perhaps it is covertly reliant on a modal primitive in the first place.

Obviously Lewis must go for option C with the proviso that there is nothing Pyrrhic about his polemical victory. We must go for option A, with B and a Pyrrhic C as fallback positions.

Is Plenitude clear? Not really. As we have seen, it is curiously incomplete. We are entitled to postulate *some* alien individuals. And once we have got them we can combine or fail to combine them at will. But apart from this, Lewis gives us very little guidance. There are no constraints on our freedom to postulate apart from consistency. We must not assume any entity of which contradictions would be true. But neither is there a *license* to postulate. Lewis does not say that we can postulate anything we like so long as it is consistent – not when he is being careful that is. The official theory is officially silent on this point. To get an answer we must seek an informal gloss on this official formulations. Now in his (1983, p. xi), Lewis *does* say that so long as a God does not have to satisfy an inconsistent description to

[11] Originally published (1970).

[12] We had not noticed this passage until Lewis pointed it out to us. If our article performs no other function it will enable readers to make sense of Lewis's rather compressed remarks on this topic.

be then Gods exist in very large numbers. Which suggests we can assume anything which does not have to satisfy an inconsistent description. Taking Lewis at his word, we were able to postulate our paradoxical worlds. In other words, if Plenitude is read in the light of Lewis's informal pronouncements, it *is* clear and *does* entail the existence of spread worlds. But if we confine ourselves to the official theory, and scorn such suspect remarks as mere 'gas', then the theory does not seem to be clear enough to afford an answer. To save Lewis's theory from contradiction it must be deprived of clarity. To paraphrase Lewis himself, 'this is one way to regain consistency. It is not a good way' (POW, p. 191). Nor does it help to cite Lewis's explicit rejection of Saturation. This simply shows that even Lewis can contradict himself (pages xi and 16 of his *Philosophical Papers* being at odds) and it still leaves us in the dark about *which* alien worlds and alien beings we are entitled to postulate.

Thus if Lewis is to avoid the disastrous consequences we have tried to fasten upon him, he must eat at least some of his words. We hope the 'Introduction' to his *Philosophical Papers* makes a healthy meal. But it will be a meal in the midst of confusion. We can't show that spread worlds follow from Plenitude. But he can't show that they *don't* follow either. We just don't know which aliens we are allowed to posit. Plenitude can be cleared of overt inconsistency, but only because it is not clear what it implies. Furthermore the *adequacy* of Plenitude is called into question. Lewis's theory is supposed to 'get the facts of modality right', to do justice to our modal intuitions (POW, pp. 154). Among these intuitions is the thought that every coherent supposition is possible or at least *logically* possible. Of course, if our argument is correct, this claim must be rejected (so long as we retain S5 that is). However, a decent attempt to reconstruct our modal intuitions in the light of philosophical theory will come as close to this naive claim as it possibly can. But it now looks as if the official version of Plenitude falls a long way short of this ideal. And what is worse, we don't know *how far* short it falls. We can assume some alien entities, but apart from that, all we know is that we can't assume any that will induce contradictions in Plenitude. If Lewis's theory cannot do justice to our modal intuitions, we at least ought to know how much justice it is not doing.

Our claim then is that Plenitude in itself is unclear, but that if we interpret it in the light of Lewis's informal remarks it entails the existence of spread worlds. Some of our critics disagree. Plenitude *is* clear, or, at any rate, clear enough. It is just that we have not understood it properly. If we understood it we would see that it does not imply the spread worlds. This is not simply because the principle preserves a prudent reserve about which alien worlds we are entitled to posit. On the contrary, Plenitude, rightly understood gives us a clear conception of the worlds we are licensed to suppose, and the spreadworlds do not correspond to this conception. Indeed some of our critics seem to think that Plenitude *prohibits* such modal monstrosities. And it does this (presumably) without the aid of modal primitives.

Our basic trick is to devise world-descriptions which are internally consistent but which cannot be allowed to describe genuine worlds. What is wrong with these descriptions? According to Richard Miller, 'spread world descriptions are cooked up by adding absolutely unrestricted modal statements to genuine world-descriptions. But this is not the approved way that new world descriptions are cooked.' Modal statements (or at least *de dicto* necessities) 'are not part of the description of any world – they are part of the description of all worlds.' Spread world descriptions are cooked up by adding to the descriptions of *bona fide* worlds, statements which falsely describe *all* worlds (Miller, 1990, Private communication). (Miller should say statements which *entail* such statements.) The answer, presumably, is to specify that world-descriptions do not describe worlds if they entail absolutely unrestricted modal statements which are false (i.e. statements which misrepresent logical space).

Now if this suggestion is intended as a 'new wrinkle' in the definition of Plenitude, a restriction designed to deal with our problem cases, we have a ready reply. Fair enough, we say, such a restriction would probably do the trick. But it would also employ a primitive modality. And primitive modalities must be avoided by Lewis on pain of losing out in the philosophical price war. But perhaps that is not what Miller means. Perhaps he thinks that his restriction is *already* implicit in Plenitude?

This also seems to be what John Bigelow is driving at. The Lutherworld, he claims, is not (or would not be) a sum of worldmates! True, there is a possible world answering to the non-modal part of the description. But in attributing modal properties to its denizens we are implicitly quantifying over other worlds. The description therefore is not a description of one world, but perhaps of all of them. Furthermore the modal parts of this description are false since they are inconsistent with Plenitude (Bigelow, 1989, Private communication).

Finally Greg Restall: 'Plenitude is not of the form – all first-order consistent sentences are true in some possible worlds – even if meaning-postulates are allowed. Because, first-order consistent sentences can talk trans-world in the way Lewis described, Lewis is careful to allow only certain first-order sentences as truly world-describing, and these are given by combinatorially recombining the "stuff" in worlds to "make" other worlds. This is a process of telling you what kinds of sentences are really world-describing. These are certainly not all of the first-order consistent sentences (even given meaning postulates).' (Restall, 1993, Private communication).

Bigelow and Restall then are making two (related) claims: (a) that Plenitude does not license us to postulate worlds whose descriptions implicitly quantify over other worlds; and (b) that by specifying that worlds are 'made' by recombining the 'stuff' in others worlds, Lewis gives us to understand which kinds of sentences are really world-describing. The sentences describing the spread worlds are not of the right kind. Why not? Because 'they talk about worlds other than themselves'.

Before answering these opponents we had better explain our own attitude to the modalities. Since we are critics of Lewis we are not committed to his reductive analysis. Necessity cannot be reduced to truth in all worlds nor can possibility be reduced to truth in some. There is more to modality than that. But we don't want to deny that *if* there were possible worlds of the kind Lewis supposes, *then* possible truths would be true in some of them whilst necessary truths would be true in all.[13] So it is not quite correct to say that modal propositions implicitly quantify over possible worlds. Rather, *given Lewis's metaphysical framework*, they do so quantify. Or at least they *imply* propositions which quantify over worlds. Thus we are quite willing to admit that *in the context of Lewis's theory*, to postulate the Lutherworld is to make large claims about logical space. And the same goes for Plantinga's world as well.

Does Plenitude license us to posit worlds whose descriptions quantify over other worlds? Well, it can hardly *prohibit* such worlds or world descriptions. Although Lewis's ambition is to take the modality out of modality, to do this he must be able to translate plausible modal claims into his own idiom. This means that he must be able to talk about the essential properties of the entities he postulates. And this cannot be done in a non-trivial way without the aid of counterparts and hence of other worlds. Furthermore, in his *persona* as philosopher of mind, he is a functionalist who believes that beliefs and desires are characterized by the causal role they play in the explanation of the subject's behaviour.[14] It is only because of a complex set of dispositions that we can be said to have minds at all. And dispositions too require counterparts.

But perhaps this is not what our opponents are getting at. The point is rather, that the Principle of Plenitude only allows us to postulate possible worlds *one at a time*, so to speak, although once we've got them, the modal truths will re-emerge. There is a world (and hence a true world-description) containing Rebecca and her dispositions. But she has those dispositions (such as being a good seamstress) only because Plenitude (or perhaps just Recombination) has licensed us to endow her with a family of counterparts. In postulating the counterparts, we are not entitled to assume the dispositions to begin with. Similarly, if Charles is good at sketching, rather than someone who simply produces a sequence of good sketches, this is because of the fine sketches produced by his counterparts under different conditions. But in postulating these counterparts we are not entitled to take his skill at sketching as a datum so as to derive the counterparts automatically. These must be 'manufactured' piece by piece through the mechanism of Recombination.

So the idea is that we are only allowed to postulate new worlds *if the entities involved are shorn of their modal properties*, though if we postulate enough of them we get the modal properties back later on. But if this suggestion is to be of any help to Lewis, our opponents must do two things. (i) They must show that Plenitude *as currently formulated* does not allow us to postulate modally charged worlds. (Note the careful wording. Plenitude must not *allow* or *imply the existence of* modally

[13] It is because they subscribe to this conditional that fictionalists about possible worlds are entitled to make use of them in elucidating the modal concepts. If your grasp of modal notions is a little shaky then the myth of possible worlds may be quite helpful. ('It's like this. Imagine there is a whole universe corresponding to every possibility ...') But such elucidations are not explanations. For a fictionalist, the modalities remain obstinately primitive.

[14] See for instance Lewis (1966, pp. 99–107).

charged worlds. But neither can it *prohibit* or *deny the existence of* such worlds. It has to be agnostic.) And (ii) they must show that this formulation does not employ a primitive modality.

Does Lewis *say* that the modal properties of an entity do not survive the process of duplication? Not in so many words. Indeed he could not say this as part of his official theory without making use of the modalities. But perhaps this is somehow implied or hinted at in what he does say? Let us first consider Lewis's examples of things that get duplicated. The list includes dragons, unicorns, puddles, protons, and talking heads (both detached and undetached). His favorite possibilia are talking donkeys though whether they are aliens or arrived at by means of recombination he does not say. However they are definitely candidates for reduplication given that they exist somewhere in logical space. Now it seems to us that talking heads and talking donkeys are not really *talking* unless they are thinking heads and thinking donkeys. Tape-recorders and parrots don't talk or do so only by courtesy. Unless the head is disposed to respond in a reasonably intelligent way, it is not really a talking head. And the same goes for the talking donkey. So unless the head and the donkey are reduplicated together with their dispositions we haven't really reduplicated a talking head or a talking donkey. But dispositions require counterparts. And counterparts require worlds to house them. Hence the Principle of Recombination allows us – perhaps requires us – to postulate worlds in batches. You can't duplicate a talking head without its counterparts – not if it is to be genuinely talkative that is.

But perhaps this talk of talking heads is just loose talk. Perhaps Lewis subscribes to a 'thin' conception of talking (according to which tape-recorders and parrots measure up) or perhaps he would be willing to concede that the talkativeness of the talking head is not one of the characteristics that its duplicate automatically inherits. Two things are duplicates iff (1) they have exactly the same perfectly natural properties, and (2) their parts can be put into correspondence in such a way that corresponding parts have exactly the same perfectly natural properties, and stand in the same perfectly natural relations (POW, p. 61). Natural properties are in some sense basic and explanatory in the way that gerrymandered and disjunctive properties are not. (Lewis is an admirer though not an adherent of D.M. Armstrong's theory of universals.) If talkativeness is not a natural property then the duplicates of the talking head will not be automatically talkative and hence possessed of counterparts. If dispositional, or more generally, modal properties are not natural, then the Principle of Recombination only allows us to posit one world at a time. Bigelow and Restall would appear to be right. We do not have a blanket license to assume worlds whose descriptions implicitly quantify over other worlds. But all this depends on the thesis that modal or dispositional properties are non-basic. And this claim seems to be (a) difficult to state without making use of modal concepts, and (b) false. Why can't we state the claim without recourse to modal concepts? We have answered that one already. Why is it false? Well, there is an elegant argument due to James Franklin (1986) designed to show that dispositional properties cannot be reduced to categorical properties in every case. Imagine a Democritean world in which the gross properties of matter are to be explained by the micro-properties of the atoms. In particular, the dispositional properties of gross things are to be explained by the categorical properties – the shapes – of their smallest components. Democritus' model looks like a paradigm of the way the modal can be reduced to the categorical. Indeed, the popularity of the Democritean paradigm is one reason why dispositions fell into disrepute in the seventeenth and eighteenth centuries. But wait! The Democritean explanation is only going to work if the atoms are *rigid*, that is, if their shapes do not degrade under impact or stress. But rigidity is a disposition. Hence dispositions have not been done away with after all. They are ultimate, basic or, to use Lewis's terminology, *natural*. (We may remark that *atomicity* which is also vital if the Democritean explanations are going to work is also a modal property: things are atomic if they *cannot* be broken up.) But if dispositions are basic to the creatures of Democritus' fancy, how much more so to the apparently ultimate particles that science has discovered! These look more like bundles of dispositions than anything else.

But even if Recombination did not permit us to assume the existence of modally charged worlds, there is more to Plenitude than Recombination. Sometimes, at least, we are entitled to assume alien worlds and alien entities, a privilege of which Lewis avails himself. He thinks that Gods are possible and hence that there are some worlds where a God exists. But the defining characteristics of Godhead are omnipotence (or at least almightiness), omniscience, and moral perfection. An omnipotent being is one which can ... well, it is not quite clear *what* omnipotence amounts to, but whatever it is, it is certainly a dispositional or modal property of a fairly spectacular kind. What is more, a God must not only possess these modal properties but possess them necessarily. To posit a God, therefore, is to posit a being with counterparts, and hence to quantify over other worlds besides the one the God inhabits. Indeed, to posit a God is to quantify over *all* worlds, since if a God is possible, then every world either contains a counterpart who is omnipotent, omniscient etc. or contains no counterpart of God. Since, we may presume, Lewis only assumes those entities that are licensed by Plenitude, the principle implies the existence of entities whose nature is such that they cannot be posited without quantifying over all possible worlds. Plenitude is a more relaxed – or perhaps more ontologically reckless – principle than Bigelow and Restall suppose.

12.10 Conclusion

Our original aim was to embarrass Lewis by positing worlds which undermined the Principle of Plenitude and led to paradox and contradiction. Because these spreadworlds *followed* from Plenitude, the principle would have to be amended and this would be difficult to do without recourse to a modal primitive. Once Lewis was saddled with a modal primitive, his theory would lose its main selling point (one might say its *raison d'etre*) and he would lose out in the philosophical price war. Lewis replied that Plenitude did *not* imply the spreadworlds and hence that he had no need for a modal primitive. We insisted that Plenitude does so imply the spreadworlds *if* it is interpreted in the light of Lewis's informal remarks. If these remarks are dismissed however, the theory becomes unacceptably vague. (We don't know which worlds, and in particular which alien worlds, we are entitled to posit.) Friends to Lewis (or at least, critics of our argument) claimed that the Principle of Plenitude *does* supply us with a clear idea of what worlds we are entitled to believe in and that the spreadworlds do not correspond to this idea. We argued that this allegedly clear version of Plenitude relies on the thesis that dispositional properties are not natural. This is not only false but cannot be stated without the aid of a modal primitive. Besides Lewis himself subscribes to worlds which cannot be posited without quantifying over other worlds.

This is not quite the knockout polemical victory we had originally hoped for. Lewis's theory survives though its value is diminished. One of the selling points of a good theory is clarity. And Lewis theory is not only vague on a crucial point but difficult to clarify without lapsing into contradiction or resorting to a modal primitive. Still, we would like to do better. Like many philosophers, we can't help thinking that Lewis's theory is *outrageous* (a sentiment expressed by Lycan with his highly colored talk of 'rape-and-loot' modal realism). However useful possible worlds may be, however brilliant Lewis's defense, this theory can't be right. (Hence the incredulous stares.) What's more, it ought to be possible to *disprove* it – or at least to make it so unattractive that it won't find any takers. But what we end up with is just one more argument that the price is wrong.[15]

References

Armstrong, D.M. 1989. *A combinatorial theory of possibility*. Cambridge: Cambridge University Press.

Boolos, G. 1985. Nominalistic platonism. *Philosophical Review* 94: 327–344.

Franklin, James. 1986. Are dispositions reducible to categorical properties? *The Philosophical Quarterly* 36: 62–64.

Haack, S. 1978. *Philosophy of logics*. Cambridge: Cambridge University Press.

Lewis, D.K. 1966. An argument for the identity theory. *Journal of Philosophy* 63(1): 17–25.

Lewis, D.K. 1968. Counterpart theory and quantified modal logic. *Journal of Philosophy* 65: 113–126. Reprinted in Lewis, D.K. 1983. *Philosophical papers,* vol. 1. Oxford: Oxford University Press.

Lewis, D.K. 1970. Anselm and actuality. *Nous* 4: 175–188. Reprinted in Lewis, D.K. 1983. *Philosophical papers*, vol. 1. Oxford: Oxford University Press.

Lewis, D.K. 1983. *Philosophical papers*, vol. 1. Oxford: Oxford University Press.

Lewis, D.K. 1986. *On the plurality of worlds*. Oxford: Blackwell.

Lewis, D.K. 1991. *Parts of classes*. Oxford: Blackwell.

[15] Our thanks to all those who have talked or corresponded with us about this matter, Mike Thrush, Jo Asscher, Ken Perszyck, Pavel Tichy, Denis Robinson, Richard Miller, Peter Milne, John Bigelow, Alvin Plantinga, Greg Restall, Chris Mortensen, Peter Forrest, David Armstrong and above all David Lewis. The paper was much improved as the result of a highly critical session at the 1993 AAP Conference in Adelaide attended by several of the above.

Luther, M. 1525/1957. *On the bondage of the will*. Trans. J.I. Packer and O.R. Johnston. Cambridge: James Clarke & Co.

Lycan, W.G. 1979. The trouble with possible worlds. In *The possible and the actual*, ed. M.J. Loux, 274–316. Ithaca: Cornell.

Lycan, W.G. 1988. Review of *the plurality of worlds*. *Journal of Philosophy* 85(1): 42–47.

Mackie, J.L. 1982. *The miracle of theism*. Oxford: Oxford University Press.

Miller, R.B. 1989. Dog bites man: A defence of modal realism. *Australasian Journal of Philosophy* 67(4): 476–478.

Pigden, C.R. 1989. Logic and the autonomy of ethics. *Australasian Journal of Philosophy* 67 (2): 127–151.

Pigden, C.R. 1990. Ought-implies-can: Luther. Erasmus and R.M. Hare. *Sophia* 29: 2–30.

Plantinga, A. 1974. *The nature of necessity*. Oxford: Oxford University Press.

Plantinga, A. 1987. Two concepts of modality: Modal realism and modal reductionism. In *Philosophical perspectives*, Metaphysics, vol. 1, ed. J. Tomberlin, 189–231. Atascadero: Ridgeview.

Quine, W.V.O. 1961. *From a logical point of view*, 2nd ed. Cambridge, MA: Harvard University Press.

Chapter 13
Defending Quine on Ontological Commitment*

Emily Gill

13.1 Ontological Commitment

A (meaningful) sentence is ontologically committed to the objects required for its truth. A theory is a set of sentences, so a theory is committed to the objects required for the truth of all its sentences. Thus, ontological commitments are part of the truth conditions of sentences. In particular, if a sentence S is committed to an object of kind K, then the existence of Ks is necessary for the truth of S. For example, the sentence U 'There are unicorns in my garden' is true only if unicorns exist. Unicorns are required for the truth of U, so U implies that unicorns exist, and if there are no unicorns, U is false.

Deciding what our theories are committed to is important. It enables us to test whether the theories that we affirm are consistent with our beliefs, in particular, with our ontological convictions, as well as enabling us to compare the commitments of different theories.

13.2 Quine's Approach

The most well-known and general response to this demand is that of W.V.O. Quine. It can be summarised succinctly:

> The idea is that – roughly – one tells what a theory says there is by putting it in predicate calculus notation, and asking what kinds of things are required as values of its variables if theorems beginning '(∃x) ...' are to be true (Haack 1978, p. 45).

---

* *In this paper I defend a Quinean approach to the ontological commitments of theories against some neo-Carnapian alternatives.*

E. Gill (✉)
Department of Philosophy, University of Otago, Dunedin, New Zealand
e-mail: emilyruth.gill@gmail.com

J. Maclaurin (ed.), *Rationis Defensor: Essays in Honour of Colin Cheyne*, Studies in History and Philosophy of Science 28, DOI 10.1007/978-94-007-3983-3_13,

> The idea is that, in order to determine the ontological commitments of a theory, one must first transform or regiment it into a more perspicuous notation, one in which its commitments are clear. [...] A theory is committed to just those entities that are required as values of its bound variables when the theory is expressed in primitive notation on any interpretation that renders its theorems and axioms true (Macdonald 2005, pp. 44–5).

The key virtues of the Quinean method lie in its simplicity. The problem is that the theories we believe or consider are expressed in natural language, but it is often unclear or controversial what the commitments of theories in natural language are. For example, whether or not our theories require the existence of numbers, or universals, or tropes, for their truth is not the sort of thing that is obvious to a competent language user. On the face of it, our theories and beliefs are consistent with various metaphysical views. Quine adopts a Tarskian view, on which the objects a sentence requires for its truth are those it attempts to refer to. So our problem is that it is unclear how we should understand the referential structure of sentences and theories in natural language.

Where natural language is unclear in some way, a standard technique is to regiment or 'tidy up' our sentences to make the possible interpretations clear. This is the basis of the Quinean approach, and he proposes a canonical notation (CN) as the language of regimentation. CN is based on the language of first order predicate calculus with identity, and with names eliminated in favour of bound variables. The existential demands of sentences in CN are explicitly defined, as the quantifiers are interpreted objectually:

The universal quantifier (∀x) ... is read 'every *object* x is such that ....'
The particular or existential quantifier (∃x) ... is read 'at least one *object* x is such that...'.

The truth conditions for quantified sentences are specified in terms of what is required from a domain of objects in order to satisfy the sentence. A universal generalisation is true if every object in the domain of the theory satisfies the sentence following the quantifier. An existential generalisation is true if at least one object does so.

Quine's criterion of commitment for theories applies to theories expressed in CN:

> *(Q) A theory is committed to those and only those entities to which the bound variables of the theory must be capable of referring in order that the affirmations made in the theory be true* (Quine 1961, pp. 13–14 – Quine's italics).

Since (Q) follows from the specified interpretation of the notation, it is trivially correct. A theory is committed to the objects it attempts to refer to, and CN is designed so that there is one and only one mechanism of attempted reference. Thus given a sentence or theory in CN we can say exactly what entities it requires for its truth, or more carefully, we can say for any domain of objects whether or not the theory is true on that domain.

Quine's proposal is that we should adopt CN as the language in which we regiment natural language theories when discussing their ontological commitments.

The advantage of rendering theories in CN should be clear; in this notation their commitments are clear and unambiguous. Since the theories we are interested are those about the world, the domain in question is simply all the entities that there are. By regimenting a theory in CN, we specify exactly what objects there must be to make the theory true.

For example:

(1) A table is red.

Translated into CN in the most obvious way, we get:

(2) (∃x) (x is a table and x is red)

And applying the objectual interpretation of the quantifier, we can state the truth conditions and ontological commitments of this sentence:

(TC2) (2) is true if and only if there is (exists) an object, x, such that x is a table and x is red.

So (2) is committed to just one object, and if we accept it as an adequate formalisation of (1), then we can say (1) is also committed to no more than one object.

It is worth noting that this is not to say any more than the regimented theory itself says. To say that the sentence (∃x) (x is black & x is a swan) requires the existence of a black swan for its truth is simply to say that the sentence is true of any domain containing an object that is black and is a swan; it does not tell us anything about what it means to be black, or to be a swan, or to be an object. There is nothing of metaphysical significance about the language of CN; its advantage consists simply its ontological transparency. There might seem to be a regress problem here. If we express the commitments of a sentence in one language by expressing it in another, do we now need a further language in which to specify the commitments of this language? No, because the language of regimentation is not a metalanguage, more an ontologically sanitised version or part of the natural: that part in which attempted reference is achieved by means of quantification.

13.2.1 Paraphrase and Theory Choice

If we can agree on how to render a theory in CN then we can agree about its ontological commitments (at least those that are made explicit.) But it is quite possible that we will not agree. Regimentation is not a mechanical process, since the commitments of sentences in natural language are underdetermined by the way they are used. So it is useful to see Quine's technique as a way of clarifying ontological disputes. We may agree that some sentence in natural language is true, but disagree about what it refers to, or requires for its truth. For example, I might agree that (1) is true just in case there is a red table, but argue that to be red, a table must instantiate *the property* of being red. What the sentence actually requires for its truth is a table and a property. I disagree that (2) is the best way to understand the ontological implications of (1). I can make my view clear by offering a paraphrase of (1):

(3) A table instantiates the property of being red.
(4) $(\exists x)(\exists y)$(x is a table and y is the property of being red and x instantiates y)

Now we can treat (2) and (4) as two alternative interpretations of (1), or just as two different theories. So which is right? There is no magic bullet here: all we can do is consider which of the two theories is better. The fact that we can express an interpretation in CN – or in any language – clearly doesn't show that interpretation is correct. But because commitment is explicit in CN, it is supposed to make it easier to evaluate the alternatives. On this view, paraphrase is simply a way of clarifying, revising and improving one's theory in the light of ontological considerations. So questions about what our theories are committed to can be cashed out as questions of a more familiar kind, about which of the alternative theories is better. Paraphrase is often used to avoid commitment to unwanted entities by eliminating reference to them (so that our formal theory doesn't quantify over them.) But there is no reason why it can't be used to increase the commitments of a theory, if we think that by doing so we get a better overall theory. So Quine's method doesn't pick out a unique interpretation as correct.

However, Quine does argue that regimentation in CN shows some ontological interpretations to be unacceptable. Firstly, the use of CN imposes some standards on the sorts of things we can reasonably treat as existing objects. This is because we have to be able to give adequate identity and individuation conditions for objects (Quine's other ontological doctrine.) This is a minimal standard; it cannot decide between two or more competing ontological interpretations which meet its demands. However it can be used to evaluate whether an ontological interpretation is or is not viable.

Secondly, according to Quine, the one and only way a theory can be ontologically committed to objects of some kind – abstracta, for example – is to attempt to refer to abstract objects. The only way to show a theory is not so committed is thus to argue that the theory need not be understood in this way; that there is a viable way to interpret the theory so that it does not attempt to refer to abstracta. In doing this, Quine demands that we produce a regimented theory that is satisfied by a domain without abstract objects; a theory that does not quantify over abstracta. This is where Quine's method finds it teeth, for our best scientific theories are awash with apparent references to abstract objects such as numbers and classes. Since we should believe our best scientific theories, if we want to deny commitment to abstract objects we should be prepared to regiment our best scientific theories so that they do not quantify over abstracta. If we can't do this – and Quine thinks we can't – we ought to accept commitment to abstracta. At least, we ought to unless we are prepared to give up believing our best scientific theories, and Quine doesn't think you should do *that*.

In the face of this, there are two ways the would-be nominalist might go. She might, like Quine, accept that the best way of discussing the ontological implications of theories is by regimenting them in CN. Then she must keep looking for an acceptable interpretation that avoids quantification over abstracta, or accept that given her current theoretical commitments she is also committed to abstract objects of some kind. Alternatively, the would-be nominalist might reject Quine's method on the grounds that she does not accept that abstracta exist, and thus a criterion that says our best theories are committed to such must bewrong. If she takes this option, then I suggest she should offer an alternative criterion.

Before looking at two alternatives, I note the appealing features of (Q).

1. It provides a useful methodological framework in which to conduct ontological disputes. We want a method that is neutral between competing ontological theories, but allows us to compare these carefully. By offering an interpretation of a theory in an ontologically explicit notation, we can see exactly what commitments it assigns to our original theory. In stipulating CN we say nothing about what sorts of individuals can be the values of its variables – they may be particulars, universals, tropes, sets, numbers, possibilia, or anything else. Furthermore, Quine's method does not assume any systematic relationship between types of expression in natural language and ontological categories, as a Fregean approach does. So we do not assume that terms always refer to objects and predicates always refer to properties. Rather, we use CN to represent all the attempted references or commitments of a sentence in a standard way, with bound variables.
2. It provides a rigorous way to test whether an ontological claim is consistent with a prior theory. If you say there are no properties, for example, you must demonstrate that this is consistent with the theories you believe, by showing how to render these in CN without quantifying over properties.
3. It captures the initial idea that ontological commitments are a part of truth conditions. It is based on a straightforward account of the relationship between truth, and what exists. For Quine, the claim that a sentence is true is an ontological claim, for it entails that the objects it is committed to exist.

13.3 Hofweber

Now I'm going to tell you about the first challenge to Quine's criterion.

Thomas Hofweber thinks that CN is inadequate for representing the ontological commitments of natural language, and as a result it commits us to too much. He argues that quantification in natural language is ambiguous; it has two semantic functions which can come apart. But in CN, there is only one kind of quantifier, so we can't express everything we can say with natural language in CN.

Quantification in natural language can have the objectual function identified by Quine, which carries commitment. Hofweber calls this external quantification. But he thinks quantifiers also have another role, as placeholders for other expressions. He calls this internal quantification. For example, if (5) is true:

(5) Fred admires Sherlock Holmes.

Then it follows that (6) is true:

(6) Fred admires someone.

An obvious way to render this in CN is (7):

(7) ($\exists$x) ($\exists$y) (x Fredises and x admires y).

If we accept (7) as rendering (6) we say that (6) is committed to an object, such that Fred admires that object. But this is wrong. For in this case, the object of Fred's admiration is not an object in the world, but a fictional character.

This shows that quantification in natural language does not always carry commitment. Although the two roles of the quantifier often go together, they can come apart. So Hofweber suggests a new criterion of commitment. He says,

> (H) The methodology for settling questions about ontological commitment will have to be to see what statements with external quantifiers in them we accept and believe to be true (Hofweber 2005, p. 277).

This approach clearly has advantages for the nominalist. Consider the following argument that properties exist:

(8) Fido is a dog.
(9) Therefore, Fido has the property of being a dog.
(10) Therefore, there is a property that Fido has, namely, being a dog.
(11) Therefore, there are properties (among them, being a dog.)

The argument seems to be valid – so it seems to show 'Fido is a dog' is committed to properties. Moreover, it suggests that paraphrase will not help us avoid this. You might think that (9) or (10) could be paraphrased as (8), so that (9) and (10) are not committed to properties. But if (9) and (10) follow from (8), the paraphrase doesn't seem to help.

Hofweber's approach avoids all these difficulties. We can simply say that the quantification in (10) and (11) is internal, and thus does not commit us to anything. And here, this seems to be the right kind of response. For we surely don't think that the fact we can express (8) in a pretentious way, by talking about properties, is evidence that properties exist, or that (8) requires a property for its truth.

Hofweber thinks his approach applies more widely than this, because he thinks the distinction between internal and external quantification can be applied to whole areas of discourse. Whether or not we are committed to properties depends on whether property-talk is external talk about objects in the world, or internal talk based on features of our language like the equivalence between (8) and (9) in the argument above, and this is supposed to be an empirical question about the function of language. If property-talk is internal, we are not committed to properties and there is no need to eliminate such talk by paraphrase.

13.3.1 Objections to Hofweber

What exactly is involved in extending the idea of internal quantification to whole areas of discourse? External quantification is standard objectual quantification. Internal quantification is a form of substitutional quantification. When we read a quantifier substitutionally, the truth conditions are quite different to those we get on an objectual reading. On a substitutional reading, 'Fred admires someone' is true iff there is a name whose substitution for 'someone' results in a true sentence. On an objectual reading, it is true iff there is some object that is admired by Fred. So 'Fred admires someone' may be true on a substitutional interpretation, but false on an objectual reading.

The truth conditions for substitutional quantification don't answer the question about ontological commitment, or what is required for the truth of the sentence – they just postpone it. A particular substitutional quantification is true if there is a true substitution instance. The interesting ontological question seems to be: what objects does the true substitution instance require for its truth? So where substitution instances are available, we don't *need* to appeal to a substitutional reading of the quantifier. For example, we need not accept the simple argument for properties because we can deny that (9) is part of our best theory. If someone suggests that it is, we can offer (8) as a paraphrase. (8) seems to imply (9) in ordinary situations, since whenever (8) is assertable so is (9). But (8) doesn't imply (9) once we represent them in CN.

What about cases where substitution instances are not available? These are the really hard cases for the nominalist, because these are the cases where paraphrases are not readily forthcoming. Moreover, in these cases the substitutional reading can deliver the wrong truth-values. One example is where we quantify over infinitely many objects, like the natural numbers.

(12) Every number has a unique successor.

If we read this substitutionally, we get something like:

if any expression that names a number is substituted for Ω in the following schema there is exactly one expression that names a number whose substitution for Σ yields a true sentence:

(Sch) Σ is the unique successor of Ω.

(A schema is like an incomplete or 'gappy' sentence. The Greek letters just stand in the gaps where expressions of a specified type can be inserted.)

The problem with the substitutional reading is that it makes (12) come out false, since infinitely many numbers do not have names. Yet (12) is true, and it seems to be true of numbers – all numbers – whether or not they have names.

Other difficult cases are those where we are in ignorance:

(13) There are properties we haven't discovered yet.

is very likely true, but since we can't yet name these properties, we can't say it is true *because* there is a true substitution instance.

Here, the ontological question seems to be permanently postponed. Hofweber doesn't discuss such examples, but given that his is a neo-Carnapian view, it might be that he would try to account for them in terms of the possibility of generating linguistic expressions. Given any number, we can name its successor, even if we can't name every number. And presumably, we can always generate a new predicate and form a name of a property from this. If this is the right way to understand Hofweber, the truth of (12) and (13) turns out to depend on the possibility of constructing new expressions in a language.

The problem with this solution is that it undermines the important link between truth and what exists. If we take ontological questions seriously, we want the truth-value of (12) and (13) to depend on what there is, not on what our future language might be like. Of course we can always invent new expressions or languages if they will be useful, but the ontological concern is with the contribution the world makes to the truth of our claims. (12) and (13) look like non-modal claims about what exists, not claims about possible extensions to our language.

This also highlights the epistemic burden that Hofweber's account places on our claims about language. To decide whether or not we are committed to properties we have to decide whether property-*talk* is external, referring talk, or internal, non-committing talk. Hofweber doesn't say much about how we are supposed to decide this. But the original problem was that it is not clear whether talk of properties should be construed as attempting to refer to properties or not. In the simple example, our intuition is that 'Fido is a dog' doesn't refer to properties, but what about 'the chemical properties of elements depend on the number of electrons in their outer shell'? Here, at least some people do think they are really talking about properties. So we might just end up arguing about whether property talk is internal or external, and Hofweber doesn't tell us how to settle this.

The Quinean method, however, provides a way forward. We can test our intuitions by giving alternative interpretations of the referential structure of property talk. It makes no difference whether quantification in natural language has more than one kind of significance, or is used in different ways. The aim in using CN is to clear up just such confusions or ambiguities in natural language. The Quinean will treat internal or substitutional quantification as a figure of speech, the kind of thing we aim to eliminate in expressing a theory in CN. If we say property talk is not referring to properties at all, then apparent references to properties will be treated in the same way. To challenge the adequacy of CN, Hofweber has to show more than just that quantification or naming in natural language is not always committing. He must show there is reason to think we cannot express the ontological implications of our theories in CN, and he has not established this. We should not see this focus on objectual quantification as arbitrary or unmotivated; it reflects the importance of genuine reification in theorising about objects. Objectual quantification, along with identity conditions for objects, imposes robust standards for the positing of objects.

13.4 Truthmakers

A more radical criticism of (Q) claims that Quine's own approach is based on a fallacious view about the relationship between true sentences and the world. The claim is that Quine is attempting to derive a metaphysical theory from premises about language, and in particular, the structure of true sentences. Radical truthmaker theorists think this is wrong-headed, because it assumes there is a systematic relationship between the structure of a true sentence and the "structure" of its truthmaker – the bit of the world in virtue of which the sentence is true. The radical claim is that there is no such reliable systematic correspondence between sentence-structure and the structure of (potential) truthmakers. So the Quinean approach, which is based on the idea that the commitments of a sentence or theory can be identified with its referential structure, may give us the wrong results. Quineans ask what must our quantifiers range over in order for our theories to be true? Truthmaker theorists ask more simply: what must exist in order to make our theories true?

This looks like an attractive approach for two reasons. It suggests we can stop worrying about finding paraphrases for sentences which appear to refer to unwanted objects. And it seems to shift the focus to what is really important – the question of what the world is like, rather than what we have to say in order to describe it.

For example, Heather Dyke says

> there is no need to think that a truthmaker is structurally isomorphic to the truth it makes true. ... If truthmakers are not structurally isomorphic to the truths they make true, then there is no need to fear that one will be committed to an entity answering to every referring term in a sentence one takes to be true. The truthmaking project undermines the paraphrase project. Once we recognize this we will realize there is no need to find a paraphrase for sentences with apparently problematic ontological commitments. (Dyke 2008, p. 85)

A similar view is expressed by Ross Cameron:

> Do we need an ontology of numbers to make mathematical truths true? Whether or not we are ontologically committed to numbers depends solely on whether we need them as truthmakers: the dispensability of number talk is neither here nor there. (Cameron 2008, p. 11)

And Cameron explicitly states a criterion of commitment:

> (C) I hold that the ontological commitments of a theory are just those things that must exist to *make true* the sentences of that theory. (Cameron 2008, p. 4)

And this fits well with the original account of commitment.

13.4.1 Advantages for Truthmakers

We can see how the nominalist might employ this approach. For example, consider

(14) There are anatomical features shared by spiders and insects.

This will pose a problem for a Quinean who does not want to be committed to properties like anatomical features, for the sentence quantifies over such things. So the Quinean must look for a paraphrase, without the unwanted commitments, that can replace this sentence.

Cameron's approach apparently avoids this demand. All we need to ask is what we need to make (14) true. For example, (14) would be true in virtue of the fact that spiders have exoskeletons and compound eyes and insects have exoskeletons and compound eyes. In this case, we don't need properties to make (14) true, so (14) is not committed to properties.

13.4.2 Problems for Truthmakers

13.4.2.1 Evaluating Truthmakers

Cameron's approach works well here because we can identify a suitable truthmaker for (10). But again, the hard cases are those where it's difficult to see what *could* make some sentence true, if it's not the properties it appears to refer to. For example:

(15) Pressure is proportional to temperature at a fixed volume.
(16) The chemical properties of elements depend on the number of electrons in their outer shell.
(17) Half of the species on earth are species of beetle.

These are the sorts of the cases that really drive the Quinean to accept the existence of classes. They seem to be about properties or kinds, rather than the particulars that instantiate these. So it is very hard to see how talk of properties or classes can be paraphrased away. According to Cameron, these sentences are only committed to properties or classes if we need such things to make them true. So do we? This is a very hard question. One response is that all we need to make (15) true is all the gases, since (assume) if (15) is true of anything it is true of these. This is just to treat (15) as a generalisation about gases. But then it seems our nominalist is unable to distinguish between law-like generalisations like (15) and accidental generalisations, like

(18) All the coins in my pocket are silver.

The truthmaker for (18) is all the coins in my pocket. But whereas (18) is only contingently true – there might have been a copper coin in my pocket, so (18) might have been false – the same doesn't seem to apply to (16). It isn't just that (16) is true because, as a matter of happy coincidence, it is true of all gases, but it could just as easily have been false. (16) is supposed to be nomologically true: it says you won't find a gas anywhere that (16) is not true of.

But maybe this isn't such a big problem as it seems. Truthmaker theorists have to account for various modal truths, and many want to do this without invoking modal truthmakers or properties. One effective strategy employed is to suggest that modal truths can be explained in part by the application conditions of the concepts involved. For example, consider the statue Goliath and the lump of clay, Lumpl, which constitutes it. A traditional argument for the existence of two distinct entities here is the fact that there are sentences which are true of Goliath and not of Lumpl, or vice versa. For example, Goliath could survive the loss of his nose, but Lumpl could not survive this loss. Therefore, Goliath has a property – the property of possibly surviving the loss of his nose – which Lumpl does not have. Therefore, (by the indiscernibility of identicals) Goliath and Lumpl are distinct objects.

Heil (2005, p. 188) explains this by pointing out that Goliath is an object picked out under a certain concept or sortal, that of being a statue. Lumpl is an object – apparently the very same object – picked out under another sortal, that of being a lump of clay. Our concept of a statue allows that statues may survive the loss of various parts yet continue to count as the same statue. Lumps of clay, on the other hand, seem to be identified with their parts, so by definition, a lump of clay cannot survive the loss of a part. So 'Goliath (the statue) could survive the loss of his nose' is true in virtue of the fact that *statues are the sort of thing that can survive the loss of a part*. The truthmaker for this sentence need involve no modal properties, since what makes it true is simply the application conditions of our concept of a statue. Similarly, 'Lumpl could not survive the loss of a part' is true in virtue of the fact that *lumps are identical to their parts*, and need not involve a modal property.

This does seem like the right sort of story to me. But we should note that this seems to imply that the minimal truthmakers for the modal truths here are concepts and their applications conditions. That is, we don't need to invoke any entity in the world to account for or guarantee the truth of these sentences, so long as we understand "Goliath" and "Lumpl" as associated with the sortals "is a statue" and "is a lump of clay" respectively. So all we *need* to make these modal sentences is true is our concepts, and thus according to (C) all we need be committed to, when we assert them as true, is our own concepts of statues and lumps, or the application conditions of these sortals.

In other words, this strategy is similar to the way we think of analytic truth. You might argue that it is analytically true that statues can survive the loss of their parts and lumps cannot. That is, if you understand what 'statue' and 'lump' mean you can immediately see this claim is true. Could we apply a similar strategy to nomological truths? 'The pressure of a gas is proportional to its temperature at a fixed volume' is, let's assume, a nomological statement. It doesn't seem that we can simply say that it is part of the concept of being a gas that it obeys this law. The sentence is not analytically true: it seems to be something we discovered about gases, and we can surely imagine discovering an exotic gas that did not obey the law.

The truthmaker nominalist could bite the bullet and go for a Humean account of laws as simple regularities. But should we accept this response, and accept that his proposed truthmaker is really adequate to guarantee the truth of (16)? How can we try to decide this? The Quinean method provides one way of doing so. If the nominalist argues for a Humean view of laws, he can render all the law-like statements of our theories into such generalisations, he can clearly state the Humean version of these theories. The advantage of providing such a paraphrase is to test or demonstrate that this interpretation is consistent or acceptable. Some aspects or implications of the original theory will have to be given up – but maybe new advantages will emerge.

On the other hand, if we can't give such a paraphrase – or our paraphrase results in an unacceptably poor theory – it's still unclear whether the posited truthmakers are adequate. At best, we have a standoff. The truthmaker nominalist may maintain his claim that nothing more is needed to make the theory true, but the Quinean can maintain her claim abstract objects are needed to satisfy the theory.

13.4.2.2 Structure

The fundamental objection from radical truthmaker theorists is that Quine's approach affords too much significance to the apparent references in our sentences. Radicals deny that there is any systematic way to identify the commitments of a theory by examining its referential structure. But you might think that there must be some cases where it is clear that the commitments or truthmakers of a sentence can be identified with the objects it attempts to refer to: sentences of the form 'there are Ks' or 'a exists'. Armstrong (2004, p. 23) says the truthmaker for a sentence of the form 'a exists' is always a. And this coincides with the objectual interpretation of the quantifier. 'a exists' is equivalent to 'there is an object that is identical to a'. In these cases, the truthmaker approach and the Quinean approach seem to converge. But it is precisely such cases that Cameron wants to use his criterion to avoid:

> (The proposition) '<x exists> might be made true by something other than x, and hence … '*a* exists' might be true according to some theory without *a* being an ontological commitment of that theory.' (Cameron 2008, p. 4)

This does seem like a radical view, partly because it goes against the consensus, but also because it seems to undermine our grip on what 'exists' means. For if the sentence '*a* exists' might be made true by something other than *a*, then '*a* exists' might be true even though *a* does not exist. How can this be? This appears to be a straightforward contradiction.

Cameron's answer is that we need only be committed to the things we admit into our ontology as truthmakers. Other things can be truly said to exist, although they are not elements of our ontology – they are not, in our view, among the entities that constitute the fundamental furniture of the world. For example, say that there are no composite objects amongst the fundamental constituents of the world: there are only simples. According to Cameron, sentences that appear to refer to composite objects can still be true, if they are made true by arrangements of simples. So 'the Taj Mahal exists' is true – literally true – in virtue of the simples arranged Taj Mahalwise. But if 'the Taj Mahal exists' is true, then (by disquotation) the Taj Mahal exists. But this does not entail that the Taj Mahal really exists, only that it is true to say 'the Taj Mahal exists' (because the sentence has a truthmaker.) So on Cameron's view, this sentence is true but is not ontologically committed to the Taj Mahal, since all we need to make it true is the simples that compose the Taj Mahal.

This approach turns out to have some similarities to Hofweber's. Quantifications or existence claims in natural language aren't always ontologically committing because we need only be committed to the things we need to make them true. So does Cameron's distinction do better than Hofweber's? To tell the difference between what really exists and what can merely be truly said to exist, we have to decide on what makes true the existence claims. Again, this seems somewhat circular, since what we are trying to decide is what sentences require for their truth.

Cameron does think his approach has a key advantage: it allows us to accept sentences that quantify over problematic entities as literally true without accepting that the entities in question exist. For example, he says his solution to the puzzle above will allow nihilists (who believe in no composite objects) and universalists (who believe in loads of composite objects) to stop fighting. They were fighting because they thought the following were claims contradictory:

(a) 'The Taj Mahal exists' is literally true.
(b) All that really exists are simples.

But on Cameron's approach, the two turn out to be consistent, for simples arranged Taj-Mahalwise are enough to make true the sentence 'the Taj Mahal exists'.

But the agreement here seems to be superficial. It depends on accepting the following claims:

(i) 'The Taj Mahal exists' is not committed to a composite object.
(ii) A sentence is literally true just in case it has a truthmaker.

But the serious universalist claims

(c) Composite objects really exist.

And this really does contradict (b).

Furthermore, presumably the nihilist will still want to say that 'The Taj Mahal really exists' is false, since this contradicts (b). So on Cameron's approach, the nihilist says:

(a) 'The Taj Mahal exists' is literally true.
(d) 'The Taj Mahal really exists' is false because all that really exists are simples.

Again, Cameron's view that existence claims in natural language need not be treated as committing is just the sort of thing the Quinean approach is designed to address. Contra what the neo-Carnapians sometimes suggest, the Quinean does not read her commitments directly from the apparent references of natural language. On the contrary, she actively develops ontological interpretations of theories. Of course, we do 'read' the commitments of such interpretations in CN from the bound variables, but this is because such regimentations are designed to make commitments explicit based on the specified interpretation of the quantifiers. To show that the Quinean approach commits what Dyke (2008) calls the 'Representational Fallacy', it must be argued that translation into CN somehow forces us to preserve commitment to objects we don't need to make our theories true. Cameron, like Hofweber, would have to show that assertions of 'mere' existence cannot be eliminated in favour of sentences that only assert the existence of objects that really exist. And again, this is what he has not done.

13.5 Conclusion

The neo-Carnapian solutions assume that we are supposed to 'translate' natural language quantifiers using '$\exists$'. But as Hofweber and Cameron argue, natural language expressions like 'there is' or 'exists' need not always be committing. The Quinean approach gives us a way of making use of this insight. Where we want to deny that a particular use of such an expression is ontologically committing, we can replace the sentence in question with one that clearly shows what the sentence requires for its truth. It is a mistake to think of this a matter of discovering what the original sentence means or says, still less as a way of working out what there is. Rendering a theory in CN is not a mechanical process, nor is it a matter of translating from one language to another. It is a way of picking out a particular interpretation of the ontology of a theory. There will be many alternative ways to do this, and it is misleading to think that we can apply some further, independent standard for the correctness of a regimentation. On the contrary, regimentation here is just more theorising. Given the clarity and rigour of (Q) compared to the alternatives I have considered, I propose that we should retain CN (or some similar first-order language) as the notation for expressing ontological claims, interpretations, and disagreements.

Acknowledgements I owe great thanks to Colin Cheyne for his time and patience in helping me to develop and express the views in this paper.

References

Armstrong, D. 2004. *Truth and truthmakers*. Cambridge: Cambridge University Press.
Cameron, R. 2008. Truthmakers and ontological commitment. *Philosophical Studies* 140: 1–18.
Dyke, H. 2008. *Metaphysics and the representational fallacy*. New York: Routledge.
Haack, S. 1978. *The philosophy of logics*. Cambridge: Cambridge University Press.
Heil, J. 2005. *From an ontological point of view*. Oxford: Oxford University Press.
Hofweber, T. 2005. A puzzle about ontology. *Noûs* 39: 256–283.
Macdonald, C. 2005. *Varieties of things: Foundations of contemporary metaphysics*. Malden: Blackwell.
Quine, W. 1961. On what there is. In *From a logical point of view*, 2nd ed. New York: Harper & Row.

Chapter 14
The Scandal of Semantic Platonism

Vladimír Svoboda

14.1 Introduction

The title of this short article, which I am dedicating to a distinguished Dunedian philosopher, echoes the title of an article written about 20 years ago by another outstanding philosopher and logician who spent most of his academic career in Dunedin – Pavel Tichý. In an article entitled *The Scandal of Linguistics* (Tichý 1992), Tichý articulates a harsh criticism about linguistics. Linguistics, he says, is supposed to be the scientific study of language. And language is a system of signs that allows us to make what we have on our minds accessible to our neighbours, thanks to the fact that the signs mean something. But if you ask a linguist what meaning is, the likely reaction is not only that he will not give any straight answer, but he will consider the question as somewhat improper. Tichý's point is interesting and his way of presenting the issue is amusing. In the background of his argument one can recognize Tichý's conviction that the logico-semantic theory he promotes in fact offers what the linguists fail to provide – namely an explanation of the nature of meaning.[1]

In this article, I want to suggest that the account of meaning that Tichý, his followers, and many other philosophers of the same bent favour can be challenged in a way similar to that in which Tichý challenged linguists – namely, that proponents of the doctrine tend to neglect questions that concern the very foundations of their theory of language. Put very simply, Tichý and his philosophical fellows offer an account of meaning that is underpinned by a strong version of Platonism, in which entities from the realm of Platonic objects serve as

[1] A comprehensive account of Tichý's views is presented in Tichý (1988).

V. Svoboda (✉)
Institute of Philosophy, Academy of Sciences of the Czech Republic, Jilská, Prague 1, 11 000, Czech Republic
e-mail: svoboda@site.cas.cz

J. Maclaurin (ed.), *Rationis Defensor: Essays in Honour of Colin Cheyne*, Studies in History and Philosophy of Science 28, DOI 10.1007/978-94-007-3983-3_14,

a kind of unexplained explainers; the philosophers, however, don't aspire to provide any elucidations either concerning organization of the realm, or concerning our grasp of its constituents. Colin Cheyne challenged different aspects of this kind of methodological strategy at several places – though mostly in a different context.[2] Here I don't aspire to make a deep and thorough criticism of the Platonist position. I just want to put forward some rather naive questions that the friends of Platonic entities who would like to provide a full-fledged theory of the functioning of our language should (try to) answer. I suggest that the fact that they tend to overlook questions of this kind might be deemed "scandalous".

14.2 Plato and Platonism

Before I put forward the questions that form the core of this article, a short excursion into the roots of Platonism will perhaps be worth undertaking. Quite naturally, this excursion will lead us to the philosopher who gave the doctrine its name – Plato (427–347 b.c.). In many places of his works – mostly the Socratic dialogs – Plato presents deliberations concerning the entities that are supposed to be outside of our material, changeable world. He usually calls the entities ideas (εἶδος). Considerations that could be characterized as contributions to the general theory of ideas can be found only in a few of Plato's writings,[3] but the pursuit of the nature of particular ideas (or concepts) is a typical theme of Socratic dialogues (let us note that the pursuit mostly does not yield anything close to a satisfactory outcome).

In the dialogues Plato usually focuses on rather abstract evaluative concepts/ideas associated with various characteristics of human behaviour such as (the ideas of) *temperance*, *beauty*, *virtue* or *justice*. Ideas associated with terms of more mundane vocabulary are addressed only minimally. In some places, Plato outlines a kind of a general account of the idea of idea and discusses some related problems. In the dialogue *Parmenides* – which is usually classified among Plato's later works – we can find a number of passages that are worth noting. I will mention just two of them. In the first one Socrates mentions his misgivings concerning ideas of plain things:

> ... visible things like these [hair, mud, dirt] are such as they appear to us, and I am afraid that there would be an absurdity in assuming any idea of them, although I sometimes get disturbed, and begin to think that there is nothing without an idea; but then again, when I have taken up this position, I run away, because I am afraid that I may *fall* into a bottomless pit of nonsense, and perish; and so I return to the ideas of which I was just now speaking [the just, the beautiful, the good], and occupy myself with them (1892, pp. 49–50).

---

[2] See Cheyne (1999, 2001).

[3] Crucial discussions concerning the nature of ideas can be found in *Phaedrus*, *Republic*, *Timaeus* and *Parmenides*.

Parmenides warns Socrates not to despise even the meanest things, but he immediately proceeds by pointing out problems that the theory of ideas faces. After a discussion of some questions that leave Socrates perplexed, he concludes:

> These, Socrates, [. . .] are a few, and only a few of the difficulties in which we are involved if ideas really are and we determine each one of them to be an absolute unity. He who hears what may be said against them will deny the very existence of them—and even if they do exist, he will say that they must of necessity be unknown to man; and he will seem to have reason on his side, and as we were remarking just now, will be very difficult to convince; a man must be gifted with very considerable ability before he can learn that everything has a class and an absolute essence; and still more remarkable will he be who discovers all these things for himself, and having thoroughly investigated them is able to teach them to others (1892, p. 55).

It is not entirely clear whether the note about the "very considerable ability" needed for overcoming the problems related to the theory of ideas is to be read as Parmenides' (or Plato's) irony, or as a serious hope that a true genius might be able to put things right. What seems quite clear is that Plato himself was struggling with the problems that his present-day followers promoting the doctrine of Platonism and in particular Platonic semantics tend to neglect. Formulation of most of the problems may be traced within Plato's texts and there exists an extensive literature dealing with them on different levels (c.f. e.g. Ross 1951; Fine 1993), but in spite of this it may be useful to present some of them in an illustrative form.

14.3 Constructions

Before we get to the questions that I will be raising, I will sketch the basic idea of the doctrine to which my questions will pose a challenge. The central idea of the philosophical doctrine is the idea that language is essentially a code – linguistic expressions are selected to encode abstract entities from the Platonic realm. Thus we can say that a linguistic expression is meaningful if and only if it is associated with a certain entity from the Platonic heaven. In the case of Tichý's elaboration of the doctrine, the Platonic entity has a character of an abstract nontemporal (Platonic) procedure that is called a *construction*.

> A construction as such is a non-linguistic object. It is best looked upon as a cognitive itinerary: it starts at some given objects and leads, by way of some operations, to another object. A linguistic expression is a typographic or audial diagram of such a journey. The way words are combined into the expression depict the way objects and operations are combined into the corresponding construction (Tichý 1992, p. 79).

The specific nature of different kinds of constructions is not important in the present context.[4] We should, however, spend some time with the most fundamental

---

[4] I am ready to admit that Tichý's conception of constructions is ingenious and is, in many respects, a revolutionary step in the development of intensional semantics.

of these entities – constructions called trivializations. They are essential as they serve as the basic input of more complex constructions. The somewhat puzzling feature of these constructions is that their input is by definition identical with their product. Thus trivialization is a (or the?) 'step' consisting in leaving object as it is: The construction of trivialization takes an object, say *X*, and generates *the same object* (this construction is written '^{0}X').

The fact that any object can be trivialized is something of a straightforward confirmation of Socrates' urging that "there is nothing without an idea". Thus, presumably, there are (out there in the realm of Platonic entities) not only things[5] like number 16, but also the trivial construction of the number, namely $^{0}16$ and also the trivial construction of the construction – $^{00}16$. It may not be easy to understand why the Platonist heaven is so abundant with entities that are so difficult to discern, but it is not shocking – heaven is not supposed to be parsimonious. What is somewhat surprising is that trivialization also transports, at a very cheap cost, spatiotemporal individuals, or more precisely their abstract surrogates, into the 'heaven' as the construction provides an abstract counterpart for any concrete object. Let us take the famous horse Phar Lap as an example. We can be certain that, in the Platonic heaven, there is an object ^{0}*Phar Lap* that constructs this distinctive animal.[6] If we take any other English term like "horse", "fast", or "final lap" we can be sure that ^{0}*horse*, ^{0}*fast*, and ^{0}*final lap* are present in the Platonist heaven. We can, of course, be sure of that only provided that we can be sure of the fact that objects to be (re)constructed in the trivial way *are available*. It must be remembered: any construction *takes an object, and generates the same object*. If there is nothing to be taken as the input, then nothing can be constructed as the output. On the other hand, according to the theory the same object can be constructed in an endless number of ways.

14.4 Platonism and Meaningfulness

Any theory based on the assumption that expressions of natural language are meaningful due to the fact that they encode Platonic entities faces, of course, a number of questions. Here I will mostly omit the more general ontological and epistemological issues. I will instead focus on questions that challenge the view that a theory of language based on Platonist principles is fruitful as an explanatory tool that reveals how our natural languages work. I will try to show, using a few examples, that the theory fails to offer a satisfactory account of meaningfulness and

[5] In the broadest sense of the word.

[6] This claim is somewhat problematic as individuals, from Tichý's perspective, are just metaphysical surrogates of full-blooded (material) objects – bare 'hangers' on which properties can be 'hanged', and so within this theory, Phar Lap is, strictly speaking not an animal. I, however, do not want to dwell on the metaphysics of individuals as it would lead us into quite abstruse theorizing.

suggest that adherents of the theory don't in fact seem to have the ambition to provide anything close to a well-developed account of meaning.

As I said, linguistic expressions are supposed to be meaningful thanks to the fact that they are associated with entities from the Platonic heaven (they encode them). If we apply this thesis to simple expressions like "beauty," "bateau", "bead", "beakch" we can depict the situation by the following highly simplified scheme:

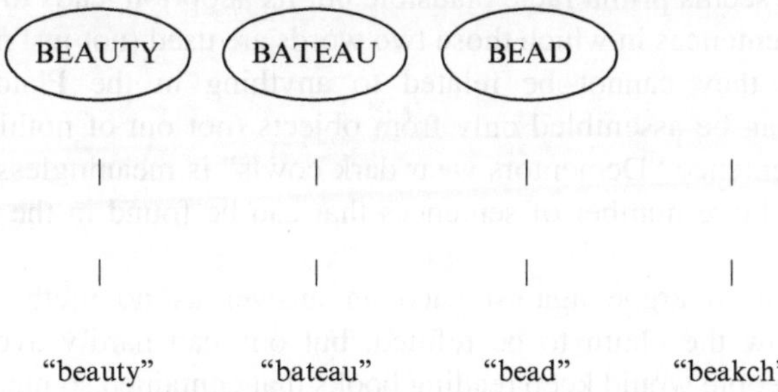

As is perhaps obvious, the entities belonging to the Platonic realm are depicted by ellipses. The picture suggests that while the words "beauty", "bateau", and "bead" are meaningful, the word "beakch" is not – it does not succeed in picking out any object inhabiting the Platonic realm.[7] Though words, phrases and sentences can change their meanings, gain one or lose one, there seem to be only two options – a given expression at a given stage of evolution of the natural language in question either succeeds in picking out a certain object, or not, as nothing like a 'near miss' counts in case of the picking out of an object.[8]

Let us now consider three simple expressions: "wizard", "dementor", and "nargles".[9] The natural question to ask here is:

Q. Are the words "wizard", "dementor", and "nargles" successful in picking out certain Platonic objects?

Hardly any adherent of Tichý´s doctrine will vacillate over the answer to the question whether the idea (concept) of wizard(hood) picks up a Platonic object. They will all agree that it does, and that the expression ^{0}*wizard* represents a 'successful' trivial construction of the object. I, however, presume that answering the same question in case of the expressions ^{0}*dementor* or ^{0}*nargle* will not be so straightforward. The answer requires deciding whether the ideas of dementor(hood) and nargle(hood) are to be found in the Platonic realm.

[7] The scheme is simplified (also) in the sense that at least in the case of the homonymic word "bead" there should be a number of distinct objects associated with the word.

[8] This assumption is perhaps not necessary, in principle we might think of stronger and weaker ties between an expression and its meaning, but so far I have not heard any proposals aiming in this direction from the camp of proponents of Platonic entities.

[9] For those who have not read books about Harry Potter, I should note that wizards, dementors and nargles appear in the books by J.K. Rowling.

I can think of four different answers that adherents of semantic Platonism could offer if the question is limited to the words "dementor" and "nargle". Let us consider them in the sequel:

(a) No, the words do not pick out any Platonic objects – no entities associated with the expressions "dementor" and "nargle" are present in the Platonic realm.

This answer seems prima facie plausible but its adoption leads to a problematic consequence: sentences in which those two words are used (not just mentioned) are meaningless – they cannot be related to anything in the Platonic realm as constructions can be assembled only from objects (not out of nothing). Thus, for example, the sentence "Dementors wear dark cowls" is meaningless and the same holds true of a large number of sentences that can be found in the books by J.K. Rowling.

It is difficult to argue against such an answer as no method is available that would allow the claim to be refuted, but one can hardly avoid wondering why so many people would keep reading books that contained so many nonsensical sentences. Moreover, there seems to be quite a significant difference between sentences speaking about dementors and sentences (or rather 'sentences') like the following one: "Degrubregrugs wear dark cowls". Thus, I suggest that that the answer (a) is not quite satisfactory, though I can imagine that some adherents of semantic Platonism would be willing to defend it.

We can, of course, imagine a complementary answer:

(b) Yes, the words do pick out Platonic objects – entities to be associated with the words "dementor" and "nargle" (as well as their trivial constructions) are present in the Platonic realm.

This answer, however, does not seem less problematic than the previous one. If we appreciate the fact that all that we learn about nargles from the Harry Potter series is that they hide in mistletoe[10] (and though we learn much more about dementors they are also quite mysterious creatures), it seems that the idea of dementor(hood) and nargle(hood) have made it into the Platonic heaven somewhat too easily. Moreover, if the idea of nargles is present in this heaven, then we have to assume that the sentence "Nargles have legs" is true or false as the idea must be clear and definite. But we humans don't have any clue as to how we could reach a decision concerning the truth value. Moreover, we cannot rely on J.K. Rowling as a source of information. Even if she were to devote her next book to nargles and describe them as green quadruped creatures that are about one inch long, we cannot be sure that she can be trusted. Perhaps the idea of a nargle, as it is eternally present in the Platonic heaven, is incompatible with the idea of having legs and Rowling is

[10] Elsewhere we learn nothing. Frankly, even the information itself shouldn't be taken for granted, as Luna Lovegood, who is the source of the information about nargles, might not be completely trustworthy.

simply wrong – just like those who claim that snakes have legs and unicorns have no horns.[11]

Though deliberations of this kind seem to lead to very strange results, we can say, similarly as before, that the answer (b) is hardly refutable. The only thing we can do is to call attention to some consequences of its adoption that are hard to swallow. If suspicious things like nargles have their ideal prototype in the realm of Platonic entities, then it seems that perhaps any expression might succeed in picking out some Platonic object. In such a case it was, perhaps, premature to reject the 'sentence' "Degrubregrugs wear dark cowls" as meaningless as there is no serious reason to deprive degrubregrugs or degrubregrugshood of presence in the realm of Platonic entities. The only difference between nargles and degrubregrugs is that the latter do not appear in Rowling's books, but only in this article. I may be wrong but I suppose that not many adherents of semantic Platonism would be willing to adopt such a profligate account of Platonic entities.

The third possible reaction to our question is less bold. It may be formulated as follows:

(c) It is unclear whether the words "dementor" and "nargle" pick out Platonic objects – it is necessary to find out whether they do.

This answer naturally evokes the question *how* could we find out whether they do if they do, and that they don't if they don't. We could undertake empirical research hoping that, if we find some dementors or nargles in the actual world, it would serve as a kind of confirmation that the ideal patterns are out there (at least as much as the presence of actual horses indicates that their ideal prototype is out there). Quite obviously, there is little likelihood that we will find some real dementors or nargles, so this strategy is not too promising. On the other hand, it is clear that no empirical research can provide even the slightest evidence to the effect that a certain idea is *not* present in the Platonic heaven.

Somebody might suggest that the best strategy is to ask the one who apparently has the best idea of dementors and nargles – J.K. Rowling. But, in fact, we don't have any serious reason to think that the writer has a privileged insight into the realm of Platonic entities. We might conjecture that when she started writing about the strange creatures (and about wizards) she must have had a certain idea of what she was talking about, but this is not quite realistic. It is likely that J.K. Rowling does not have any readymade answers to questions like: "How many legs do nargles have?" or "Can dementors sit down?" (though she would probably readily answer the question "Do dementors tell jokes to their grannies?" negatively, as a dementor joking with his granny seems inconceivable).

In general the situation is quite hopeless – it seems foolish to search for an answer to a question if we don't have any clue as to how we could find it. And it is worth noting that the fact that we – humans – are supposed to have an insight

[11] We will return to the problem of incompatibility of ideas later.

into the realm (as otherwise we could not mean anything) does not really help in cases like this one.

The last of the obvious possible reactions to our initial question is the following one:

(d) It is unclear whether the words "dementor" and "nargle" pick out Platonic objects – but this need not bother us as this issue is unimportant.

This answer suggests, quite unequivocally, that we could (or should) adopt a similarly agnostic position with regard to any other expression as well. Thus, we might conclude that even with simple expressions like "beauty" or "bateau" or complex expressions like "Wizards wear dark cowls" it is unimportant whether they manage to pick out any Platonic object (i.e., it is unimportant whether they are meaningful). This last approach, which might be called *semantic agnosticism*, quite clearly undermines the original conception of meaningfulness that adherents of semantic Platonism adopt. If we don't need to worry about particular objects from the Platonic realm, we probably also don't need to worry about the existence of the realm as such and the assumption that there is such a realm cannot play a central role in our semantic theory. We can therefore conclude that adopting the answer (d) would be a bad choice for anybody who takes the Platonic theory of language seriously.

14.5 Setup of the Platonic Realm

The simplified discussion presented in the above paragraphs indicates, I believe, how problematic the Platonist picture of natural language is – even on the level of simple expressions. Things, however, become even more problematic when we move to the level of compound expressions. Let us, for the sake of illustration, consider expressions like "pork bateau", "triangular circle", or "braised Tuesday". Are these phrases meaningful? For the Platonist, once again, the answer depends on the setup of the Platonic realm. It seems quite clear that the idea of triangularity, as well as the idea of a circle, are among the Platonic objects. But do these ideas combine into the more complex idea of a triangular circle? Can such an idea be found in the eternal realm?

Some adherents of semantic Platonism are convinced that they can. Pavel Materna, for example, claims that the expression "triangular circle" picks out a full-fledged construction that constructs a set whose elements are triangular circles, i.e. the empty set.[12] Thus, as concerns its meaning, the expression is perfectly fine and probably also the expressions "pork bateau" and "braised Tuesday" are similarly meaningful. Even the sentence "Pork bateaus ride triangular circles on braised

[12] See Materna (1998). Materna mentions the case of round squares.

Tuesdays" probably comes out as meaningful under this approach. Its proponents may say that it offers (as all other meaningful expressions) an abstract itinerary that we have to follow if we want to grasp what the sentence says (to get a mental grasp of the Platonic object that the sentence picks out). The itinerary starts with taking bateaus and pork (trivialized) and composing them together into a compound construction, and then taking triangular objects and circles and composing them together, etc.

If we recall Socrates, who was afraid that he might "*fall* into a bottomless pit of nonsense, and perish" should he concede that even visible things like hair or mud have their ideas, we can appreciate the fearlessness of his present days followers. I, however, dare conjecture that not all of them would be willing to admit that, as soon as we can compose some words in a grammatical way (and the respective constructions fit together as concerns their types), we are licensed to presume that the Platonic realm contains an object that makes the resulting compound expression meaningful.

But those who favour this more cautious approach are in an uneasy position. To differentiate the meaningful and meaningless grammatical constructions they must face the uneasy task of providing some illuminating insights into the structure of the Platonic realm that would reveal the organization of the grand domain. Thus, they should somehow escape from the Platonic cave. In the end, they should be able to answer questions like: Are there some principal or superior ideas and some ideas that are inferior? Are some objects in the Platonic realm dependent on others or are they all self-reliant? What determines how ideas combine into complex ideas?

Questions like these, however, are not likely to intrigue many present day analytic philosophers. They are too heavily reminiscent of the idle disputes of scholastic metaphysics. But I am afraid that those Platonists who aspire to provide a plausible theory of meaning cannot avoid such questions if they want their theory to be adopted by analytically-oriented philosophers. Of course, they may adopt a defeatist strategy and take recourse to the position of Platonic semantic agnosticism, but such a position is not likely to gain them many supporters.

14.6 Conclusion

I don't want to insist that Platonistic theories of (structured) meanings cannot provide valuable insights into the logical structure of our language. The realm of Platonist entities may, in my view, be taken as a useful theoretical posit which can serve as a fruitful explanatory tool of semantic analysis. If it is understood as such a posit, it wouldn't make much sense to raise deep or naive philosophical questions about the nature of the entities inhabiting the realm and our ability to penetrate its vast territory. In such a case, questions similar to those that I raised here would be beside the point. But if somebody maintains that a serious philosophical theory of language is to be built on the foundations of a strong Platonism, then he or she should confront questions of this sort head-on.[13]

References

Cheyne, C. 1999. Problems with profligate platonism. *Philosophia Mathematica* 7(2): 164–177.

Cheyne, C. 2001. *Knowledge, cause, and abstract objects*. Dordrecht: Kluwer Academic Publishers.

Fine, G. 1993. *On ideas*. Oxford: Clarendon.

Materna, P. 1998. *Concepts and objects*, Acta Philosophica Fennica, vol. 63. Helsinki: Societas Philosophica Fennica.

Plato. 1892. *The dialogues of Plato*, Parmenides, Theaetetus, Sophist, Statesman, Philebus, vol. 4. London: Oxford University Press.

Ross, W.D. 1951. *Plato's theory of ideas*. Oxford: Clarendon.

Svoboda, V. 2008. Nesnesitelná lehkost platonismu. In *Organon F – Príloha*, ed. M. Zouhar, 198–208. Bratislava: Jednotliviny, všeobecniny, významy.

Tichý, P. 1988. *The foundations of Frege's logic*. Berlin/New York: de Gruyter.

Tichý, P. 1992. The scandal of linguistics. *From the logical point of view* 1: 70–80. Reprinted in Svoboda, V., B. Jespersen, and C. Cheyne. eds. 2004. *Pavel Tichý's collected papers in logic and philosophy*, 775–788. Prague: Filosofia, University of Otago Press.

[13] This paper is an elaboration of my article Svoboda (2008). Preparation of the paper was supported by projects 401/07/0904 and 401/10/1279 of the Grant Agency of the Czech Republic. I wish to thank Jarda Peregrin for his helpful comments on a draft of this paper.

Part IV
In Logic

Chapter 15
A Neglected Reply to Prior's Dilemma*

Jc Beall

15.1 Introduction

Consider the long-debated principle concerning *is* and *ought*:

IOP. There is no valid deduction of an *ought* (or 'ethical') statement from and *is* (or 'factual') statement.

One very broad motivation for IOP comes from the thought (however vague) that *ought* statements are 'funny' in a way that *is* statements aren't. But whatever the motivation, IOP has long faced a short and sharp challenge from Arthur Prior (1960). Letting $\mathbb{O}$ be our unary *ought* operator, Prior's now-well-known dilemma runs thus:

- Let A be any *factual* statement, and $\mathbb{O}B$ any *ought* statement.[1]
- Dilemma: either the disjunction of A and $\mathbb{O}B$ is factual or ethical.
- Case 1: $A \vee \mathbb{O}B$ is ethical. We may validly derive an *ought* from an *is*.

  1. A
  2. $\therefore A \vee \mathbb{O}B$

---

* I am delighted to contribute to this volume in honor of Colin Cheyne. What I've always enjoyed about Colin is his willingness to explore ideas, but also his no-nonsense approach to finding his own views on the matter at hand. My hope is that this paper will give Colin the opportunity to voice his own response to Prior's dilemma. The debate will undoubtedly be better for Colin's input

[1] Throughout, I will assume that these categories are exhaustive and exclusive, just to simplify discussion.

J. Beall (✉)
Departments of Philosophy, University of Connecticut, Storrs, CT, USA

Departments of Philosophy, University of Otago, Dunedin, New Zealand
e-mail: jc.beall@uconn.edu.

J. Maclaurin (ed.), *Rationis Defensor: Essays in Honour of Colin Cheyne*, 203
Studies in History and Philosophy of Science 28, DOI 10.1007/978-94-007-3983-3_15,

- Case 2: $A \vee \mathbb{O}B$ is factual. We may validly derive an *ought* from an *is*.

  3. $A \vee \mathbb{O}B$.
  4. $\neg A$.
  5. $\therefore \mathbb{O}B$.

- Either way, IOP fails.

On the flat-footed account of *ought* statements – namely, as statements that contain (use) the *ought* operator (Jackson 1974; Prior 1960) – the dilemma has appeared to be particularly challenging: one is stuck with the first horn (viz., Case 1), which relies only on the rule of Addition, namely,

$$A \vdash A \vee B$$

that is, that A implies A ∨ B for all A and B. And Addition, if any rule, *is* surely not subject to rejection – or so the standard thought goes.

My aim, in this paper, is to highlight a neglected response to Prior's dilemma (specifically, the first horn). The response arises naturally from two thoughts (however vague). The first thought is the above flat-footed method of individuating *ought* statements (advocated by Jackson and, indeed, Prior): an *ought* statement is any statement that uses the *ought* operator. A slogan for this individuation criterion is this: any dose of 'oughtiness' in a statement makes the entire statement '*oughty*'. (More soberly: any use of '*ought*' in A renders A an *ought* statement.) But, now, combine this thought with the (however vague) thought that *ought* statements are 'funny' in important ways that 'factual' statements are not – for example, though details aside, that they, unlike (let us suppose) factual statements, needn't always be true or false. Combining these two – not uncommon – thoughts (viz., flat-footed individuation and 'funniness' of 'oughtiness') motivates a neglected but natural reply to Prior's dilemma (viz., the first horn): it is precisely the rejection of Addition which jumps out in this context. If, for example, *ought* statements are decidedly 'funny' vis-a-vis *is* statements, why think that 'adding' an *ought* to an *is* should be a *logically valid* step? Our standard boolean operators are built for standard (say, 'factual') discourse; and using them (the standard operators) to combine funny or odd sentences with the standard ones does nothing but produce compound funniness – or so a natural thought goes.

This line of thinking, I suggest, naturally motivates a so-called *Weak Kleene* framework (Beall 2010; Kleene 1952): combining factual sentences with 'funniness' (in this case, ethical statements) results in funniness.[2] But while Prior's dilemma has prompted many interesting and sophisticated replies that defend one

[2] Weak Kleene logic is typically associated with Bochvar's interpretation of it (Bochvar 1938) – where meaningless discourse is involved. It will be plain from my proposal for the *ought* operator (see Sect. 15.3) that, whatever 'funny' may ultimately mean with respect to 'ought' (a matter for theories to debate), it does not mean meaningless.

version or another of IOP, none has replied along these lines.[3] In what follows, I briefly sketch a framework for this sort of reply to Prior's dilemma – to the first horn, which is the only relevant horn given the assumed flat-footed approach to individuating *ought* statements.

15.2 Weak Kleene

The Weak Kleene (or WK) approach to the boolean operators runs as follows. Let $V = \{1, .5, 0\}$ be our set of semantic values. Our stock A of atomic sentences p_i are interpreted by valuations v: $\mathbb{A} \rightarrow V$ which are (total) functions from the atomics into V. In turn, we extend the valuations v: $S \rightarrow V$ to cover all (boolean-made) sentences via the following conditions:

$\neg$	
1	0
.5	.5
0	1

$\vee$	1	.5	0
1	1	.5	1
.5	.5	.5	.5
0	1	.5	0

$\wedge$	1	.5	0
1	1	.5	0
.5	.5	.5	.5
0	0	.5	0

Observe that the boolean connectives behave perfectly classically if and only if all subsentences are treated classically: the entire compound takes the 'non-standard' or 'funny-business' status (in the formal picture, value 0.5) if any part of the sentence has that status; other*wise*, the compound takes a classical value.

15.2.1 Validity

Towards validity, we define *truth in a model* (or truth on a valuation) in terms of value 1: sentence *A is true on v iff v(A)* $= 1$. In turn, we may say that a *set X* of sentences is true on v iff all of its members are true on v. Finally, we define the WK-validity relation $\vdash$ along standard lines:

- $X \vdash A$ iff every model on which X is true is one on which A is true.

For reasons given above, this logic is a proper sublogic of classical logic: anything WK-valid is classically valid, but there are classically valid arguments that are not WK-valid. An important example of a classically valid argument (form) that fails to be WK-valid is Addition: arbitrary A fails to imply $A \vee B$. A counterexample is any valuation v according to which $v(A) = 1$ but $v(B) = .5$.

[3] Pigden's recent book (Pigden 2010) is the now-classic source for Prior's dilemma and latest replies.

15.3 Adding Ought

What we're interested in is our *ought* operator. A simple *broad* error theorist about ethical discourse – who maintains that all *ought* statements are one and all *untrue* – could easily treat $\mathbb{O}$ as a constant function such that $v(\mathbb{O}A) = .5$ iff $v(A) \in \{1, .5, 0\}$. (In other words: $\mathbb{O}A$ winds up with value 0.5 regardless of the value of A.) While this makes sense within a broadly error-theoretic approach to ethical discourse, I shall suggest a more neutral option – one that treats $\mathbb{O}$ in a supervaluational fashion.

In short, we move to a point-based – or, if you want, world-based or situation-based – framework, with a non-empty set W of points and a binary 'accessibility' relation R on W (i.e., subset of $W \times W$). We define our standard boolean connectives as per the tables in Sect. 15.2 but now relativised to points, so that if at point x there's no 'funniness' involved in any part of A, then $v_x(A) \in \{1, 0\}$, but otherw*ise* $v_x(A) = .5$, and so on.

Turning to $\mathbb{O}$, we give the following (supervaluational) conditions:

$$v_x(\mathbb{O}A) = \begin{cases} 1 \text{ iff } v_x(A) = 1 \text{ for all } y \text{ such that } Rxy; \\ 0 \text{ iff } v_x(A) = 0 \text{ for all } y \text{ such that } Rxy; \\ 1/2 \text{ otherwise.} \end{cases}$$

Unlike in error-theoretic accounts, this approach allows *ought* statements – that is, sentences in which $\mathbb{O}$ occurs – to take any range of our semantic values: true, false, or 'funny' (as it were). Indeed, the proposal enjoys the freedom enjoyed by standard intensional treatments of *ought*, where details of $\mathbb{O}$'s logical behaviour is subject to details of R, a matter I leave for debate among differing theorists.

15.3.1 Validity

Towards defining validity, it's convenient to add a designated point @ in our models. (This isn't required, but does streamline things.) Specifically, let models be structures $M = (W, R, @, v)$ where $W \neq \emptyset$ and $v : W \times S \rightarrow \{1, .5, 0\}$ and R are as above. In turn, define *truth in a model* thus: A is *true-in-M* iff $v_@(A) = 1$, and similarly for sets $X \subseteq S$ of sentences. Define validity thus:

- $X \vdash A$ iff every model on which X is true is one on which A is true.

As one expects of an *ought* operator, $\mathbb{O}$ fails to 'release', that is, $\mathbb{O}A \nvdash A$; the easiest sort of countermodel is one wherein $v_@(A) = 0$ but the only points that @ 'sees' with respect to ought-accessibility relation are ones whereat A is true.

As always in such accessibility-based intensional frameworks, fiddling with constraints on R results in different logical behaviour for $\mathbb{O}$. But, again, this is a matter for different theories of *ought* to debate. My concern here is only with Prior's dilemma, to which I now return.

15.4 Prior's Dilemma

The reply to Prior's dilemma, in short, is that it turns on a mistake: Addition is *invalid*. A simple countermodel: let $W = \{@, y\}$ and let @ access both points. In turn, let $v_@(A) = 1$ and $v_y(A) \in \{.5, 0\}$, in which case $v_@(\mathbb{O}A) = .5$. Hence, this is a model in which A is true but $A \vee \mathbb{O}A$ untrue.

Of course, the thought, assumed by Prior and others in the debate, that Addition *is* valid is not unreasonable: when we're doing science or other *ought*-less inquiry, Addition may well be free of counterexamples. But throw in 'funny' discourse – in particular, ethical discourse – and the standard logical operators cannot but produce a 'funny' compound.

15.5 Closing Remarks

Prior gave a simple, flat-footed challenge to advocates of IOP. The challenge deserves a simple, flat-footed response. By my lights, the foregoing proposal fits the bill. *Ought* statements are ones that contain (use) the *ought* operator. Such statements are 'funny' in ways that standard *is* (i.e., factual) statements are not. But our basic boolean operators (viz., negation, disjunction, and conjunction) are largely built to handle standard discourse – factual, no-funniness discourse. One may indeed step from factual statements to ethical statements by way of boolean operators: disjunction, as Prior noted, is one such step. But what Prior overlooked is that the step is free of logical backing: the step can take one from truth to untruth.[4]

References

Beall, Jc. 2010. *Logic: The basics*. Oxford: Routledge.
Bochvar, D.A. 1938. Ob odnom trechznacnom iscislenii i ego primenenii k analizu paradoksov klassiceskogo rassirennogo funkcional'nogo iscislenija, *Matematiceskij Sbornik* 4 (46): 287–308.

[4] Thanks very much to Charles Pigden and Michael McLeod for prompting and, in turn, discussing these ideas. This paper emerged in the lively logic reading group in the Otago University Philosophy Department, where, in addition to Pigden and McLeod, Hannah Clark-Younger and Juan Manuel Gomez also gave early feedback. For useful comments I thank Dominic Hyde, Scott Lehmann, Greg Restall, David Ripley, Marcus Rossberg, Gill Russell, and Lionel Shapiro.

[English translation: Bochvar, D.A. On a three-valued logical calculus and its application to the analysis of the paradoxes of the classical extended functional calculus. *History and Philosophy of Logic* 2: 87–112.]

Jackson, F. 1974. Defining the autonomy of ethics. *Philosophical Review* 83(1): 88–96.

Kleene, S.C. 1952. *Introduction to metamathematics*. Amsterdam: North-Holland.

Pigden, C.R. (ed.). 2010. *Hume on is and ought*. New York: Palgrave Macmillan.

Prior, A.N. 1960. The autonomy of ethics. *Australasian Journal of Philosophy* 38: 199–206. Reprinted in Pigden, C.R., ed. 2010. *Hume on is and ought*. New York: Palgrave Macmillan.

Chapter 16
Mathematical and Empirical Concepts

Pavel Materna

16.1 Introduction

In his (2002, p. 135) Buzaglo formulates his theory of *concept expansion.*

Buzaglo adduces convincing examples of forced expansions and is, in my opinion, successful in showing what he promised to show. His work inspired me in the following sense: Buzaglo (as well as (Manders 1989)) shows the way in which it is rational even for a realist to consider 'development of concepts', and documents the theory by numerous examples from the area of mathematics. A natural question arises: in which way can the phenomenon of expanding mathematical concepts influence empirical concepts? But at the same time a more general question can be formulated: in which way do the mathematical concepts influence empirical concepts?

What I want to show in the present paper can be described as follows:

The problem articulated by Buzaglo deserves some semantic refinements. Following explications are needed:

What is meaning? (In particular: *What are concepts?*)
What are questions? (Or, equivalently: *Semantics of interrogative sentences.*)
Further, a useful notion will be the notion of *problem.*

Taking over the notion of *conceptual system* from Materna (2004) and using Tichý's Transparent intensional logic (TIL) I can try to solve the problem of the relation between *mathematical and empirical concepts* (not only for the case of expanding some mathematical concepts).

P. Materna (✉)
Department of Philosophy, Masaryk University, Brno, Czech republic
e-mail: maternapavel@seznam.cz

J. Maclaurin (ed.), *Rationis Defensor: Essays in Honour of Colin Cheyne*,
Studies in History and Philosophy of Science 28, DOI 10.1007/978-94-007-3983-3_16,

16.2 Structured Meanings

Whoever wants to use the term *meaning* for solving some problem has to make a choice: the term is used in many ways, and to solve the given problem we have to choose one of them (or, indeed, to propose various, probably pairwise distinct solutions of the problem.)

Instead of presenting just here a definition of my choice I will describe some features that meaning should possess. (I cannot accept any theory for which meaning is *use*.)[1]

First, meaning should roughly correspond to Frege's vague idea of *Sinn*: an expression denotes an object (if any) due to possessing a meaning; thus understanding an expression E involves knowing the meaning of E.[2]

Second, meaning should not be simple: it should be *structured*. That set-theoretical explications of meaning are essentially unsatisfactory has been suspected by more philosophers. Remarkable hints can be found in Bolzano's (1837), Tichý's (1968, 1969) are probably the first explicit declarations of a procedural character of meaning, David Lewis' "General Semantics" (1972), Cresswell's attempts from (1975) to (1985), Bealer's (1982), Chierchia's (1989), Moschovakis' (1994) are examples of exhibiting dissatisfaction with set-theoretical conceptions, which do not enable us to explain how a structured expression can express an unstructured meaning. A special remark concerns Carnap's (not fully successful) *intensional isomorphism* (1947) and Church's endeavour to define synonymy in a sufficiently fine-grained way (final version 1993).[3] (That meaning should be structured is a desirable requirement not only because of a necessary harmony with structured expressions: in general there are more equivalent expressions which denote one and the same object and therefore more meanings leading to the same object. This is easily intelligible if meaning is structured).

Third, we would like to accept Church's revision of Frege's semantic schema. This revision can be found in Church's (1956) and consists in the following claim: ***The*** *meaning of an expression E is* ***a*** *concept of the denotation of E*. So any meaningful expression possesses just one meaning (*The*), and the object (if any) denoted by the given expression can be denoted by various expressions and so reached by various concepts (*a*). So concepts are meanings: the meaning of any non-indexical expression is a concept. (See Materna 1998, p. 2004).

Fourth, what does it mean *to be structured*? Let us examine some attempts at exemplifying structured objects.[4]

Sets as boundaries between members and non-members are simple. To contain members is surely not the same as to be structured. Therefore, *functions* as

---

[1] See the chapter, 'What should a nice theory of meaning look like' in Newton-Smith (1981/1996, pp. 162–164), where the author argues that meaning should be non-holistic and structured.

[2] Knowing the meaning of E does not involve knowing the denotation of E.

[3] See also Anderson (1998).

[4] See also Tichý (1995).

mappings are simple as well. Therefore, *intensions* as functions (from possible worlds) cannot be meanings if meaning has to be structured. This is a very natural constraint: if intensions were meanings then no subexpressions of an expression E would find its semantic counterpart in the intension that would be the meaning of E and the meaning of all expressions equivalent to E would be the same.

What about Cresswell's (and Kaplan's) *tuples*? Here it seems that at least some correspondence between the parts of a structured expressions and the 'parts' of the respective tuple can be stated. Yet there is nothing in the definition of tuples which would specify the role that the particular components of the given tuple should play as components of the meaning.[5]

What remains? The ingenious move to solve our problem has been made by Tichý (1968/2004), where *algorithm-like abstract procedures,* modeled in 1968 by Turing machines and defined later as *constructions,* have been shown to possess a structure. An abstract procedure *is a sequence of steps ('instructions'), being itself an executable instruction*. It cannot be reduced to a set of instructions. Indeed, let **A** be the algorithm[6] and $\{\mathbf{i_1},\ldots,\mathbf{i_k}\}$ be the set of its instructions. Obviously $\mathbf{A} \neq \{\mathbf{i_1},\ldots,\mathbf{i_k}\}$. (Also: $\mathbf{A} \neq <\mathbf{i_1},\ldots,\mathbf{i_k}>$, *pace* Cresswell).

Example. Consider following definitions of primes:

(a) *natural numbers greater than 1 divisible just by themselves and 1*
(b) *natural numbers having just two factors*

The procedures making up the meanings of (a), (b), respectively, can be roughly described as follows:

(a) Is what you examine a natural number?

No: → no prime
Yes: → continue: Is it greater than 1?

No: → no prime
Yes: → continue: is it divisible just by itself and 1?

No: → no prime
Yes: → prime

(b) Is what you examine a natural number?

No: → no prime

Yes: → continue: Does the set of its factors contain just two members?

No: → no prime
Yes: → prime[7]

[5] As for a detailed criticism of Cresswell's theory in this respect see Tichý (1994, 2004) and Jespersen (2003).

[6] Later we will show that abstract procedures as defined in TIL are not necessarily algorithms in the sense of effective procedures.

[7] More such steps can be, of course, derived from a more fine-grained analysis of (b).

The procedures sub (a) and (b) are different. This is an example of two distinct concepts identifying the same object (the set of primes) and making up the unambiguous meaning of (a), (b), respectively.

16.3 Questions

Setting aside various technical means of logical analysis of interrogative sentences, i.e., s. c. *erotetic logics,*[8] the basic approach to questions offered by TIL is formulated in Tichý (1978) as concerns empirical questions. Expanding this approach to the case of non-empirical, typically mathematical questions we can present following principles:

Analyzing interrogative sentences we have to distinguish two factors: on the one hand there is a logical object, be it an intension (case empirical questions) or a construction (case mathematical questions), which I would call *semantic core*, on the other hand a *pragmatic factor*: the *interrogative attitude* is what distinguishes interrogative sentences from expressions that share the semantic core with the former and are not interrogative sentences.

The interrogative attitude consists just in asking. If the semantic core is a construction the respective attitude asks *what is constructed*, if it is an intension then the attitude asks *what the value of the intension is* in the actual world-time. In both cases answering the question means to find the *solution* (if any). In the mathematical case the solution is found *a priori*, in the empirical case the intension (a condition of finding solution) is found *a priori* (only knowledge of the respective language is necessary), the solution proper is a result of empirical, *a posteriori* methods. See Chap. 4.

Examples.

Interrogative sentence	Semantic core[1]
Is 2^3 greater than 3^2?	$[^0 >[^0\text{exp}\ ^02\ ^03]\ [^0\text{exp}\ ^03\ ^02]]$
What is the square root of – 4?	$[^0\surd\ ^0{-4}]$
Is Jupiter the greatest planet?	The proposition that Jupiter is the greatest planet
What is the age of the Universe?	The age of the Universe
Why is wood burnable?	The cause of the fact that wood is burnable

[1] The first two examples exemplify the way TIL would use. (S. c. 'constructions'.)

[8] A good overview can be found in Harrah (2002). The first two examples exemplify the way TIL would use. (S. c. 'constructions'.)

Remark: In the mathematical cases the semantic core is the meaning of the interrogative sentence. In the empirical cases the construction that constructs the semantic core is the meaning of the interrogative sentence.

16.4 Problems

In general *problems* can be viewed as abstract procedures.[10] When we choose constructions as explications of abstract procedures we can see that only *constructions that do not contain free variables* can be said to be unambiguous *problems*: Let C be a construction containing n free variables. If $n = 1$ then the respective problem is λx C. If $n > 1$, then there are more problems determined by C, each of them without any occurrences of free variables.[11]

Example. Let C be the construction underlying the expression
(x – y). The constructions underlying the following λ-expressions are the problems determined by C:

$$\lambda xy(x - y),\ \lambda yx(x - y),\ \lambda x\lambda y(x - y),\ \lambda y\lambda x(x - y).$$

Similarly, one of the problems determined by the construction underlying the expression

$$\forall n(n{>}2 \supset \neg(a^n + b^n = c^n)),$$

where a, b, c, n are variables ranging over natural numbers, is Fermat's Last Theorem, viz.

$$\forall n\,\forall abc\,(n{>}2 \supset \neg\,(a^n + b^n = c^n)).$$

The construction underlying this last expression is not effective but – as we know already – an equivalent (more than complicated) expression expresses an effective (mechanically provable) procedure.[12] An effective procedure given by the construction underlying the expression

$$(n{>}2 \supset \neg\,(a^n + b^n = c^n))$$

---

[10] See Materna (2009).

[11] Cf. Church (1956), Introduction, where Church explains the distinction between form and function.

[12] So we can see that it happens that among *equivalent* expressions some express an effective procedure unlike the rest. See Duží, Materna (2004).

is, e.g., the construction underlying the λ-expression

$$\lambda n\lambda abc(n>2 \supset \neg(a^n + b^n = c^n))$$

(the function constructed by it is primitive recursive).

Evidently, every mathematical question is a problem.[13] As for empirical questions, the construction that constructs the semantic core is a problem as well. Now we can generalize the notion of *solution* applying it to problems.

16.4.1 Non-empirical (Mathematical) Problems

If the problem C is an effective procedure then the *solution$_1$* of C is the object (function, number etc.) – if any – constructed by C.

Otherwise, if there is a procedure C′ that is effective and equivalent[14] to C, then the *solution$_2$* of C is C′. (Thus the solution$_1$ of C = the solution$_1$ of C′).

Otherwise, there is no solution to C.

16.4.2 Empirical Problems

First of all, we probably suspect (and rightly so) that no empirical problem is an effective procedure.[15] Let C be an empirical problem. Such a procedure constructs an intension, i.e., a function from possible worlds. Intensions are not recursive functions, so there is no algorithm that would calculate the value of the given intension in the actual world (no algorithm can decide, which among all possible worlds the actual one is). Thus the *solution$_3$* of C is the value of such a not computable function in the actual world-time. In other words, we cannot get the solution$_3$ of an empirical problem using only *a priori* methods.

[13] If we try (as we do) to give a *general* definition of problem we have to ignore such connotations as 'being difficult' or 'not being solved yet' etc. So even the construction underlying the expression 1 + 1 = 2 is a problem according to our definitions.

[14] It is obviously clear that two procedures are equivalent iff they construct one and the same object.

[15] See Materna (2008).

16.5 Concepts[16]

I have already suggested that in TIL concepts (as possible meanings of expressions) are structured. They are constructions but not every construction is a concept.

First of all, concepts are – similarly as problems – *closed* constructions, i.e., constructions not containing free variables.

Remark: This constraint is in harmony with our intuition: a concept should unambiguously identify an object. As soon as some free variable is present no object can be unambiguously identified. Thus the meaning of an expression that contains some indexicals (mostly pronouns) is an open construction, where the indexical is represented by a free variable whose valuation is a matter of some context. Consider the expression

my house.

This 'my' is an indexical that behaves in the respective procedure (construction) as a variable. We have concepts of the property *being a house* or of the 'individual concept' (Church) *the oldest house in Prague* but no object is unambiguously identified by what is expressed by *my house*. As soon as we learn that that expression has been uttered by the King of France, and consider therefore the expression

The house of the KING OF FRANCE

we can state that this expression expresses a concept of the individual role (an intension that associates every world and time with at most one individual). (Remember, of course, that in some worlds and times no individual plays this role.)

Second, considering constructions that differ just by λ-bound variables we surely do not want to say that they represent distinct concepts. The constructions underlying the λ-expressions (variables ranging over real numbers)

$$\lambda x\,(x{>}0),\, \lambda y\,(y{>}0),\, \lambda z\,(z{>}0),\, \lambda z_1\,(z_1{>}0),\, \ldots$$

represent one and the same concept (*real numbers greater than 0*).

Also, such constructions of functions that differ just by the degree of η-expansion do not represent different concepts. Consider λ-expressions

$$\text{Bel},\ \lambda w\,(\text{Bel}\,w),\ \lambda w \lambda t\,((\text{Bel}\,w)t),\ \lambda w \lambda t\, \lambda xp(((\text{Bel}\,w)t)xp)$$

(Bel(ieve) an intension that links every world with a chronology of the relation linking individuals *x* with propositions believed by *x*, *w* ranging over possible worlds, *t* ranging over time moments, *x* ranging over individuals, *p* over propositions)

[16] A systematic exposition of a procedural theory of concepts can be found in Materna (1998, 2004).

The procedure given by the constructions that are meanings of these expressions is one and the same concept (*believe*).

Summarizing we can define *concepts* as closed constructions modulo α and η – conversion.

Since however the same constraints can be applied to problems we can see that *concepts are problems and vice versa: the only distinction is pragmatic: every construction can be viewed either as a procedure that identifies (in the better case) some object (in the broadest sense), or else as a procedure that defines a problem of identifying an object.*

16.6 Conceptual Systems

Every theory of (abstract) procedures has to admit some *elementary, no more decomposable* procedures; otherwise an infinite chain of definitions would be detrimental to it. For the above outlined procedural theory of concepts such an elementary procedure is what is defined as *simple concept*. To explain we have to define one of the kinds of construction, called *trivialization*.

Let X be any object/construction. Then ^{0}X is a *construction* called *trivialization*. It takes X and let it be without any change.[17]

Now let X be an object that is not a construction. In this case ^{0}X is a *simple concept*.

Concepts are (abstract) procedures, so simple concepts are procedures as well: it's an extreme case of most simple procedure.

Examples. (a) ^{0}Planet is the simple concept of the property (*being a*) *planet*.
(b) ^{0}0 is the simple concept of zero.
(c) ^{0}Prime_number is the simple concept of the set of prime numbers.

Any object whatsoever that is named in the given language can be given *via* a simple concept. *Let 'X' be a name of an object.*[18] *Then* ^{0}*X is the simple concept of the object X.*

Here two points have to be emphasized:

(a) The 'way to the denotation', which a concept should be if defined procedurally is extremely poor in the case of simple concepts: they take ('mention') the object and let it be without any manipulation. In the case of concepts of infinite mathematical objects the respective procedure is not even feasible, consisting in presentation of some case of actual infinity (cf. the concept ^{0}Prime_number, where the procedure that identifies particular prime numbers is reduced to presenting the whole infinite set).

---

[17] We can say that ^{0}X *mentions* X.

[18] We will omit the self-evident parameter "in the given language".

(b) Considering a given language **L** we need the category of simple concepts without assuming that there is a fix set of simple concepts and all concepts encoded by **L** can be derived from such a set. Instead we can conceive of particular simple concepts as being simple *with respect to* a *conceptual system.*

The general idea of a conceptual system can be articulated as follows: We have to consider '*conceptual resources*' whenever we intend to solve some problem. Since a *finite* set of such resources is a necessary condition of successfully solving a problem we can and must choose a finite set of such concepts that will be given as basic (i.e., not decomposable any more) and that can be combined with other basic or compound concepts creating an infinite set of compound concepts.

It is the role of such a basic finite set of resources what is played by a finite set of *simple* concepts. This set will be called the set of *primitive concepts* of a *conceptual system (c.s.) S*, $\mathbf{Pr_S}$.

Then we have at our disposal *types* within a ramified hierarchy (see, e.g. Tichý (1988)). The atomic types can be called *preconcepts* of the given c.s.

Further, for any type an infinite set of *variables* is at our disposal. Since constructions (and thus concepts) are extra-linguistic procedures the variables are extra-linguistic procedures as well. They construct objects dependently on a total function known as *valuation*. Therefore, we will say that the variables, as well as constructions that contain (free) variables, *v*-construct objects, where *v* is the parameter of valuation. Besides, the letters used for variables ($x, y, z, x_1, x_2, \ldots,$) are just names of variables.

Finally, an infinite set of *derived concepts* of a c.s. *S*, $\mathbf{Der_S}$, is unambiguously determined by the respective $\mathbf{PR_S}$ due to the way the particular kinds of construction are defined.

Example[19]**.** A most simple c.s. S for arithmetic of natural numbers can be defined as follows:

Simple types: Atomic types[20]: ο (*truth-values*, {**T**, **F**}), ν (*natural numbers*)

Functional types: sets of partial functions $(\alpha\beta_1 \ldots \beta_m)$, where α is the type of the value and $\beta_1, \ldots, \beta_m$, respectively, are types of the arguments

Higher types: they make it possible to not only use but also mention constructions (and so concepts): the latter become objects *sui generis.*

For any of the infinitely many types infinitely many variables are at our disposal. Their names can be arbitrary letters, preferably from the second half of alphabet. The type of the objects they range over must be however explicitly given. For example, if the variable *p* happens to range over classes of naturals

[19] Needless to stress that our exemplification will not follow the real history of, e.g., mathematics. Some historically real examples can be found for example in Manders (1989) or in Buzaglo (2002).

[20] So *preconcepts*.

we must write $p \rightarrow (\text{o}\nu)$ (because classes and relations are represented by characteristic functions).

PR$_S$ will be

$$\{^0 0, {}^0\text{Succ}, {}^0+, {}^0., {}^0=, {}^0>, {}^0\neg, {}^0\supset\}.$$

(Types are respectively ν, (νν), (ννν), (ννν), (oνν), (oνν), (oo), (ooo).)

Concerning **Der**$_S$, it contains each concept that can be composed of the members of **PR**$_S$. Setting aside concepts that construct objects of higher types (e.g. other concepts) we have to state an important fact: among the members of **Der**$_S$ there are infinitely many concepts that construct truth-values (and are thus meanings of arithmetic sentences). Traditionally the expressions of this kind (i.e., sentences) are not connected with *concepts*. The reason should be that concepts are neither true nor false. (That this reason is untenable is shown in Author (2004, pp. 49 and 50).) We accept however Church's proposal (1956) according to which *every* meaningful expression expresses a concept. Thus the concepts constructing truth-values are members of **Der**$_S$. But then some special features of such concepts – let us call them *o-concepts*[21]– must be emphasized. We have stated that the idea of conceptual systems consists in defining 'conceptual resources' necessary for solving some problems. A conceptual system should offer such resources, it itself however cannot decide about accepting or rejecting the results of applying respective concepts. The members of **Der**$_S$ can be seen as a *list of problems that can be formulated* if the respective **PR**$_S$ is given. Now the o-concepts can be seen either as yes-no problems or as constructions of the solution. The way **Der**$_S$ has been defined excludes the possibility of deciding in advance which problems have been already solved. Thus we cannot apply a consistency criterion to this set. **Der**$_S$ contains every o-concept that can be defined in terms of **PR**$_S$, so it contains o-concepts together with their negations: to adduce very simple example, the sentence

$$3\,.\,5 > 2\,.\,6$$

can be viewed just as a claim that it is true, or as expressing the problem (in this case really *very simple*) whether it is true. The negation of this sentence is as well either a (false) claim that 3 . 5 is not greater than 2 . 6, or as expressing the problem whether this inequality holds. Accepting the first interpretation of both sentences we have to admit that **Der**$_S$ is in a sense inconsistent. Accepting the second interpretation we need not admit such a conclusion: we have got simply two dual o-problems.

Now the resources given by our c.s. S make it possible to decide for the infinite class of very simple problems which of the two dual problems is solved by *yes* and which by *no*. The class of such problems could be called *Childrens' Arithmetic*

[21] An analogous class of concepts can be defined in the case of empirical concepts: the latter construct *propositions* and are meanings of empirical sentences. See below.

Problems (ChAP). Observe now what happens if a *quantifier* is added to $\mathbf{PR_S}$. The resulting $\mathbf{PR_{S1}}$ is

$$\{{}^0 0,\ {}^0\text{Succ},\ {}^0{+},\ {}^0{\cdot},\ {}^0{=},\ {}^0{>},\ {}^0{\neg},\ {}^0{\supset},\ {}^0\forall\}$$

where the type for $\forall$ is $(o(o\nu))$.

Remark: The choice of quantifier is arbitrary. If the universal quantifier is a member of $\mathbf{PR_{S1}}$ then the existential quantifier can be constructed in $\mathbf{Der_{S1}}$ and *vice versa*.□

The c.s. S_1 offers something more than *ChAP*. As an example of formulating a not as trivial problem as we are used in the case of *ChAP* we can adduce the following pair of sentences expressing dual problems:

(a)
$$\forall xy \neg \forall z \neg x = z + y$$

(b)
$$\neg (\forall xy \neg \forall z \neg x = z + y)$$

Now the arithmetic using S_1 decides that (a) does not hold, since it fails to for any pairs x, y of naturals such that $y > x$. Therefore, (b) is true.

Well, but if I own 3 euros and borrow 5 euros, how many euros do I own now? This practical question should find its counterpart in arithmetic: does a z exist such that $3 = z + 5$? Using S_1 arithmetic answers that no such z exists. Time for a new change!

This time a new *preconcept* replaces the type ν. Let it be signified by ι (*integers*) and involve naturals + negative integers. For S_2 we have preconcepts o, ι, and $\mathbf{PR_{S2}}$ is

$$\{{}^0 0, {}^0\text{Succ}, {}^0{+}, {}^0{\cdot}, {}^0{=}, {}^0{>}, {}^0{\neg}, {}^0{\supset}, {}^0\forall\}$$

with types respectively ι, $(\iota\iota)$, $(\iota\iota\iota)$, $(\iota\iota\iota)$, $(o\iota\iota)$, $(o\iota\iota)$, (oo), (ooo), $(o(o\iota))$.

What did we gain? Arithmetic using S_2 makes it possible to give a more satisfactory answer to the question formulated by (a) ((b)): (a) does and (b) does not hold. Some intuitions must however be given up. For example:

On S_1 the o-concept expressed by $\forall x \neg$ Succ (0) $= x$ constructs **T**, whereas on S_2 it constructs **F**.

On S_1 the o-concept expressed by $\forall xy\ x + y \geq x$ constructs **T**, unlike on S_2.

But: as soon as we let the variables range over ν the above intuitions are preserved. The preconcept ι, being simple, not decomposable within S_2, can in some way 'remember' its 'origin' so that the mentioned 'communication' between S_1 and S_2 allows for stating that the 'old intuitions' are not lost and can be applied if the constraint "work on ν only" is heeded.

The fact that in this last paragraph so many single-quoted metaphors have been used is symptomatic. The necessity to do it is best explained in Manders (1989, p. 561):

> [c]urrent logic-based epistemological traditions can hardly assign a role to domain extensions. One cannot usefully treat domain extensions as inferences within a fixed theoretical context, nor as the strictly extramathematically motivated establishment of a new theoretical context unrelated to others. The philosophical discussion of the bases of theoretical contexts has never dealt with relations between contexts. But relations between contexts are central here. . .

The above schematic story with transition from naturals to integers can be continued and result in the transition from integers to fractions/rationals. The question given by the sentence $\forall xy\,(\neg\, y = 0 \supset \exists z\; x = z.y)$ is answered negatively on S_2. Replacing the preconcept ι by *rationals* (denoted by, say, ρ) we get a c.s. S_3 with $\mathbf{Pr}_{S3}$

$$\{{}^0 0,\ {}^0{+},\ {}^0{.},\ {}^0{=},\ {}^0{>},\ {}^0{\neg},\ {}^0{\supset},\ {}^0\forall\}$$

with respective changes of types and ^{0}Succ omitted. Now the above question is answered positively.

Let us return to S_1. Our question expressed by $\forall xy \neg \forall z \neg\, x = z + y$ has been answered negatively (which motivated the transition to S_2). We could instead formulate another question as follows (to make it more similar to the analogous question in S_2):

$$\forall xy\,(\neg\, y > x \supset \neg \forall z \neg\, x = z + y)$$

On S_1 this question is answered positively. Our dissatisfaction with the negative answer in the case of $\forall xy \neg \forall z \neg\, x = z + y$ can be refined as an only partial satisfaction with $\forall xy\,(\neg\, y > x \supset \neg \forall z \neg\, x = z + y)$: we can namely ask:

(1) *What if* $y > x$?

And in the case of S_2 and the question expressed by $\forall xy\,(\neg\, y = 0 \supset \exists z\, x = z.y)$ can motivate us to ask

(2) *What if* $y = 0$?

Consider (1). Add a primitive concept to our systems: a function *the only* (type (ν(oν)) for S_1, (ι(oι)) for S_2 , (ρ(oρ)) for S_3). This function returns the member of the given singleton and is *undefined* on any set that is not a singleton. A particular question over S_1 *what is the only z such that* $2 = z + 5$? can be generalized to a general problem[22]

$$\lambda xy \textit{ the only } z(x = z + y).$$

[22] (i.e., the meaning of the following expression)

Now the solution of this problem is a *partial* function *f1* such that it is *undefined* wherever $y > x$. Changing preconcepts, i.e., going over to S_2, we see that the expression $\lambda\, xy$ *the only* $z\ (x = z + y)$ expresses another problem (due to the change of types), this time a problem whose solution is a *total* function *f2*, whose values are the same as the values of *f1* if $y \leq x$.

Consider (2). The general problem is expressed by

$$\lambda xy \textit{ the only } z(x = z.y).$$

Again, the solution of this problem is a partial function *f3*, undefined if $y = 0$. Now we ask: Can we change preconcepts so that the expression

$\lambda\, xy$ *the only* $z\ (x = z\, .\, y)$ will express a problem whose solution will be a *total* function, say, *f4*, sharing its values with *f3* if $y \neq 0$?

The analogy with the preceding case (*f1* $\rightarrow$ *f2*) fails. Theoretically, *f3* could be "totalized" *via* introducing *per definitionem* some 'special number'/'special object', say, ∞, defined as follows:

$$\infty = \textit{the only } z\forall x(x = z.0)$$

but this would a typically *ad hoc* solution without any further useful effect (as to "render things understandable", see Manders (1989, p. 562)).

Returning to S_3 we have stated that the expression $\forall xy\ (\neg y = 0 \supset \exists z\, x = z.y)$ is a true claim of arithmetic over S_3. The question *What if* $y = 0$? did not trigger a change of S_3 analogous to the changes of S, S_1 and S_2. A change has been triggered (see however the footnote 19) by an 'extraneous' factor: incommensurability of diagonal/hypotenuse (proved already by Euclid). If what we like to call *numbers* should be associated with geometrical magnitudes then the conceptual resources of S_3 (offering the preconcept of rationals) would not suffice. Schematically a new c.s., say, S_4, with preconcept τ, *reals*, instead of ρ, *rationals*, would resolve this situation.

Our schematic story goes on. A function $\surd$ is definable in $\mathbf{Der}_{\mathbf{S4}}$[23] as follows:

$$\surd = \lambda x\ \lambda y(x = (y.y))$$

Using S_4, does it hold that $\forall x\ (x \geq 0 \supset \exists y\ y \in \surd x)$? It does, while it does not for S_1, S_2, S_3. But now we can ask:

(3) *What if* $x < 0$?

This question cannot be satisfactorily answered if arithmetic uses S_4. We know, however, that complexification, introducing complex numbers and leading thus to another conceptual system, say, S_5, makes it possible to answer the problem

[23] As well as in S_1, S_2, S_3.

expressed by $\forall x\ \exists y\ y \in \sqrt{x}$ positively: this time the problem differs from the problem expressed by the same expression within S_4 – the preconcept is no more τ but the set of complex numbers, say, κ. The result is what Manders (1989) calls 'existential closure' (or, if successful, 'model completion'), which leads to unification and useful simplification of concepts. True, something has to be sacrificed, i.e., some laws proved in arithmetic on S_4 do not hold any more, e.g., ordering of real numbers cannot be expressed.

16.7 Empirical Systems

We have seen that empirical problems/concepts can be seen as (abstract) *procedures* that construct non-trivial (= non-constant) *intensions*, i.e. functions whose value depends on the state of the world at the given time moment. Thus the type of any intension is $((\alpha\tau)\omega)$ (abbreviated as $\alpha_{\tau\omega}$), where α is an arbitrary type, τ is the set of time moments/real numbers, and ω is the collection of possible worlds. To solve an empirical problem means to find the value of the constructed intension in the reality (modeled as the actual world) at the given time, which cannot be achieved *a priori,* logically, but only empirically. Conceptual resources for solving empirical problems are therefore conceptual systems S such that $\mathbf{Pr}_S = \{C_1,\ldots,C_n\}$, where at least one concept C_i is an empirical concept.[24] Naturally, the interesting empirical c.s. contain a non-empty class of non-empirical (logical and mathematical) primitive concepts.

Remark: The role that was played by o-concepts in the non-empirical conceptual systems is played by $o_{\tau\omega}$-concepts in the empirical c.s. Any $o_{\tau\omega}$-concept (i.e., a concept that constructs a proposition) is in the respective $\mathbf{Der}_S$ accompanied with its dual counterpart: Let P be a proposition. Then if $\mathbf{Der}_S$ contains a concept of P, say, $\lambda w\, \lambda t\, [\ldots X \ldots]$, then it contains $\lambda w\, \lambda t\, \neg[\ldots X_{wt} \ldots]$ as well. The theory that works on the given system takes any of these two concepts and tries – using empirical methods (registering, measuring etc.) to verify or falsify that proposition (the result may change with time in the same world).□

In the preceding chapter we observed two ways in which a given c.s. can be changed:

A. A new primitive concept is added, *preconcepts are the same.*

For example we can have ν as a preconcept and
$\mathbf{Pr}_S = \{{}^0 0, {}^0\mathrm{Succ}, {}^0{+}, {}^0{=}, {}^0{>}, {}^0\neg, {}^0\supset\}$ and
$\mathbf{Pr}_{S'} = \{{}^0 0, {}^0\mathrm{Succ}, {}^0{+}, {}^0\cdot, {}^0{=}, {}^0{>}, {}^0\neg, {}^0\supset\}$. Ignoring the historical process of using particular systems (Peano vs. Pressburger) we can state that adding

[24] If no empirical concept is a member of $\mathbf{Pr}_S$ then no empirical concept can be derived from S, of course.

multiplication in S′ increases expressivity. (When formalized, completeness and decidability are lost, but this is another story.)

B. *Preconcepts are changed.*

See, e.g., the transition from S_1 to S_2.

Can we accept this classification of changes in the case of *empirical* c.s.?

Before answering this question we will try to refine this classification.

First of all, the case **A.** can be refined as follows:

A1. as **A.**, i.e., a primitive concept (or in general some primitive concepts) is (are) *added.*

A2. Some primitive concepts are *changed*, i.e., replaced by other concepts.[25]

Further, our intuition has it that the logico-mathematical concepts play an 'instrumental role' w.r.t. the empirical problems, and that therefore we should distinguish between changes of a c.s. S due to changes of the logico-mathematical part of $\mathbf{Pr_S}$ and changes due to changes of the empirical part of $\mathbf{Pr_S}$. We will thus separate any empirical c.s. S into its logico-mathematical part (LMP) and its empirical part (EP), considering separately $\mathbf{Pr_{LMP}}$ and $\mathbf{Pr_{EP}}$.

Now the case **A1.** can be obviously exemplified (if an analysis based on conceptual systems is applied). Let us suppose that the contemporary cosmology works using some conceptual system **S**. We know that some interesting problems can be formulated in **S** in virtue of the phenomenon called 'redshift'. A schematic (simplified) question can arise: *What is the cause of the accelerating expansion of universe?* A new primitive concept is added to $\mathbf{Pr_{EP}}$ and so to **S**, viz. *dark energy*.

Just such examples induce the fundamental question of *fruitfulness* of such empirical systems: consider the real example of adding *flogiston*, perhaps also of adding *ether*,[26] and Molière's generalizing literary example of *vis dormitiva*.

In any case what is (or should be) called *concept expansion* is not a simple addition or replacing some primitive concept: Such cases are not fully arbitrary but the necessity of their emergence is a kind of nomological necessity[27] rather than of logical or an analytical necessity.

As for **B**, it seems that the change of *preconcepts* is definable just for the $\mathbf{Pr_{LMP}}$ part of a c.s. Apparent counterexamples showing that e.g. some changing primitives of physics do change their type (see the constant length in Newton's physics vs. the Einsteinian velocity dependent length) are not convincing: the change is here not a consequence of a change of preconcepts: empirical concepts are type-theoretically conservative (at least when compared with mathematical concepts), using just truth-values, individuals, times/reals, possible worlds as atomic types.

---

[25] For example Newton's concept of length is replaced by Einstein's concept.

[26] But the famous Michelson-Morley's experiment proves an indirect usefulness of adding the concept *ether*.

[27] (which can be detected by studying History of Science)

Thus we can consider a conceptual system **S** as given by $\mathbf{Pr}_{\mathbf{EP}}$ and by various $\mathbf{Pr}_{\mathbf{LMP}}$. The topic of *concept expansion* in Manders' or Buzaglo's sense can be reduced to changes in the $\mathbf{Pr}_{\mathbf{LMP}}$ of the given c.s.

So the idea is that some conceptual system S used by an empirical theory ***T*** and containing conceptual resources for solving some problems of ***T*** is no more able to answer some questions that can be formulated within S (+ the respective 'common part' of the given natural language). In particular let a member C of $\mathbf{Der}_{\mathbf{EP}}$ construct a proposition **p** and a member C′[28] construct the proposition $\lambda w\, \lambda t\, [{}^0\neg\ {}^0\mathbf{p}_{wt}]$. Now saying that the "conceptual system S ... is no more able" etc. means that ***T*** using S cannot decide whether **p** or $\lambda w\, \lambda t\, [{}^0\neg\ {}^0\mathbf{p}_{wt}]$ holds. It is the empirical case of *concept expansion* when the decision is made possible by a change in $\mathbf{Der}_{\mathbf{LMP}}$ such that some preconcepts of $\mathbf{Pr}_{\mathbf{LMP}}$ are changed so that some important properties of the relations that link mathematical objects before the change took place are preserved and some new properties arose (or better some new *relations* can be stated). So for example if general solvability of quadratic equations is a condition of solvability of an empirical problem and if the c.s. S, where this problem is a member of $\mathbf{Der}_{\mathbf{EP}}$, is connected with $\mathbf{Pr}_{\mathbf{LMP}}$ containing just reals as numbers then complexification of $\mathbf{Pr}_{\mathbf{LMP}}$, i.e., expanding mathematical operations from reals to complex numbers, sacrificing some laws holding for reals but preserving those ones that are relevant for solving the given problem means that this problem is solved due to a concept expansion. In this way even changes *of non-empirical concepts can be* (and provably often are) *relevant for solutions of empirical problems.*

This italicized statement should be suspect or rather inspiring: how come that non-empirical, in particular mathematical concepts could be relevant for solving *empirical* problems? But we should begin with our condition: "if general solvability of quadratic equations is a condition of solvability of an empirical problem..." Can a solution or solvability of a *mathematical* problem be a *condition of* solution or solvability of an *empirical* problem? And if so, in which sense?

Let us recapitulate. An empirical problem constructs a non-trivial intension. Its solution (solution$_3$, see above, 4.) is the actual value of that intension. Consider the following schematic example. I am standing on some place and want to drop a stone down a steep hill of known depth, say, 60 m. My problem is how long the stone will fall. The intension constructed by this problem is a 'temporal role', i.e., a function that associates possible worlds + times with at most one time interval. The solution 3 will be therefore a definite time interval, say, 4 s.

Now we ask: is the solution of a mathematical problem (we know, which one) a necessary condition of solving the above empirical problem? Not at all: I can use an empirical procedure, viz. measuring the time by my watch.

What about an empirical mass problem? In our case: For any length **d** to find the time interval **t** such that **t** will be the duration of the free fall of a stone dropped to a

[28] So C, C′ are 'dual' $o_{\tau\omega}$-concepts.

vertical hole whose depth is **d.** This time the respective intension is a function that associates every world and time with a function from lengths (say, in meters) to time intervals (say, in seconds). This intension is non-trivial because there is no *logical* necessity which would be responsible for the result.[29] The solution$_3$ consists in finding the real values of all such functions from **d** to **t** in the actual world. Once more, is the solution of the well-known mathematical problem a necessary condition of solving this mass empirical problem?

Surprisingly enough neither here can we say that without calculating $\sqrt{(2d/g)}$ as the (approximate) value of **t** we could not solve our empirical mass problem. We would simply have to realize the empirical step of measuring time for every instance of our problem.

Yes, but having at our disposal the actual data, viz. the depth of the hole where the stone is thrown, *and* realizing the respective calculation we need no more to make any empirical steps (measuring **t**).[30]

Our question is a special case of the more general problem: Can mathematical calculations contribute to our empirical knowledge?

In practice, there may be no doubts: our historical experience confirms that there would be nearly no progress of empirical sciences and technical practice if mathematic calculations were missing. Our question is however a theoretical one, viz. how is it possible? We surely know that mathematical claims are independent of empirical facts.

A simple analogy could be useful. Imagine that multiplication is an unknown operation. Your problem will be, e.g., how many units of the given currency 27 pieces of some commodity you have to pay if one piece costs 15 units. Is your problem unsolvable? Not at all: you will add 15 units 27 times. The solution will be realized much slower, that's all. The contribution of multiplication to our empirical knowledge obviously consists in making our solutions faster (and easier).

But of course: this *quantitative* difference between our knowledge equipped with mathematics and one doing without mathematics becomes a practically *qualitative* difference as soon as more complex problems come to being.

Our question was:

Can a solution or solvability of a *mathematical* problem be a *condition of* solution or solvability of an *empirical* problem? And if so, in which sense?

Our preliminary answer is now: Solution or solvability of a *mathematical* problem is in a sense *relevant* for solution or solvability of an *empirical* problem but it is not a necessary condition thereof.

---

[29] A kind of *nomological* necessity is certainly present.

[30] See a highly interesting paper (1995, 2004) by Pavel Tichý, especially pp. 183–184 (883–884 in 2004). We will return to this problem in the final part of the present paper.

16.8 Concepts and Expanded Concepts in Mathematics

Considering our examples from 6 we can summarize w.r.t. particular concepts. Thus the transition

- from S_1 to S_2 led to expanding *subtraction* (type (ννν) → type (ιιι)),
- from S_2 to S_3 led to expanding *division* (type (ιιι) → type (ρρρ)),
- from S_3 to S_4 led to expanding *square root* (type (ρρ) → type (κκ)).

The problem with concept expansion gets its practical shape as soon as we become aware of the fact that mostly (or perhaps always) the concept before the expansion is expressed by the same expression as the result of expansion: we speak about subtraction, about division, about square root indiscriminately whether we talk about the not-yet-expanded or the (an) expanded stage. Thus an impression arises that there is one concept here, independently of the type of the function it constructs.

Notice that something similar happens not only in arithmetic; we speak also about a universal (existential) *quantifier*, be it over numbers, individuals etc., or take such a set-theoretical concept as *cardinality*: be the class whose cardinality we consider a class of numbers, of individuals, of classes of individuals, of propositions etc. etc., or take *singularizer*. This general phenomenon can be captured by the notion *type-theoretical polymorphism.*

Intermezzo: Type-theoretical polymorphism

Let a concept C construct a function *f* of the type $(\alpha\beta_1 \ldots \beta_m)$. C is said to be *type-theoretically polymorphous* iff at least some of α, β_i can be replaced by other types and the resulting function *f′* returns the same value as *f* would return if the respective substitutions were realized.

Examples. Here we use the standard TIL base of atomic types:

o (truth-values **T**, **F**),
ι (individuals),
τ (time moments/real numbers),
ω (possible worlds)

Let the concept expressed by ∃ construct

1. a function of type (o(oι)) and
2. a function of type (o(oτ)).

Let x range over ι and y over τ. Compare concepts expressed by

1′. $\exists x\ (x = \text{Saturn v } x = \text{Jupiter v } x = \text{Mars})$ and
2′. $\exists y\ (y = 2 \text{ v } y = 3 \text{ v } x = 5)$.

Clearly, the procedure given by ∃ is indifferent w.r.t. the kind (type) of the particular procedural steps.

A little less banal case, i.e., a case where also the type of α is replaced, is the case of the *singularizer* the. This is a function that returns the only member of a singleton and is undefined on each set that is not a singleton. Thus the schematic type is (α(oα)), where α is an arbitrary type. So consider again two cases (here we use the type ν of natural numbers):

1. the function constructed is of type (ν(oν)) (as in "the smallest prime") and
2. the function constructed is of type ((oν)(o(oν))) (as in "the greatest class of factors of 60").

Again all what is relevant for the result of applying the function is whether the argument is or is not a singleton.

(Notice:

1. The biggest planet

   $\lambda w\, \lambda t\; {}^0\text{the}\; \lambda x\, [{}^0{=}\; x\, [{}^0B_{wt}\, {}^0P_{wt}]]$[31] α is here ι.

2. The most well-known proposition

   Here ^{0}the is of the type $((o_{\tau\omega})(o(o_{\tau\omega})))$, α is $o_{\tau\omega}$.

In both cases the function the selects an object or is undefined dependently on whether the respective class (argument) is or is not a singleton in the given world + time).

End of Intermezzo

The case of expanded concepts is not identical with the case of type-theoretical polymorphism: the concept C′ after expansion accords with the concept C before expansion as far as the function *f*′ constructed by C′ shares arguments with the function *f* constructed by C. Beyond this common area *f*′ returns values where *f* was undefined. The analogy with the polymorphism ends here. Yet independently of this difference the problem with "one term for distinct concepts" remains to be an unpleasant fact, which is not always taken into account. (Take the 'box' and 'diamond' symbol in modal logics.)

At least three options suggest themselves (see also Buzaglo):

(a) The results of expanding a concept C make up a sequence or rather a tree whose branches consist of particular distinct concepts that are however united by a kind of necessity.
(b) The whole 'tree of expansion' is one concept.
(c) The particular results of expanding are simply distinct concepts. No intrinsic (i.e., not pragmatic) relation between them can be defined.

A specific ('intercontextual') necessity connecting particular results of expanding strongly impugns accepting the option (c). The kind of necessity that connects the 'stages' has been described by Manders as well as by Buzaglo. For example the idea of forced or strongly (or absolutely) forced expansion in the latter case is not a pragmatic idea; the definitions determine the respective relations independently of the 'human factor'. How far we can call the resulting necessity a kind of logical or analytic necessity is an interesting question not to be answered here.

In any case it is not recommendable to accept the option (c).

[31] Again a 'construction'.

Some well founded doubts about the option (b) can be found in Buzaglo's book (2002, pp. 86 and 87). Be aware that it would be on rare occasions only when we could say that the procedure of expanding terminated and we have got *the* concept.

It seems that we should accept the option (a) according to which every 'stage' of expanding is a separate concept (a common trait with (c)), which is connected with the other 'stages' by a well definable necessary link. (Perhaps the term 'cluster' would characterize the collection of the particular results of expanding the given concept.)

16.9 Empirical Expanded Concepts

Recapitulating: Conceptual systems serve as resources for formulating and possible solution of problems (in a given area). They are unambiguously determined by the set **Pr** of primitive (simple) concepts.

Consider the following situation:

The set $\mathbf{Pr_S}$ is $\{C_1,\ldots,C_k, E_1,\ldots,E_m\}$, where C_i, $1 \leq i \leq k$, are logico-mathematical concepts and E_i, $1 \leq i \leq m$, are empirical concepts. Let us sequester the LM part from the EMP part and consider two sets:

$$\mathbf{Pr_{LM}} = \{ C_1, \ldots, C_k\} \text{ and } \mathbf{Pr_{EMP}} = \{E_1, \ldots, E_m\}.$$

Let t_j, $j \geq 1$, be variables ranging over mathematical/logical objects. Suppose that at least some members of $\mathbf{Pr_{EMP}}$ contain (bound) variables from among t_j.

Now consider another conceptual system S' defined as follows:

$\mathbf{Pr_{LM\ S'}}$ differs from $\mathbf{Pr_{LM}}$: at least some of C_i are replaced by concepts that are their expanded versions. (Some new preconcepts come into being.) This change induces the respective change in $\mathbf{Pr_{EMP}}$ in virtue of our assumption that the variables t_j whose range now changed,[32] occur in the members of $\mathbf{Pr_{EMP}}$. We can ask: Will some values of some (partly) measurable features of S increase in S′?

Consider the following features:

Area of conceptually definable objects,

expressivity.

As for area: Will the system S′ be able to define more non-mathematical objects than S?

Let us first ask the following question: Consider just some logico-mathematical conceptual system S and define a system S′ differing from S just by replacing at least some of the members of $\mathbf{Pr_S}$ by expanded concepts. Will the system S′ be able to define more *mathematical* objects than S? The positive answer is trivial. Among mathematical objects definable in S there were at least some partial functions. They remain to be definable in S′ but in S′ the functions that are constructed by expanded concepts (that decrease partiality) are definable too.

[32] (Better: which have now to be replaced by other variables with a changed range)

In the case of non-mathematical objects we must recapitulate that by non-mathematical objects we mean *intensions*. Now we will set aside trivial (=constant) intensions, thus considering just non-trivial intensions.

Our question will be answered on theoretical level. Some historical support can be sought.

Theoretically, having at our disposal some *new* functions due to the effect of expansion in $\mathbf{Pr_{LM}}$ we can obtain *new* intensions as constructed in $\mathbf{Pr_{EMP}}$ and define new intensions in $\mathbf{Der_S}$. That this claim is as obvious as it might seem banal can be shown by an artificial example. Let α, β, γ be respectively types (preconcepts) *integers, rationals, reals.* Let $\text{Temperature}_{1,2,3}$ be an intension of type $(\alpha\iota)_{\tau\omega}$, $(\beta\iota)_{\tau\omega}$, $(\gamma\iota)_{\tau\omega}$, respectively. This is a 'classical' case of *intensions that are distinct due to distinctness of the mathematical components of their concepts, which in their turn differ due to concept expansion based on change of preconcepts.*

Now we can solve, of course, the problem whether Temperature_i differs from Temperature_j in the sense of (a) or of (b) (end of Chap. 8), which is probably not very important here and now. What may be of some interest is perhaps the following fact: whereas a further expansion of γ is theoretically possible and has been realized historically (transition to complex numbers) there is nothing in the case of Temperature what would require such an expansion. On the other hand, this further expansion proved to be useful ('necessary') in mathematical physics,[33] which is in the last analysis an empirical science. The concept of *twistor* (for which the mentioned expansion proved to be relevant) is not a purely mathematical concept: in some possible worlds the space-time possesses other properties than in our world.[34]

As for expressivity: Since the mathematical concepts extend the set of problems that can be formulated in the given conceptual system the expressivity of the conceptual system increases with adding mathematical concepts.

Both these answers, i.e., the one concerning the area, the other concerning the expressivity, hold even if new mathematical concepts are added without being products of the concept expansion.

Now we can try to generalize our analyses and characterize the specific nature of mathematical concepts.

16.10 The Mad Tailor

In one of his essays St. Lem compares mathematics to a mad tailor. The latter fabricates all and sundry products without taking into account any order and builds up in this way a steadily growing storage. Sometimes a customer comes and calls for a commodity. Mostly he gets it from the storage.

---

[33] See Manders (1989, p. 559).

[34] Notice that this formulation is equivalent to saying that "space-time might possess other properties".

Now imagine that our mad tailor can offer a following 'commodity'[35]:

$$\mathrm{D'}, \mathrm{T'}/(\tau\iota), \mathrm{G'}/\tau, \mathrm{The}/(\tau(o\tau)),$$

(a) $$\exists x(\mathrm{T'}x = \mathrm{The\_}y(y = +\surd((2\mathrm{D'}x)/\mathrm{G'}))),$$

(b) $$\lambda\ tdg\ \forall x(tx = \mathrm{The\_}y(y = +\surd((2dx)/g))), d, t \to (\tau\iota), g \to \tau$$

For a suitable choice of D′, T′, G′ the commodity (a) constructs the truth-value **T**: the class of real numbers satisfying (a) is not empty. The construction (concept) (b) constructs the class of such possible interpretations of D′, T′, G′, for which it holds that the respective equality holds for any individuals x.

Investigating the relation between time (T) and depth (D) in the case of free fall the 'customer' Galileo detected a regularity, whose simplified record is since then taught in every normal school.[36] Using our types and explicitly stating the empirical character of the respective physical constants we can write

$$\mathrm{D}, \mathrm{T}/(\tau\iota)_{\tau\omega}, \mathrm{G}/\tau_{\tau\omega}$$

and apply β-reduction to (b) as follows:

$$((\lambda tdg\forall x(tx = \mathrm{The\_}y\ (y = +\surd((2dx)/g))))\mathrm{T}_{wt}\mathrm{D}_{wt}\mathrm{G}_{wt})$$

$$\forall x(\mathrm{T}_{wt}x = \mathrm{The\_}y(y = +\surd((2\mathrm{D}_{wt}x)/\mathrm{G}_{wt})))$$

Notice that the presence of modal and temporal parameters w, t causes that the mentioned β-reduction is no mathematical law. Binding the free variables we get

$$\lambda w\lambda t\ \forall x(\mathrm{T}_{wt}x = \mathrm{The\_}y(y = +\surd((2\mathrm{D}_{wt}x)/\mathrm{G}_{wt}))).$$

The respective concept constructs the class of possible worlds-times where Galileo's formula holds. The resulting class of possible worlds is clearly a proper subclass of ω.

Remark: On the hypothesis that natural laws are indeed not (logically) necessary but at least eternal we can $\lambda w\lambda t$ above replace by $\lambda w\forall t$.

[35] The respective constructions would be written in another way, of course.

[36] The following text is, of course, no attempt at a historical reconstruction of Galileo's discovery.

Thus the purely schematic (i.e., not historical) description of the situation when Galileo, the customer, asked the mad tailor to offer him a commodity looks as follows: There was a *problem* to be solved (an *empirical* problem that is): to find a regularity that would enable us to determine for any material object X the time taken by X when freely falling from a point A to the point B for any distance D from A to B. One of the conceptual systems being at disposal at that time contained empirical concepts necessary to solving that problem, but mere description, in terms of those concepts, of the values that could be measured during dropping objects, could not bring about the regularity sought by Galileo. This regularity concerned the relation between the physical constants D, T and G, or rather between their values D_{wt}, T_{wt} and G_{wt} in the given world-time. Since the resistance of the atmosphere or other background could be abstracted from and (therefore) the kind of the falling object became irrelevant the regularity has been reduced to the relation between the numerical values of $D_{wt}x$ and $T_{wt}x$, and the commodity given (sold?) by the mad tailor was at Galileo's disposal.

So what the role of mathematical concepts in solving empirical problems does consist in? In which sense do they contribute to empirical knowledge?

Empirical statements are of two kinds. *Purely contingent* statements reduce to stating particular events or other particularities. Most of the human communication consists in exchanging statements of this kind. By contrast the *law-like statements* are indeed contingent but they record some *regularities* so that we can state that a kind of (not logical but 'nomological') necessity is connected with them: we can suppose that there are some other than actual worlds that share with the actual world regularities ('natural laws') of this sort, only we know that the class of such worlds is just a proper subclass of the class of possible worlds, otherwise the natural laws could be discovered by purely mathematical calculations.

It is this area of regularities, of empirical law-like statements where mathematical concepts are indispensable. Here the mere registration of facts, a statistical generalization does not suffice.

Let us finally quote a pregnant characteristic of the role of mathematics in empirical disciplines:

> ...while plotting the values of the function against its arguments Galileo was not doing mathematics. He was just taking down what was dictated to him by nature.... But Galileo not only *identified* the free-fall function. He also noted that there is quite a straightforward way of *calculating* the values of the function from its arguments. Given an argument, all one needs to do is multiply it by two, divide the result by 9.7, and then take a square root. It was when he made this second discovery, a discovery concerning a complex involving functions and numbers, that he was doing mathematics (Pavel Tichý 2004, 883–884).

Acknowledgment The present paper has been supported by the Grant Agency of Czech Republic, project No 401/07/0451.

References

Anderson, C.A. 1998. Alonzo Church's contributions to philosophy and intensional logic. *The Bulletin of Symbolic Logic* 4(2): 129–171.

Bealer, G. 1982. *Quality and concept*. Oxford: Clarendon.

Bolzano, B. 1837. *Wissenschaftslehre*, vol. I, II. Sulzbach: v. Seidel.

Buzaglo, M. 2002. *The logic of concept expansion*. Cambridge: Cambridge University Press.

Carnap, R. 1947. *Meaning and necessity*. Chicago: Chicago University Press.

Chierchia, G. 1989. Anaphora and attitudes *de se*. In *Contextual expressions*, ed. J. van Benthem and P. van Emde Boas. Dordrecht: Reidel.

Church, A. 1956. *Introduction to mathematical logic*. Princeton: Princeton University Press.

Church, A. 1993. A revised formulation of the logic of sense and denotation. Alternative (1). *Noûs* 27: 141–157.

Cresswell, M.J. 1975. Hyperintensional logic. *Studia Logica* 34: 25–38.

Cresswell, M.J. 1985. *Structured meanings*. Cambridge: MIT Press.

Duží, M., and P. Materna. 2004. A procedural theory of concepts and the problem of the synthetic *a priori*. *Korean Journal of Logic* 7: 1–22.

Harrah, D. 2002. The logic of questions. In *Handbook of philosophical logic*, vol. 8, ed. D. Gabbay and F. Guenthner, 1–60. Dordrecht/Boston/London: Kluwer Academic Press.

Jespersen, B. 2003. Why the tuple theory of structured propositions isn't a theory of structured propositions. *Philosophia* 31: 171–183.

Lewis, D. 1972. General semantics. In *Semantics of natural language*, ed. D. Davidson and G. Harman, 169–218. Dordrecht: Reidel.

Manders, K. 1989. Domain extension and the philosophy of mathematics. *Journal of Philosophy* LXXXVI(10): 553–562.

Materna, P. 1998. *Concepts and objects*, vol. 63. Helsinky: Acta Philosophica Fennica.

Materna, P. 2004. *Conceptual systems*. Berlin: Logos.

Materna, P. 2008. The notion of problem, intuitionism and partiality. *Logic and Logical Philosophy* 17: 287–303.

Materna, P. 2009. Concepts and recipes. *Acta Analytica* 24: 69–90.

Moschovakis, Y.N. 1994. Sense and denotation as algorithm and value. In *Lecture notes in logic*, vol. 2, ed. J. Väänänen and J. Oikkonen, 210–249. Berlin: Springer.

Newton-Smith, W.H. 1981/1996. *The rationality of science*. London/New York: Routledge.

Tichý, P.1968. Smysl a procedura. *Filosofický časopis* 16: 222–232. Translated as Sense and procedure in Tichý, P. 2004. *Collected papers in logic and philosophy*, ed. V. Svoboda, B. Jespersen and C. Cheyne, 77–92. Prague/Dunedin: Filosofia, Czech Academy of Sciences/ University of Otago Press.

Tichý, P. 1969. Intensions in terms of Turing machines. *Studia Logica* 26: 7–25. Reprinted in P. Tichý. 2004. *Collected papers in logic and philosophy*, ed. V. Svoboda, B. Jespersen and C. Cheyne, 93–109. Prague/Dunedin: Filosofia, Czech Academy of Sciences/University of Otago Press.

Tichý, P. 1978. Questions, answers and logic. *American Philosophical Quarterly* 15: 275–284. Reprinted in P. Tichý. 2004. *Collected papers in logic and philosophy*, ed. V. Svoboda, B. Jespersen and C. Cheyne, 293–304. Prague/Dunedin: Filosofia, Czech Academy of Sciences/ University of Otago Press.

Tichý, P. 1988. *The foundations of Frege's logic*. Berlin: New York, de Gruyter.

Tichý, P. 1994. The analysis of natural language. Reprinted in P. Tichý. 2004. *Collected papers in logic and philosophy*, ed. V. Svoboda, B. Jespersen and C. Cheyne, 801–841. Prague/Dunedin: Filosofia, Czech Academy of Sciences/University of Otago Press.

Tichý, P. 1995. Constructions as the subject-matter of mathematics. In *The foundational debate: Complexity and constructivity in mathematics and physics,* ed. W. DePauli-Schimanovich, E. Köhler and F. Stadler, 175–185. Dordrecht/Boston/London/Vienna: Kluwer. Reprinted in

Tichý, P. 2004. *Collected papers in logic and philosophy*, ed. V. Svoboda, B. Jespersen and C. Cheyne, 873–885. Prague/Dunedin: Filosofia, Czech Academy of Sciences/University of Otago Press.
Tichý, P. 2004. In *Collected papers in logic and philosophy*, ed. V. Svoboda, B. Jespersen, and C. Cheyne. Prague/Dunedin: Filosofia, Czech Academy of Sciences/University of Otago Press.

Chapter 17
Post-Fregean Thoughts on Propositional Unity*

Bjørn Jespersen

> *The logician makes it his business to explain how Bill, the individual, and walkerhood, the property, combine to construct the proposition that Bill walks*
>
> Tichý 1978, p. 275

17.1 Introduction

This essay sketches how a theory of *procedural semantics* may offer a solution to the problem of the *unity of the proposition*. The current revival of the notion of *structured meaning* has made the problem of propositional unity

* I am thankful to James Maclaurin for the opportunity to contribute a paper to this *Festschrift* devoted to Colin, with whom I had the pleasure of editing Tichý's 900-page-strong tome of collected papers in cooperation with Vladimír Svoboda back in 2003–2004. Versions of this paper were read at *PLM*, Stockholm, 16–18 September, 2011; *LMPS XIV*, Nancy, 19–26 July, 2011; *LOGICA 2011*, Hejnice, 20–24 June, 2011; Institute of Logic, Charles University, Prague, 29 November 2010; and Department of Analytic Philosophy Slovak Academy of Sciences, Bratislava, 12 November 2010. The audiences are cordially thanked for their comments and questions. This paper follows the structure of those talks and was composed within a very short period of time, which accounts for its character of work in progress. A much-expanded version is in preparation. The research reported herein was carried out as part of Grant Agency of the Czech Republic Project #401/10/0792 *Temporal Aspects of Knowledge and Information*.

B. Jespersen (✉)
Department of Philosophy, Institute of Logic, Czech Academy of Sciences, Prague

Department of Computer Science, VŠB-Technical University of Ostrava, Czech Republic
e-mail: bjornjespersen@gmail.com

J. Maclaurin (ed.), *Rationis Defensor: Essays in Honour of Colin Cheyne*, Studies in History and Philosophy of Science 28, DOI 10.1007/978-94-007-3983-3_17,

pressing.[1] The problem, stated in its simplest form, is how an individual *a* and a property *F* combine into the proposition *P* that *a* is an *F*. That is, how do two different kinds of objects combine into a third kind of object capable of having properties that neither of its constituents could have? Standard constraints imposed on *P* include that *P* must be capable of being true/false, being known/believed to be true/false, occurring as argument of propositional connectives (such as entailment), and figuring as the meaning of a sentence.

Section 17.2 provides background to the unity problem. It is argued that ordered *n*-tuples are neither propositions nor structures. Section 17.3 presents the theses that the key to solving the unity problem for atomic propositions consists in accounting for *predication* and that the key to accounting for predication consists in construing it as an instance of the abstract *procedure of functional application*. This procedure always occurs as an organizing principle of propositional constituents, since it organizes the entities these constituents present as function and argument without executing any functional applications. That is, the procedure is being *displayed* without being *executed*. If executed, the procedure will produce a product that is external to the procedure producing it.

Two variants of structured propositions will be considered. Structured propositions will uniformly be identified with hyperintensionally individuated logical procedures; but one kind of procedure produces truth-values while the other kind of procedure produces truth-conditions (modeled as possible-world propositions). Non-empirical structured propositions are instances of the procedure of application and occasionally also of the abstract *procedure of functional abstraction*. Empirical structured propositions are instances of the

[1] King (2007) and Gaskin (2008) deserve credit for having drawn attention to the topic of propositional unity, a theme that has been lying dormant for too long. In particular, Gaskin (2008) contributes much-welcomed historical background on the structure and unity of propositions. However, their own contributions suffer various serious drawbacks. For criticism of King (2007), see Deutsch (2008) and Hanks (2009, §4.2). I was glad to see Deutsch insist on an answer to a fundamental question that has also been bothering me, namely: what is the fundamental logical background to the theory of Russellian propositions? None, it seems so far. For criticism of Gaskin (2008), see e.g. García-Carpintero (2010) and Vallicella (2010). Basically, Gaskin embraces a form of Bradley's regress, which I think is a no-go. Hanks (ibid., p. 481) claims that King's 2007 theory "is by far the best developed theory of propositional structure on the market". I strongly beg to differ. With all due respect, King's theory is riddled with deep-seated flaws (two of which Hanks points out himself) and is no match for Tichý's. As readers of Tichý (1988) and most of (2004) will probably agree, Tichý, despite his lucid pen, is a tough read. Still the fact remains that Tichý's theory of structured meaning is chronologically prior and conceptually superior to King's. In fact, I see little prospect in a neo-Russellian approach, such as King's, to propositions and their structure and unity. I fail to see how any such theory would not be bound to fall prey to, first, the list objection and, second, the regress objection when introducing remedies. As for a perceptive account of Russell's propositions (uncluttered by the use made of them by advocates of direct reference, etc.), Pelham and Urquhart (1994) remains a personal favourite.

procedure of application and always also of the procedure of abstraction, to wit, over the values of modal and temporal variables, which non-empirical propositions lack. 'Proposition' will be used as a shorthand for 'structured proposition' when no confusion can arise.

The underlying logical framework is provided by Tichý's post-Fregean Transparent Intensional Logic (TIL), which is formally a *hyperintensional, partial, typed lambda calculus* whose syntax is furnished with a procedural ('intensional') semantics as known from programming languages where it contrasts with denotational ('extensional') semantics. TIL is an extensional logic of hyperintensions in the sense that TIL includes hyperintensions in its ontology, regimented by a ramified type hierarchy, while respecting the laws of extensional logic. The dual type theory of TIL – its simple type theory and the ramified hierarchy – will not be defined here, though, in the interest of brevity. For a one-call reference on TIL, see Duží et al. (2010a). For the most recent full-scale, in-depth application of TIL as an extensional logic of hyperintensions, see Duží and Jespersen (forthcoming).

17.2 E Pluribus Unum

Λόγος, for Plato, is the interweaving of forms, according to *Sophist* (259*e*). The Stranger, at *Sophist* (262*d*), points out that a statement, or λόγος (blurring the modern-day sentence/proposition divide), is made by combining a verb with a name, adducing as an example

'Theaetetus sits'

Merely providing two names or two verbs would constitute a mere two-item *list* and not a fully-fledged λόγος. Aristotle's 'third man', of course, lambasts the notion of interweaving forms for setting off a *regress* of mediating entities put forward to 'glue together' other entities also occurring inside a given λόγος.

Since Antiquity the two pitfalls any theory of propositional unity needs to watch out for are (i) reducing unified structure to nothing but an enumeration of entities (whether words or concrete or abstract objects) and (ii) invoking mediating entities alongside the entities they are supposed to mediate between. A list offers too little by failing to account for how the items listed are unified into one structure. A regress offers too much by way of a sprawl of increasingly ramified entities. A regress does solve the unity problem, but only temporarily, by deferring the unity problem to the next step of the regress. Either way the problem remains unresolved how, in the simplest case, Theaetetus the man and the property of sitting are 'interwoven' so as to form the λόγος that Theaetetus is sitting.

At the dawn of modern logic, Frege uses the chemical metaphor of *saturation* to differentiate between *Gegenstände* and *Eigennamen*, which are saturated, and *Funktionen* and words for them such as *Begriffswörter*, which are unsaturated.[2] Syntactically speaking, the functional application of an unsaturated predicate (*Begriffswort*) to a saturated name (*Eigenname*) yields (in *oratio recta*) a name of a truth-value (rather than of a truth-condition or a *Gedanke*). Logically speaking, the application of a property (*Begriff*, i.e. a characteristic function) to an individual yields (in *oratio recta*) a truth-value (rather than the *Gedanke* that that individual has that property). Saturation is Frege's account of the logical adhesion between a name and a predicate or between an object and a *Funktion*.

One notorious problem with Frege's account is that saturation makes little sense beyond its metaphorical appeal – which Peregrin (forthcoming, p. 5) aptly summarizes as the 'key-and-lock model'. In particular, it is not helpful to be told that when an individual saturates a property the resulting truth-value somehow contains both as constituents. This idea is most of all just a mereological mystery.[3] It would have made more sense if Frege had instead maintained that when an individual saturates a property the resulting *Gedanke* contains modes of presentation of both as constituents. Yet, another problem is exactly that it remains to this day an open, and very lively, research question whether *Gedanken* are literally structured, and if so exactly how. Frege is, however, absolutely clear about the nature of the constituents of any kind of *Sinn*, including a *Gedanke*: each constituent of a *Sinn* is invariably itself a *Sinn*. The *Gedanke* that Theaetetus sits contains as constituents (assuming that *Gedanken* are structured) at least a mode of presentation of Theaetetus and a mode of presentation of the concept/property of sitting.[4]

---

[2] I have retained Frege's original German labels to stay as close as possible to his inclusive notion of proper name (cf. 1892, p. 197, n. 10) and to the oscillations of his notion of *Funktion*. *Funktionen* oscillate between being *mereological* entities ('functions-in-intension', like rules or instructions) – wholes with discernible proper parts that can be factored out and replaced – and *set-theoretic* entities, viz. mappings or graphs ('functions-in-extension'). Cf. Tichý (1988, Ch. 2). For *gesättigte Gegenstände* and *ergänzungsbedürftige Funktionen*, see, e.g., Frege (1892) and (1923).

[3] For criticism, see Tichý (1988, Ch. 2, esp. §9 'Objects as their own constructions'). Interestingly, Westerhoff (2004, p. 376) claims that "there is a unique decomposition of every number into its prime factors, [therefore some sense can] be made of the prime numbers being actually 'in' a number, so that we might want to regard a number as composed of its prime factors". Westerhoff wisely hedges his locative metaphor by placing 'in' between scare quotes (although preceded by 'actually', as though a literal account were being given), for in fact the prime factors of a given number are stepping-stones on the path toward the number and are not located in, or 'in', the destination of that path.

[4] Klement (2001, p. 20) suggests "$G(*a)$" as symbolization of the Thought that a is an f, i.e. the Thought that object a falls under the concept (property) f. $G(\ )$ is the unsaturated sense of the unsaturated functor '$f(\ )$' and $*a$ the saturated sense of the saturated name 'a', making $G(*a)$ the saturated sense of the saturated sentence "a is an f". Both G and $*a$ are a *Sinn*, while $f(a)$ is a truth-value. As one would expect from a Fregean or Frege-style formal semantics, functional application (or as I would have it, the procedure of functional application) serves as unifier in $G(*a)$. However, it seems that $G(*a)$ leaves it ambiguous whether ( ) is to be applied to G at $*a$ or to what they present: f, a, respectively. Frege does not have a type theory to assist him in making this explicit, nor does he have functional arguments like w, t for the empirical case to indicate that ( ) applies to what G presents at what $*a$ presents.

Russell will have none of this. While trying on a string of positions throughout his *Hochleistungsperiode* during the first decade of the twentieth century, he remains adamant that a proposition may accommodate not only abstract but also concrete objects, both universals and particulars like people or numbers.[5] This brings Russellian propositions in line with what medieval logic would dub *propositiones in re*. The 1904 exchange with Frege over whether the solid rock of Mt Blanc is a constituent of the sense of 'Mt Blanc is more than 4,000 m high' sums up the clash between Fregean and Russellian propositions. Thus, for Russell at one point, the proposition that Theaetetus sits will result from applying the propositional function *Sit*(*x*) to Theaetetus the man and literally contain a human body as a constituent.

Russell, however, when confronting the unity problem head-on, admits to being puzzled by how (to use his own example) the difference relation and its two terms *A*, *B* (supposedly unified in the proposition that *A* and *B* differ) can recombine into that same proposition again after having been factored out of it (1903, §54). The answer he suggests is that the unifier of that proposition is simply the objective *fact* that the difference relation holds for *A*, *B*. This choice of unifier, pivoted on relations and properties being *instantiated*, bestows objective unity upon the proposition. But, notoriously, it rules out the possibility of false propositions. It also sets off a regress even when *A* and *B* do differ, for if the sole constituents are allegedly a binary relation and its two terms the question still remains *how* the relation hooks up with its relata (cf. the dispute over relations being internal or external). The entities mediating between the relation and *A* and between the relation and *B* will themselves require mediators; and so on.

After having provisionally declared propositions syntactic entities, Russell's last-ditch attempt is his multiple-relation theory.[6] This puts the judging subject in charge of unifying discrete objects into a unit that is judged to be true or judged to be false, thus trading objective for subjective unity. However, also when the judging subject is entrusted with unifying, say, the difference relation, *A*, *B* into a unit that can be judged to be true or false, it remains mysterious just how the subject does this. (There is also the risk of psychologism creeping back into logic and semantics, from where Frege had expunged it.) Moreover, there is Wittgenstein's damning 1913 objection that the unit judged to be true or false must be capable of being true or false on its own. Because this condition is not fulfilled by Russell's

[5] My broadly – and broad? – Fregean mind still boggles at the scenario of Jupiter literally being a (spatio-temporal) part of a proposition, despite countless attempts to get same mind around it. That's just too much gas for one proposition!

[6] Recent attempts to contain the unity problem at the syntactic level would include Peacock (2011) and Peregrin (forthcoming). Russell's multiple-relation theory of subjective unity has been reinstated by, e.g., Moltmann (2003) and Soames (2010). I am indebted to François Recanati for alerting me to the latter. For background on Russell's multiple-relation theory, see Pears (1977, e.g. p. 190), which Stefan Koller kindly suggested to me, and Stevens (2008).

theory, it is also not possible to express *factivity*, in case the judging subject *a* judges, *J*, something that is true:

$$\frac{Ja(R, A, B)}{(R, A, B)}$$

As Russell goes on to raze his universe, he eventually finds himself abandoning propositions altogether. In analytic philosophy, a general animosity toward propositions and other so-called intensional entities begins to take hold. What is more, the consequent success of set theory, and model-theoretic semantics in particular, within analytic philosophy leads arguably to a drop in interest in *complex* (composite, compound, ...) concrete and abstract objects. Set-theoretic entities seem to suffice. (Post-) Leşniewskian *mereology* continues to be a alive and well, but my understanding is that mereology is still a theory of how multiple *individuals* fuse into a single *individual*.[7] In principle, one could imagine a theory of propositions according to which a proposition is an individual just like its constituents. But this line is too remote from the direction I wish to go into, even if my mereology should detail how several *Sinn*-like objects fuse into one *Sinn*. Nonetheless, I wish to emphasize that the unity problem is not a *set-theoretic* problem: it is a *mereological* problem, since we need to figure out how several objects become *parts* forming one *whole*.[8]

One of several reasons why possible-world semantics, gaining currency from 1960 onwards, has shown itself to be such a fruitful research program is that it provides a clear-cut answer to Quine's query of the individuation of intensions: *functions from possible worlds* to entities, e.g. truth-values, are individuated up to

[7] "[T]he problem of what holds complex unities together (...), especially as applied by various logicians to the unity of judgements or propositions" briefly rears its head in Simons (1987, p. 342) in connection with copula-centric theories of unity and the alternative accounts offered by Frege, Wittgenstein, and Husserl.

[8] Fine (1994, p. 137) enumerates examples of objects that appear to him to be composed of parts: "..., a set of its members, and a proposition of its constituents". It is puzzling to see sets and (structured) propositions mentioned side by side. If one removes an element from a set, the result is a new set, with no gap in it, rather than the original set with a fresh gap in it. See Jespersen (2003, p. 177) for an objection to Braun's 'gappy tuples' along those lines: the result of removing *A* from $\langle A, B, C\rangle$ is not $\langle —, B, C\rangle$ but $\langle B, C\rangle$. Using Fine's vocabulary, TIL will qualify as *mereological pluralism*, as well as *mereological intensionalism*, for TIL embraces several different ways of combining parts into wholes (each kind of construction being one such way of constructing wholes from parts) and denies that identity of parts entails identity of wholes. Bolzano's example given in (1837, §56) of 3^5 *versus* 5^3 is a fine example of mereological intensionalism: the same parts – 3, 5, exponentiation – give rise to two different wholes, 3^5 *versus* 5^3, which translate into TIL as two different Compositions of two different numbers (243 and 125, as a matter of mathematical fact). See also Tichý (1986, esp. §2 'Gaps').

logical equivalence.[9] Possible-world intensions are mathematically precise notions and perfectly suitable for a formal semantics of large portions of natural language and artificial modal languages. Like all set-theoretic objects, possible-world propositions are devoid of structure and so elegantly circumvent the problem of propositional unity, as it simply does not arise. While a set-forming rule may well possess (at least syntactic) structure, the resulting set will lack structure; this is so even when its elements happen to be structured.

Possible-world semantics is indispensable – but also insufficient. The individuation in terms of logical equivalence has turned out to be too crude for, not least, the individuation of linguistic senses and of some attitude complements (notably the *explicit* ones, which are not logically closed). For instance, epistemic logics phrased in the world idiom have typically been haunted by logical omniscience. One sort of solution has been sought in sententialism, i.e. the fine granularity of syntactic structure. Another has been sought in what has become known as *structured meanings*. Cresswell is driven by hyperintensionality concerns, see 1975, 1985; Kaplan wishes to be able to insert individuals directly into propositions to dramatize the fact that directly referring terms pick out particular individuals rather than rules for picking out individuals:

> That's right, John himself, trapped in a proposition. (1990, p. 13.)

Kaplan's 'structured propositions' hark back to the Russell of *c.* (1903), and both Kaplan and Cresswell develop their notion of structure by means of *ordered n-tuples*.[10] Tichý (1994) and Bealer (1998) may have been among the first to criticize the tuple theory of structured propositions. Basically, tuples are neither structures nor propositions. Here follows a catalogue of the most common and most obvious objections that have surfaced over the last 20-odd years in the literature. First, there is the Benacerraf-style objection that it remains unprincipled *which* of the two tuples

$$\langle a, F\rangle, \langle F, a\rangle$$

is the proposition that a is an F. If the sequence between a, F does not matter we may just as well settle for

$$\{a, F\}$$

[9] Needless to say, the introduction of possible worlds as functional arguments (or elements of sets) is not going to impress anyone of a nominalist persuasion, unless reconstrued along syntactic lines. From a historical point of view, it is interesting to note that when Montague began to develop an intensional logic within a largely Tarskian framework, he did not introduce an independent type for possible worlds: s only occurred together with other types, as in $\langle s, t\rangle$. It seems that s as an independent type would still have been one bridge too far for Montague. For a comparison of Montague's IL with Tichý's TIL, see Duží et al. (2010a, §2.4.3 'Montague's implicit intensionalization').

[10] See Duží et al. (2010a, §1.1 'The programme in outline') for a survey.

For sure, there are theories holding that propositions are sets. Possible-world semantics is in fact one such theory, holding that propositions are sets of possible worlds, thus identifying a proposition with its satisfaction class, as one would expect from an extensionalist theory of intensional entities.[11] Constructivism identifies a (mathematical) proposition with the set of its proof-objects (constructions), such that judging a proposition to be true is tantamount to judging its set of proof-objects to be non-empty. The idea that a proposition is a set is *prima facie* absurd, though, for while the hallmark of a proposition is that it can be true or false, a set cannot be true or false. Something may be true or false *of* a set, but that is something else. The second objection is, therefore, that a set, whether ordered or unordered, is unfit as *truth*-(value-) *bearer*. 'First a, then F', or its converse, simply does not make a claim, so truth and falsity remain orthogonal notions. Consequently, neither $\langle a, F\rangle$ nor $\langle F, a\rangle$ can be *known* or *believed* to be true, and so cannot figure as complement of an attitude.[12]

Third, a tuple cannot occur as argument of a propositional connective, like entailment, or operand of a propositional operator, like box and diamond. Boxes and diamonds modalize the truth or falsity of a proposition; entailment has the (assumed) truth of one or more propositions necessitate the truth of a further proposition.

Fourth, the tuple notation gives rise to ill-formed formulae. For instance, the following is neither here nor there, so one is left wondering how the tuple theory would express something as mundane as existential generalization over individuals:

$$\frac{\langle F, a\rangle}{\langle \exists, x, \langle F, x\rangle\rangle}$$

Nor is this a *wff*:

$$\Diamond\langle F, a\rangle$$

Fifth, set theory shares with proposition theory the general theme of generating unity from multitude – *e pluribus unum*. As Cantor famously said, "A set is a Many that allows itself to be thought of as a One." But proposition theory needs, furthermore, to generate a *structured molecule* from what are ultimately *unstructured atoms*. And an ordered n-tuple is a mere enumeration of elements, itself

[11] Kripke (1963, §5.2) explains why possible-world propositions may equivalently be viewed as *sets* of possible worlds.

[12] One obvious argument, however, against $\{F, a\}$ would be that it does not generalize to reflexive relations: $\{R, a, a\}$ is the same set as $\{R, a\}$, providing the binary reflexive relation R with one argument too little. An advocate of propositions-as-tuples may propose that of $\langle a, F\rangle$, $\langle F, a\rangle$ one should simply pick one and stick with it, without invoking principles, and decide once and for all in which order Fs, Rs, as and bs and whatever other propositional components are to occur. Personally, this would leave me less than satisfied, for I very much want to know why a particular (self-styled) theory of structured meanings favours one kind of structure over any other.

devoid of structure. (A so-called model-theoretic structure is just a tuple, and this usage of 'structure' is internal to model theory.) Pointing out this much amounts to an instance of the Ancient list objection: the attempt, namely, to obtain unity by means of listing entities one after the other, without specifying how they interact so as to form a unit. It would be to no avail to add a third element *X* (a relation, an operation, or whatnot) to mediate between *a* and *F*. $\{X, a, F\}$, $\langle X, a, F\rangle$ and its permutations remain enumerations. For instance, if *Pred* (as a possible instance of *X*) is a generic predication operation and *Apply* (as a possible instance of *X'*) a generic functional-application operation then the four-tuple

$$\langle Apply, Pred, F, a\rangle$$

is just an enumeration of two operations, a property and an individual, with no indication of how they hook up so as to form the proposition that *a* is an *F*.[13]

Finally, if the tack of inserting *Apply*, *Pred* or whatever else is pursued there is a regress in the offing. If *X* is to mediate between *a* and *F* then *Y* needs to mediate between *X* and *a* and *Z* needs to mediate between *X* and *F*; and so on. This is the

[13] For discussion, see Jespersen (2003). Almog (1986) advances an extreme form of bottom-up approach to propositional constitution by blending yes, it's Almog (1986) varying domains with world-indexed propositional generation. If *a* does not belong to the domain assigned to world *W* no proposition $\langle \ldots, a, \ldots\rangle$ can be generated, and so fails to *exist*, at *W*. For discussion of Almog's brand of serious actualism, see Jespersen (2000). TIL is not particularly troubled by existence issues for propositions, whether they be possible-world propositions or propositional constructions. Of the former kind of proposition Tichý says, in criticizing Plantinga's varying domains a consequence of which is that the proposition *Pegasus exists* ('Pegasus' stipulated to denote an individual) exists at some, though not all, possible worlds:

> This is bound to sound rather unorthodox to any adherent of logical semantics. For all existing semantic systems [publishing in 1972] are constructed in such a way that the set of propositions is fixed in advance, prior to experience – that is, independently of the consideration which of the possible states of affairs may eventually turn out to be the actual one. Reality or 'the way things are' is not supposed to discriminate between propositions and entities which are not to count as propositions: this matter is settled beforehand as part of the specification of the language in hand. (...) The state of affairs which obtains determines rather which propositions (of the pre-fixed set of propositions) are true, which propositions are false, and, in some systems [such as TIL], which ones lack either truth value. (1972, p. 87/2004, p. 182.)

As for the existence conditions pertaining to propositional constructions, thanks to the thoroughgoing top-down approach of TIL (see Duží et al. (2010a, §1.2.2 'The top-down approach', §1.4.2.3 'The top-down approach to semantics revisited')) and its constant domain (see (ibid., §4.2 'Intensional essentialism')), as soon as *F* has been typed as a property of individuals and *a* has been typed as an individual, both can be Trivialized and so the propositional construction $\lambda w\lambda t\,[{}^0F_{wt}\ {}^0a]$ can be generated. Whatever cannot be Trivialized is nowhere and never an object in the universe of TIL. For instance, Sherlock Holmes cannot be Trivialized, nor can the largest prime number. TIL does not entertain so-called impossible worlds whose domains may contain fictional characters or impossible numbers. Alexander the Great was right: beyond the possible worlds there are no more worlds to conquer!

backbone of *Bradley's regress*, which is an instance of the 'third man'. The important lesson is that whatever provides propositional unity cannot itself be a constituent of the proposition it unifies. (Soames (2010, Ch. 2) may be the only recent locus where I have seen this point stated explicitly.) The question which of the constituents is the unifier is, thus, ill-posed. The problem of propositional unity thus stands: what, in the simplest case, is the logic of the interaction between a property and an individual such that a proposition rolls out at the other end?

In the light of such complications, it becomes tempting to maintain that propositions are *primitive black boxes*, since their structure, if any, remains under wraps.[14] It is a formally feasible option to introduce a domain of propositions *Prop* that are stipulated to be, for instance, structured and hyperintensionally individuated and add *Prop* to the model-theoretic frame(s) by means of which one interprets formulae. Primitive propositions may be just fine when investigating *what* propositions do (i.e. when a functional description suffices), but not, however, when investigating *how* they do it (i.e. when a structural description is required). In the next section I shall try to outline such a structural description of propositional unity.

17.3 Logical Procedures as Propositional Unifiers

The motto of this paper stems from a phase in Tichý's thinking when he is still holding to a broadly Russellian conception of propositions despite his overall Fregean leanings. Thus he holds that a, F are literally constituents of the structured and hyperintensionally individuated *construction* of the possible-world proposition that is true at all those $\langle \textit{world}, \textit{time} \rangle$ pairs at which a is an F. Post-1988 TIL sees an important change in that every constituent of a construction is now itself a construction. Only in (1988) does Tichý introduce the new construction called *Trivialization*, which is a one-step construction of a given object. A Trivialization 'calls' an object with a view to mentioning or manipulating that object. Consequently, it is constructions of a, F and not a, F themselves that are now constituents of the propositional construction that a is an F. This reorientation brings Tichý into harmony with his general Fregean approach.[15]

---

[14] An example of primitive propositions would be Thomason's 1980 p-propositions.

[15] For discussion of Frege's flitting between logical equivalence and cognitive significance (*Erkenntniswert*) as the right principle of individuation of *Sinn*, see Jespersen (2010b). Frege's inconclusive thoughts on the structure and individuation of *Gedanken* lead me to invoke him as a valuable source of inspiration rather than as an authority. Hence my thoughts on propositional unity are *post*-Fregean.

If we rephrase the motto accordingly, we end up with this:

> The logician makes it his business to explain how a construction of Bill, the individual, and a construction of walkerhood, the property, combine to construct the [possible-world] proposition that Bill walks

The intra-theoretic task becomes to provide such an explanation. Formally, TIL constructions are what their definition says they are (see below). Philosophically, it is Tichý's ambition to take Frege's intuitive notion of *Sinn* to the next level of conceptual analysis by actually providing a definition and incorporating the resulting notion into modern logical theories of meaning, including possible-world semantics. Church was already pursuing a similar project, putting forward various Alternatives, a simple type theory, and eventually also a ramified type theory, in (1993), while never taking possible worlds onboard. Tichý, Church and Frege share the same methodological point of departure that functions, not sets or relations, are conceptually prior. Extensionally, all three are interdefinable (provided functions are mere mappings), but only the concept of function comes with the notions of application and abstraction. The centerpieces of Church's λ-calculus are *application*, *abstraction*, and *variables* — with the possible addition of *constants*, if we are prepared to embrace an 'applied', or 'impure', syntax.[16] Tichý's streak of genius consists in recasting application and abstraction, variables and constants as (i) abstract logical objects rather than terms and as (ii) structured, complex, hyperintensionally individuated objects rather than 'flat' mappings. His dual type theory enables Tichý to erect a superstructure upon the basis of possible-world intensions and their extensions. Roughly, the objects in the basis are regimented by a simple type theory and the objects in the superstructure are regimented by a ramified type theory.[17]

Tichý (1988, §14 'An objectual notion of variable') defines variables as constructions, and (ibid., §15 'The five modes of forming constructions') defines another five constructions. For present purposes, we need four of those, including Trivialization. The inductive definition of *construction* is as follows.

17.3.1 Definition (Constructions)

(i) (*Variable*) Let a total valuation function v be given that associates variables $x_0, x_1, \ldots, x_n$ with a sequence *Seq* of objects $a_0, a_1, \ldots, a_n$ of type α. Then the *variable* x_n *v*(aluation)-*constructs the* n^{th} *object of Seq relative* to v.

(ii) (*Trivialization*) Let X be any object of any type. Then ^0X is the *Trivialization* of X, *constructing* X without any change of the structure of X.

---

[16] See Hindley and Seldin (1986, def. 1.1, pp. 2–3).

[17] Constructions and types, See Duží et al. (2010a, §1.3.2).

(iii) (*Composition*) Let X *v-construct* a function f of type $(\alpha\beta_1\ldots\beta_m)$. Let $Y_1,\ldots,Y_n$ *v-construct* entities $b_1,\ldots,b_n$ of types $\beta_1,\ldots,\beta_n$, respectively. Then the *Composition* $[X\ Y_1\ldots Y_n]$ *v-constructs* the value (an entity, if any, of type α) of f at $\langle b_1,\ldots,b_n\rangle$. Otherwise $[X\ Y_1\ldots Y_n]$ does not *v-construct* anything and so is *v-improper*.
(iv) (*Closure*) Let $x_1,x_2,\ldots,x_n$ be pair-wise distinct variables *v-constructing* entities of types $\beta_1,\ldots,\beta_n$. Let Y be a *construction* *v*-constructing an α-entity. Then the *λ-Closure* $[\lambda x_1\ldots x_nY]$ *v-constructs* the following function g of type $(\alpha\beta_1\ldots\beta_n)$. Let $v'(b_1/x_1,\ldots,b_m/x_n)$ be a valuation identical with v at least up to assigning objects $b_1/\beta_1,\ldots,b_n/\beta_n$ to variables $x_1,\ldots,x_n$. If Y is $v'(b_1/x_1,\ldots,b_m/x_m)$-improper, then g is undefined on $\langle b_1,\ldots,b_n\rangle$. Otherwise the value of g at $\langle b_1,\ldots,b_m\rangle$ is the α-entity $v'(b_1/x_1,\ldots,b_m/x_m)$*-constructed* by Y.
(v) Nothing is a *construction*, unless it so follows from (i) through (iv).□

For instance, the variables w, t are not words, but atomic procedures for constructing the values (worlds and times, respectively) assigned to them relative to an assignment function. Not worlds and times, but procedures producing, or constructions constructing, worlds and times are what are capable of occurring as constituents. Moreover, it is Trivializations of Theaetetus and of Bill and of the properties of sitting and walking that are constituents of the hyperpropositions, or propositional constructions, that Theaetetus sits/that Bill walks.

What goes on inside a propositional construction such as the one that a is an F? The wrong, but instructive, answer is that a set – a *characteristic function* – is *applied* to an *individual*. The set is obtained by extensionalizing the empirical property F. Empirical properties are typed as functions from worlds to functions from times to sets of individuals. A function from worlds to entities is called an *intension* (in the parlance of possible-world semantics), and a function from times to entities is called a *chronology*. Extensionalization takes the form of application of an intension to worlds to obtain a chronology and application of a chronology to times to obtain an entity, typically an extensional entity like a truth-value, an individual, a set, a relation-in-extension. The result is in this case a truth-value, and which truth-value it is depends on whether a is an element of the extension of F at the world/time pair of evaluation. The resulting truth-value is then abstracted over to obtain a truth-value-in-intension, i.e. a possible-world proposition, i.e. an empirical truth-condition.

TIL argues that the above proceedings amount to *predicating* F of a.[18] If not a, F are to be combined, but the difference relation, A, B then I would say that the matching propositional construction predicates that that relation obtains between A, B. Formally, an extensionalized binary relation-in-intension is applied to its two arguments in order to obtain a truth-value, according as A, B are so related at the $\langle$*world*, *time*$\rangle$ of evaluation

[18] See Jespersen (2005), (2008), Duží et al. (2010a, §2.4.2 'Predication as functional application'). Note that functional application provides the logic both of extensionalization and predication, unlike those theories that earmark special operators for these operations, such as Bealer (1993, 1998).

The above would, roughly, be the pre-1988 account. The post-1988 account is that what goes on inside that propositional construction is that a string of subconstructions (subprocedures) display the procedure of obtaining a set from a property (itself obtained by Trivialization) and applying that set to an individual (obtained by Trivialization) in order to obtain a truth-value. Finally, it is displayed how that truth-value is to abstracted over.

I am convinced that the notion of *predication* holds the key to solving the unity problem. The inspiration is, surprisingly, due to Davidson, otherwise not known as a card-carrying friend of propositions:

> [W]hat the problem of predication is concerned with is none other than an example of what is often called the unity of the proposition. (2005, p. 87.)

This seems to me like a very promising point of departure. A function/argument logic such as Frege's can model interlocking logical structures in terms of functional dependencies, by way of how the value of one function becomes the argument of another function, or how a function applicable to some particular argument is handed down by another function. I adhere to the Fregean tenet that every sentence contains at least one functor. In 'Theaetetus sits' or 'Bill walks', the predicate denotes a function and the name denotes a matching argument. Either sentence expresses a procedure whose product is the possible-world proposition that is true at every $\langle$*world*, *time*$\rangle$ at which Theaetetus sits/Bill walks.

For an example of the logically simplest kind, consider the *mathematical proposition* that 1 is a prime. That proposition translates into TIL as a construction of a truth-value, so the kind of construction we need is Composition. To get started, Trivialize 1 to obtain the atom $^0 1$ and *Prime* to obtain the atom $^0 Prime$.[19] $^0 Prime$ constructs the set of primes, without specifying how primes are defined. Likewise, $^0 1$ constructs the number 1, without specifying how 1 is defined. This sort of semantic analysis of '1', 'is a prime', as based on Trivializations, is a *literal* one of '1 is a prime'.[20] The relevant Composition is

$$[^0 Prime\ ^0 1]$$

Its constituents are all of its molecular and atomic *subconstructions*, altogether three in number:[21]

---

[19] See Duží et al. (2010a, def. 2.17, p. 247).

[20] See Duží et al. (2010a, def. 1.10, p. 105). A TIL solution (too) close to the blackbox approach would consist in the Trivialization ^{0}*Truth_value*, which Trivializes one of the two truth-values **T**, **F**, according as 1 is a prime.

[21] See Duží et al. (2010a, def. 1.3, p. 46) and Duží et al. (2010b). The latter wrongly omits the (open) subconstructions $\lambda t\ [^0 Happy_{wt}\ ^0 Tadeusz]$ and $\lambda t\ [[[^0 Hope\ w]t]\ ^0 Katarzyna\ ^0[\lambda w' \lambda t'\ [^0 Happy_{w't'}\ ^0 Tadeusz]]]$ in the respective decompositions of the Closures matching "Tadeusz is happy" and "Katarzyna hopes that Tadeusz is happy".

$$[^0Prime\ ^01], {}^0Prime, {}^01$$

Note that Composition is not a stand-alone constituent, but only occurs together with the two subconstructions constructing arguments for functional application. The welcome consequence is that Composition, *qua* constructional unifier, is not a constituent alongside the constructional constituents it unifies. Composition unifies the two Trivializations into a mathematical proposition, but does not operate on them: it operates on what they construct, namely a function and a number. If we want to indicate that Composition is to operate on these two Trivializations we must write

$$[^{00}Prime\ ^{00}1]$$

This is the procedure of applying ^0Prime to ^{01}. Only this is an ill-typed Composition, for ^0Prime is not a function and so cannot be applied to anything. Nor is ^{01} a number. $^{00}Prime$ and $^{00}1$ are one level too high up in the type hierarchy.

Russell's approach is somewhat different: he has the difference relation figure both as propositional constituent and as propositional unifier. The upshot is a Bradley-style regress. In TIL, by contrast, the products of the subprocedures ^0Prime, ^{01} are arranged as function and argument, resp., *without executing* the Composition. $[^0Prime\ ^01]$ *displays* a procedure and not a product (a truth-value).

By accounting for unity in terms of predication and for predication in terms of the procedure of functional application (Composition), the central philosophical question becomes: *how does a procedure provide unity?* The crux of the answer is that it does so by detailing and displaying what operations to apply to what types of objects, in what order, to obtain what type of object.[22] Because TIL reifies procedures into Platonic, third-realm entities, *procedures* count as static objects, unlike *processes*, which are empirical and dynamic realizations of procedures that require agents to be unfolded at particular worlds and times.

Once we turn to constructions of truth-*conditions* (possible-world propositions) rather than of truth-*values*, things get slightly more complicated. It is a central tenet of TIL that the logical form of an empirical proposition must, canonically, be a Closure abstracting over the values of world and time variables, according to this schema:

$$\lambda w \lambda t[\ldots w \ldots t \ldots]$$

The body $[\ldots w \ldots t \ldots]$ is a Composition, and the product of that Composition (here, a truth-value) is abstracted over to construct a possible-world intension (here, a possible-world proposition).

[22] What underlies the intuitive difference between *displaying* and *executing* a procedure is the defined difference between *mentioning* and *using* constructions. See Duží et al. (2010a, §2.6 'Three kinds of context').

The logical forms fitting 'Theaetetus sits' and 'Bill walks' are the same:[23]

$$\lambda w \lambda t [{}^0F_{wt}\,{}^0a]$$

That is just a notationally more economic way of writing

$$\left[\lambda w \left[\lambda t \left[\left[\left[{}^0Fw\right]t\right]{}^0a\right]\right]\right]$$

Its subconstructions are nine in number:

$$\lambda w \lambda t [{}^0F_{wt}\,{}^0a], \lambda t [{}^0F_{wt}\,{}^0a], [{}^0F_{wt}\,{}^0a], [{}^0F_{wt}], [{}^0F_w], {}^0F, {}^0a, w, t$$

That set of subconstructions is a mixed bag of molecular and atomic, open and closed constructions, illustrating the various degrees of decomposition of $\lambda w \lambda t$ $[{}^0F_{wt}\,{}^0a]$.[24] Ultimately, it is a Closure that keeps these altogether nine procedures/constructions together in one unit, which constructs an object of a type that none of its component procedures constructs on their own.

One objection leveled above against the tuple theory of structured propositions was that tuples are not truth-bearers. Ostensibly, the same objection could be raised against both $[{}^0Prime\,{}^01]$ and $\lambda w \lambda t$ $[{}^0F_{wt}\,{}^0a]$, for neither Compositions nor Closures can be said to be true/false.[25] They *are* not truth-bearers; they *construct* a truth-value and a truth-condition, respectively. So whereas Frege's *Gedanken* are truth-bearers, none of Tichý's constructions are. But I tend to think that the rift does not run deep. One might simply stipulate that for a non-empirical Composition to be true/false is tantamount to constructing a truth-value, while constructing a truth-condition is what makes an empirical Closure capable of being true/false.[26]

Up till now I have considered only *atomic* propositions. *Predication*, so my thesis goes, is what unifies them. What about *molecular* propositions? Here we find

[23] This Closure is the *normal* form of all the constructions *procedurally isomorphic* to it. See Duží et al. (2010a, defs. 2.3, 2.4, pp. 154–55).

[24] See Duží et al. (2010a, def. 2.17, p. 247) for *atomic construction*, and (ibid., def. 1.4, p. 47) for *open* and *closed construction*.

[25] Cf. the discussion of and the solution to the problem of *epistemic shift* in Duží et al. (2010a, § 5.1.6 'Factivity and epistemic shift') arising for hyperintensions capable of performing as attitude complements, but not as truth-bearers.

[26] An additional task typically discharged by propositions is to be *sentential senses*. An intuitively appealing argument from compositionality would be that the way a sentence is composed should, by and large, match the way its sense (i.e. a proposition) is composed, thus pushing the problems of propositional structure and unity to the forefront. See Duží et al. (2010a, §2.1 'A logical method of semantic analysis'), which is strongly inspired by Carnap's *principle of subject-matter*. See also the opening pages of Frege (1923). However, the notion of sentential sense deserves separate treatment. Another topic I am disregarding for now is the logic of how propositions *determine* truth-conditions and truth-values. An answer to this question is beyond the tuple theory of structured propositions, since sets cannot determine anything.

ourselves unifying already unified atomic propositions into propositions whose unity is not pivoted on predication. The topic is a large one, and I shall only just sketch the outline of my answer for now. Here are three cases:[27]

17.3.2 Extensional Connective *(*Conjunction*)*

$$\lambda w \lambda t \left[{}^0 \wedge \left[\lambda w \lambda t \left[{}^0F_{wt}{}^0a\right]_{wt}\right] \left[\lambda w \lambda t \left[{}^0G_{wt}{}^0b\right]_{wt}\right]\right]$$

This Closure constructs a possible-world proposition by abstracting over the truth-value obtained by applying $\wedge$ to the two truth-values obtained by extensionalizing two possible-world propositions. The difference from the atomic case is that the truth-value the leftmost pair of Closures operates on has not been obtained by applying a set to an individual, but by applying a three-place extensional connective to two extensionalized possible-world propositions (i.e. truth-values).

17.3.3 Intensional/Modal Connective/Relation-in-Extension *(*Entailment*)*

$$\left[{}^0\Rightarrow \left[\lambda w \lambda t \left[{}^0F_{wt}{}^0a\right]\right] \left[\lambda w \lambda t \left[{}^0G_{wt}{}^0b\right]\right]\right]$$

This Composition constructs a truth-value by applying $\Rightarrow$ to two possible-world propositions.

17.3.4 Hyperintensional/Constructional Operator *(*Relation-in-Intension Between an Epistemic Agent and a Propositional Construction*)*

$$\lambda w \lambda t \left[{}^0 Know_{wt}\, {}^0b\, \mathbf{{}^0}\left[\lambda w' \lambda t' \left[{}^0F_{w't'}{}^0a\right]\right]\right]$$

This Closure constructs a possible-world proposition by abstracting over the truth-value obtained by applying an extensionalized two-place relation-in-intension to an individual and a propositional construction. (Notice the bold-faced Trivialization of that Closure.)

[27] For *entailment*, see Duží et al. (2010a, p. 109); for *hyperintensional/constructional attitudes*, see (ibid., Ch. 5, esp. §5.1.2.2 'Hyperintensional propositional attitudes *de dicto* and *de re*').

Finally, I briefly consider two special cases: sentences without (explicit) verbs and one-word sentences. Frege famously departs from Ancient and Scholastic subject/predicate logic, in that not the copula but the predicate is the guide to the 'heart' of a proposition. The grammatical copula simply bears no logical weight for Frege. Hence the analysis of the Hungarian sentence ('Budapest is beautiful')

'Budapest szép'

is straightforward. Its analysis is exactly the same as for 'Theaetetus sits' and 'Bill walks'.

Gaskin (2008, pp. 30–33) discusses what to make of one-word sentences, such as those describing the weather. (Grammatically, those sentences consist of one verb in third person singular and lack an explicit subject.) The TIL analysis would appear to be straightforward. Consider the Czech sentence 'Prší' ('It rains'). On a literal analysis the implicit parameters bearing on when and where it is raining are suppressed. Accordingly, we pair 'Prší' off with the Closure

$$\lambda w \lambda t[{}^0 Pr\check{s}\acute{\imath}_{wt}]$$

to model the modal and temporal fact that when it is raining it is only raining contingently and not eternally. *Prší* is typed as a possible-world proposition, so $[{}^0 Pr\check{s}\acute{\imath}_{wt}]$ constructs a truth-value, just like $[{}^0 F_{wt}\ {}^0 a]$ does. The difference between those two Compositions is that $[{}^0 Pr\check{s}\acute{\imath}_{wt}]$ skips one step of functional application; the step, namely, of applying F_{wt} to a. However,

$$\lambda w \lambda t[{}^0 Pr\check{s}\acute{\imath}_{wt}] \approx_{\eta} {}^0 Pr\check{s}\acute{\imath}$$

Redux and *contractum* are structurally distinct, have a different number of constituents, and also have two different unifiers – a Closure and a Trivialization, respectively. Yet they are η-equivalent, which renders them *procedurally isomorphic*, hence semantically indistinguishable: either of the two can be assigned to 'Prší' as its meaning.[28] It is obvious that the Closure reveals more by way of logical structure, but a principle is required that will select the Closure and not the Trivialization as the structurally superior analysis of 'Prší'. Such a principle is as yet not available to TIL.

17.4 Conclusion and Further Work

The on-going computational turn in modern formal semantics has made the need for structured and hyperintensionally individuated meanings urgent. That turn has in part contributed to reviving age-old and time-honoured questions such as the

[28] For procedural isomorphism, see Duží et al. (2010a, def. 2.3, p. 154).

following. If *P* is a *structured proposition* then *what unifies the particular (and type-theoretically disparate) constituent components of P into one unit*? Any viable answer must offer more than just a list of objects while steering clear of a regress into a never-ending ramification of mediating entities.

The answer I adumbrated above is to the effect that *predication* (and emphatically not *instantiation*) is what keeps an atomic *P* together and that the logic of predication is the *procedure of functional application*. Some propositions will also require the *procedure of functional abstraction*. So those two procedures are the central *propositional unifiers*. Crucially, neither is a stand-alone constituent of the propositions they unify, so regress is averted. Instead they are, as it were, the perimeters of the structured propositions they unify.

The project of propositional unity and unifiers sketched above is going to be worked out within Tichýs Transparent Intensional Logic. *Much* remains to be done before arriving at an actual theory of propositional unity within the confines of the paradigm of procedural semantics. First of all, it needs to be explained in much more philosophical and technical detail how a procedure unifies its subprocedures. I assume this explanation to take on the form of an intensional mereology of abstract objects. Another urgent issue is how much structure to make explicit and how much to blackbox relative to particular analytical tasks. This question bears on a standard issue like the principle(s) of individuation of linguistic meanings, see Jespersen (2010a). The TIL notion of procedural isomorphism builds upon Carnap's intensional isomorphism and Church's synonymous isomorphism, and the exact calibration(s) of procedural isomorphism is (are) very much a live issue within the TIL community.[29] Thus the topic of *propositional unity* is one where the notions of *hyperintensional individuation* and *logical structure* converge.

References

Almog, J. 1986. Naming without necessity. *Journal of Philosophy* 83: 210–242.
Bealer, G. 1993. A solution to Frege's puzzle. *Philosophical Perspectives* 7: 17–60.
Bealer, G. 1998. Propositions. *Mind* 107: 1–32.

[29] See Duží et al. (2010a, §2.2.1 'Concepts and synonymy') and Jespersen (2010b). TIL agrees with Russell's general idea that if *a*, *b* are distinct then a proposition containing *a* and a proposition containing *b* must be distinct propositions. In TIL lingo, if $a \neq_{\iota} b$ then ${}^{0}a \neq_{*1} {}^{0}b$ (and vice versa), hence $\lambda w \lambda t\ [{}^{0}F_{wt}\ {}^{0}a] \neq_{*1} \lambda w \lambda t\ [{}^{0}F_{wt}\ {}^{0}b]$. A theme underlying the discussion of propositional individuation is the Russell-Myhill *paradox of propositions*, as per Russell (1903, §500), that the number of propositions is at least the same as the number of sets of propositions. Both Klement (2001), (2002) and Deutsch (2008) make much of this violation of Cantor's theorem. Hopefully, TIL is unassailable by the paradox in virtue of its ramified type hierarchy. But the paradox will deserve full-length discussion within the framework of TIL when time comes to further motivate one or more degrees of procedural isomorphism finer than what Alternative (1) affords. (See Salmon (2010) for a recent clash between an advocate of (A0), Salmon, and an advocate of (A1), Kripke.)

Bolzano, B. 1837. *Wissenschaftslehre*. Sulzbach: von Seidel.
Church, A. 1993. A revised formulation of the logic of sense and denotation. Alternative (1). *Nous* 27: 141–157.
Cresswell, M.J. 1975. Hyperintensional logic. *Studia Logica* 34: 25–38.
Cresswell, M.J. 1985. *Structured meanings*. Cambridge: MIT Press.
Davidson, D. 2005. *Truth and predication*. Cambridge, MA/London: The Belknap Press of Harvard University Press.
Deutsch, H. 2008. Review of King (2007). *Notre Dame Philosophical Reviews*. Retrieved from http://ndpr.nd.edu/review.cfm?id=13165.
Duží, M., B. Jespersen, and P. Materna. 2010a. *Procedural semantics for hyperintensional logic. Foundations and applications of transparent intensional logic. Logic, epistemology, and the unity of science*, vol. 17. Dordrecht: Springer.
Duží, M., B. Jespersen, and P. Materna. 2010b. The *logos* of semantic structure. In *Philosophy of language and linguistics. Vol. I: The formal turn*, ed. P. Stalmaszczyk, 85–101. Frankfurt am Main: Ontos-Verlag.
Duží, M., and B. Jespersen. Forthcoming. Transparent quantification into hyperpropositional contexts *de re*. *Logique et Analyse*.
Fine, K. 1994. Compounds and aggregates. *Nous* 28: 137–158.
Frege, G. 1892. Über Begriff und Gegenstand. *Vierteljahrschrift für wissenschaftliche Philosophie* 16: 192–205.
Frege, G. 1923. Gedankengefüge. *Beiträge zur Philosophie des deutschen Idealismus* 3: 36–51.
García-Carpintero, M. 2010. Gaskin's ideal unity. *Dialectica* 64: 279–288.
Gaskin, R. 2008. *The unity of the proposition*. New York: Oxford University Press.
Hanks, P. 2009. Recent work on propositions. *Philosophy Compass* 4: 469–486.
Hindley, J.R., and J.P. Seldin. 1986. *Introduction to combinators and λ-calculus*. Cambridge: Cambridge University Press.
Jespersen, B. 2000. Singular propositions in two-stage theory. In *Topics in conceptual analysis and modeling*, ed. O. Majer, 196–218. Prague: Czech Academy of Sciences.
Jespersen, B. 2003. Why the tuple theory of structured propositions isn't a theory of structured propositions. *Philosophia* 31: 171–183.
Jespersen, B. 2005. Explicit intensionalization, anti-actualism, and how Smith's murderer might not have murdered Smith. *Dialectica* 59: 285–314.
Jespersen, B. 2008. Predication and extensionalization. *Journal of Philosophical Logic* 37: 479–499.
Jespersen, B. 2010a. How hyper are hyperpropositions? *Language and Linguistics Compass* 4: 96–106.
Jespersen, B. 2010b. Hyperintensions and procedural isomorphism: Alternative (½). In *The analytical way. Proceedings of the 6th European congress of analytic philosophy*, ed. T. Czarnecki, K. Kijania-Placek, O. Poller, and J. Woleński, 299–320. London: College Publications.
Kaplan, D. 1990. Dthat. In *Syntax and semantics*, vol. 9, ed. P. Cole. New York: Academic.
King, J.C. 2007. *The nature and structure of content*. New York: Oxford University Press.
Klement, K.C. 2001. Russell's paradox in appendix B of the *Principles of Mathematics*: Was Frege's response adequate? *History and Philosophy of Logic* 22: 13–28.
Klement, K.C. 2002. *Frege and the logic of sense and reference*. New York/London: Routledge.
Kripke, S. 1963. Semantical analysis modal logic I: Normal modal propositional calculi. *Zeitschrift für mathematische Logik und Grundlagen der Mathematik* 9: 67–96.
Moltmann, F. 2003. Propositional attitudes without propositions. *Synthese* 135: 77–118.
Peacock, H. 2011. Is there a problem about propositional unity? *Dialectica* 65: 393–418.
Pears, D. 1977. The relation between Wittgensttein's picture theory of propositions and Russell's theories of judgement. *Philosophical Review* 86: 177–196.
Pelham, J., and A. Urquhart. 1994. Russellian propositions. In *Logic, methodology and philosophy of science IX*, ed. D. Prawitz, B. Skyrms, and D. Westerståhl, 307–326. Burlington: Elsevier.

Peregrin, J. Forthcoming. There is no such thing as predication. *Conceptus*.
Russell, B. 1903. *Principles of mathematics*. New York: Norton.
Salmon, N. 2010. Lambda in sentences with designations: An ode to complex predication. *Journal of Philosophy* 107: 445–468.
Simons, P. 1987. *Parts*. Oxford: Clarendon.
Soames, S. 2010. *What is meaning?* Princeton: Princeton University Press.
Stevens, G. 2008. Russell and the unity of the proposition. *Philosophy Compass* 3: 491–506.
Thomason, R. 1980. A model theory for propositional attitudes. *Linguistics and Philosophy* 4: 47–70.
Tichý, P. 1972. Plantinga on essence: A few questions. *The Philosophical Review* 81: 82–93. Reprinted in (2004, pp. 177–188).
Tichý, P. 1978. Questions, answers, and logic. *American Philosophical Quarterly* 15: 275–284. Reprinted in (2004, pp. 295–304).
Tichý, P. 1986. Constructions. *Philosophy of Science* 53: 514–534. Reprinted in (2004, pp. 601–621).
Tichý, P. 1988. *The foundations of Frege's logic*. Berlin: DeGruyter.
Tichý, P. 1994. The analysis of natural language. *From the Logical Point of View* 3: 42–80. Reprinted in (2004, pp. 803–841).
Tichý, P. 2004. In *Collected papers in logic and philosophy*, ed. V. Svoboda, B. Jespersen, and C. Cheyne. Prague/Dunedin: Filosofia, Czech Academy of Sciences/University of Otago Press.
Vallicella, W.F. 2010. Gaskin on the unity of the proposition. *Dialectica* 64: 265–277.
Westerhoff, J. 2004. A taxonomy of composition operations. *Logique et Analyse* 185–88: 375–393.

Chapter 18
Best-Path Theorem Proving: Compiling Derivations

Martin Frické

18.1 Introduction

Increasingly, computers answer our questions in mathematics. However, mere answers are not enough in the general case. We need the actual justification, and we need explanatory insight. We want truths but we also want reasons. We want to 'Trust but verify'. And so we should. As Mill tells us in *On Liberty*, to do otherwise would make mathematics a superstition

> [were it not so]... the true opinion abides in the mind, but abides as a prejudice, a belief independent of, and proof against, argument – this is not the way in which truth ought to be held by a rational being. This is not knowing the truth. Truth, thus held, is but one superstition the more, accidentally clinging to the words which enunciate a truth (Mill 1869).

In addition to justification, we would like some understanding; that is, some explanation or insight as to what is going on. As Wolfram writes

> At some level the purpose of a proof is to establish that something is true. But in the practice of modern mathematics proofs have taken on a broader role; indeed they have become the primary framework for the vast majority of mathematical thinking and discourse. And from this perspective [some computer proofs and some proofs generated by automatic theorem proving] are quite unsatisfactory. For while they make it easy at a formal level to check that certain statements are true, they do little at a more conceptual level to illuminate why this might be so. ... [Some general proofs] can be arbitrarily long, and can be quite devoid of what might be considered meaningful structure (Wolfram 2002, p. 1156).

Another example is provided by some of Zalta's, work on computational metaphysics (Zalta 2009; Zalta et al. 2011). Some years ago Edward Zalta published some engaging results and theorems on the ontological argument, and similar.

M. Frické (✉)
School of Information Resources and Library Science, University of Arizona, Tucson, AZ, USA
e-mail: mfricke@u.arizona.edu

J. Maclaurin (ed.), *Rationis Defensor: Essays in Honour of Colin Cheyne*, Studies in History and Philosophy of Science 28, DOI 10.1007/978-94-007-3983-3_18,

Recently these theorems were given to an automated theorem prover to proof check, and the theorem prover came up with new proofs that showed that some of the original premises and steps were redundant – truly an interesting result. However, the course of this discovery was not as quick or a smooth as it might have been. The researchers did not understand the raw output of the theorem prover in a way that the output could be seen as derivations that could be compared to the original derivations. They had to work by hand to convert the output of the theorem prover into perspicuous derivations that would be useful to them. They should not have to do this. The theorem prover should do it, or there should be plug-in filters that do the desired transformations.

So here is a general problem. There can be the need for computers to transform their internal reasoning and their internal processes of proof to output forms that are suited to the particular cognitive and epistemological needs of humans. Certainly in some empirical areas, for example rule-based expert systems, computer programs do provide a full rationale, but this is not so common with logic, mathematics, and theorem proving.

This paper is going to focus on real reasoning in logic. Often computers use algorithms and data structures that are different to those natural to humans. A computer may operate on full graph theoretic structures, such as networks and trees, whereas humans are more comfortable with linear reasoning or paths. In cases like these, it is not so much that the computer has to explain what it is doing, rather it should produce for us separately proper stand alone justification that we can follow and be convinced by.

There is an example subject area central to philosophy and logic. First order predicate calculus (FOPC) is, or may be, a universal logic, perhaps even *the* deductive logic for mathematics and all formal reasoning (Lindstrom 1969; Manzano 1996; Quine 1970). And there are different styles of theorem-provers for FOPC (Robinson and Voronkov 2001). These work in different ways. Some use linear methods and heuristics. These might have a fact queue and a problem or interest queue and cycle back and forth between the two. Such methods typically produce proofs or output that is accessible to humans. Other more common automatic theorem-provers use non-linear methods for their data structures and algorithms. For example, many employ resolution, which rests on unification, Skolem functions, and a reduced connective set (Robinson and Voronkov 2001). Yet others use tableaux (D'Agostino et al. 1999). All of these latter are of themselves somewhat opaque to humans.

18.2 Theorem Proving in FOPC

In the case of FOPC, what is it that humans understand and can be convinced by? What is it that human logicians, or mathematicians, would like?

Humans reason with a full connective set, with a full 'natural deduction' rule set for that full connective set, or something similar, and they reason linearly.

The problem, then, is this. Take any ordinary rule set used by students, professors, and other humans, to write and justify their FOPC derivations; then, produce a mechanical or computational deriver or automated theorem prover for FOPC which outputs theorems in the favored rule set, any rule set. The output theorems should have a quality to them; they should be similar to or better than those produced, say, by a Professor of Logic.

First thoughts about this in terms of linear derivations are intimidating. With most rule sets: any statement that has a derivation has infinitely many different derivations; for any length of derivation there are infinitely many different derivations of that length; for any line of a derivation there are infinitely many possible next lines; for any line in the body of a derivation there are infinitely many immediate predecessor lines; if the problem is conceived of as an artificial intelligence search problem, then there is no proper evaluation function to indicate the distance of the current state from the goal state; there is not even any general way of knowing whether the goal state can be reached; theoremhood is not decidable; and so on.... There seems no obvious way of controlling the search and terminating gracefully. Certainly some counter moves can be made here. For example, possibilities can be reduced if intelligent use is made of subformula relations between the conclusion and premises, as is done, for instance with 'normal' proofs in natural deduction and intercalation calculi (Sieg and Byrnes 1998) or with 'analytic' tableaux (Smullyan 1968). Nevertheless, this seems to be a problem deserving of respect.

18.3 Tableau Theorem Proving

Since the publication of Raymond Smullyan's book in 1968, tableaux have increasingly become more common (D'Agostino et al. 1999; Smullyan 1968; Fitting 1996, 1998; Howson 1997; Jeffrey 1967). Tableaux are an interesting case to study. Mostly they are decompositional (which lends itself to designing algorithms that terminate). Decompositionality is attractive. For any particular formula, generally there is only one rule that applies (which works on the main connective or negation of the main connective). And that rule breaks the formula into shorter subformulas (which eventually themselves get decomposed if they contain suitable connectives). In a stroke many of the earlier worries are done away with. There are not infinitely many possible next steps. Some of the tableau rules are not decompositional (Fitting 1998): for example, the identity rules are not.[1] Universal instantiation is also a special case; it decomposes to an instance, but it leaves the universal formula available for a further instantiation. We know from meta-theorems about tableaux that instantiation to every closed term in every relevant branch is enough (Howson 1997; Jeffrey 1967; Bergmann et al. 1998). That will eventually close a branch and,

[1] We can omit discussion of identity from this paper; it is something of a special case and none of the techniques explained here have any distinctive insights concerning it.

by closing branches, will close a tree—*if the tree can be closed.* However, it is certainly possible for a universal instantiation to extend a branch yielding an existential formula which itself will decompose *to yield a new constant (closed term) in the relevant branch* which in turn requires that the original universal be instantiated to that new term i.e. we have a growth generator. Sometimes these feedback generators can be by-passed (a branch can close without appeal to them). Sometimes a feedback loop can be small enough to be detected and a sound conclusion made that a containing branch cannot close. But, in the general case, the loops can be bigger than any particular finite size we choose to look at. We know something like this has to happen. FOPC is undecidable, so there cannot be a mechanical decision process. Even so, tableaux are attractive.

There are still tricky decisions that range across both tableaux and the other rules sets. For example, in both, there is usually a choice of what to do, of which formula or formulas to work on, to use for extending or rewriting. In an ordinary linear proof, we always have choices (for example, of whether to conjoin line 1 with itself, or to conjoin line 1 with line 2); similarly in tableaux (almost always there will be some stages in the growth of the tableau where there is a choice among which formula to use for the next extension). There is bounded indeterminism (Fitting 1998). And different choices may well produce structurally different tableaux. For example, if you always 'split' a tree, first, rather than 'straight extend' it, you will get broad or fat trees. If, so to speak, the 'wrong' choice is made during these extensions, the eventually tableau proof is not a mistaken proof, but it may have redundant steps or inelegancies; there could be the need for optimization if a particular style of tableau is the target.

We will assume algorithms are available for closing tableaux (D'Agostino et al. 1999).[2]

18.4 Transforming Tableaux to Linear Proofs: A Construction

As we stated, we would like linear proofs. Here is a suggestion: *transform the closed tableaux into linear proofs.*

Linear proofs here include three sub-genres: Hilbert-Frege axiomatic proofs (consisting of axioms and rules of inference), Natural Deduction proofs (which have the adding and dropping of assumptions, and usually one rule for Introducing a particular connective and another rule for Eliminating that connective), and Sequent calculi (where all the sequents are unconditional truths) (Fitting 1996).

Transforming tableaux to linear proofs is not without precedent. There is the general research area of proof transformation (a good example here is 'cut-elimination' (Smullyan 1968; Fitting 1996; Gallier 1986). And also, within research

[2] Space precludes further detailed description of those in this paper, and the question of automatically closing tableaux is not the present target.

on tableaux, there has long been the awareness of the close relationship between tableaux and sequent proofs (Smullyan 1968; Beth 1969; Gentzen 1935; Hähnle 2001; Letz 1999). But here the transformation is being done in perhaps a novel way and to novel purpose. The focus is with actual running computer programs and algorithms, which will serve as automatic theorem provers and which will generate the 'best' linear proofs for any desired linear rule set.

We have not yet decided on specific tableau rules, nor on specific linear rules. Even so, this transformation can be done, from the tips of the branches, i.e. from the leaves, to the root. Here is one way to do it.

For standard proof tableaux, the tableaux are, in a graph theoretic sense, node-labeled trees (connected, acyclic graphs with one preferred node, the 'root'). Each node is labeled with a formula: there is a node for every formula. Then tableaux can be signed or unsigned (Smullyan 1968). A signed tableau has its component formulas tagged as being 'true' or as being 'false' (and an unsigned one does not). With a signed tableau, the aim is to extend all the root formulas and component formulas down a branch until there are enough atomic formulas signed true, or signed false, to guarantee that all the formulas in that branch have truth values as signed (this might not be possible if the root formulas are not simultaneously satisfiable). Unsigned tableaux do not have the signing, and then the assumption, or aim, is to get every formula to be true. Signed tableau can be converted into unsigned tableau merely by substituting the negation of a formula for any formula signed false (and unsigned can be converted into signed by signing false any formula with negation as its main connective and removing the negation).

For an argument, or tentative theorem, say

$P_1, P_2, \ldots P_n \therefore C$

where the P_i are premises (or assumptions) and the C is the conclusion or conjectured theorem. A tableau is constructed with the Ps as separate nodes signed true and the C, a node signed false. For suitable tableau rules, this will find the counter-example, if there is one.[3]

Here is a simple example, the argument

M&N ∴ N&M

can have a tableau that proceeds (Figs. 18.1, 18.2, 18.3, 18.4).

This is an unsigned tableau (and line 2 is the negated conclusion, showing the desire to make the conclusion false). This tableau has six nodes in it (two for the root formulas, then application of the rules, in two steps, generates four more).

By using an additional, or alternative, labeling system this tableau can be replaced by, or transformed into, a different equivalent one, which, in this particular case, will have only four nodes in it. The motivation for the transformation is twofold. Branches need to be closed, that is one aim. We would like to carry branch

[3] And the technique is easily adapted if the focus of interest is consistency or simultaneous satisfiability etc.

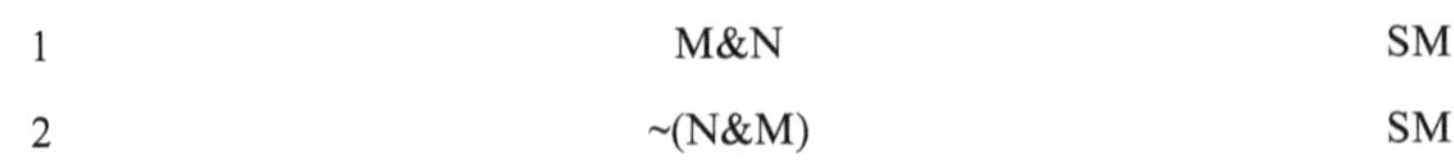

Fig. 18.1 Stage 1 of the construction of a tableau for M&N ∴ N&M

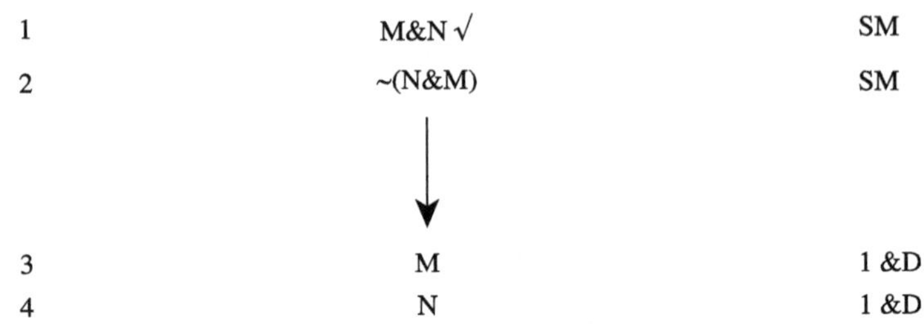

Fig. 18.2 Stage 2 of the construction of a tableau for M&N ∴ N&M

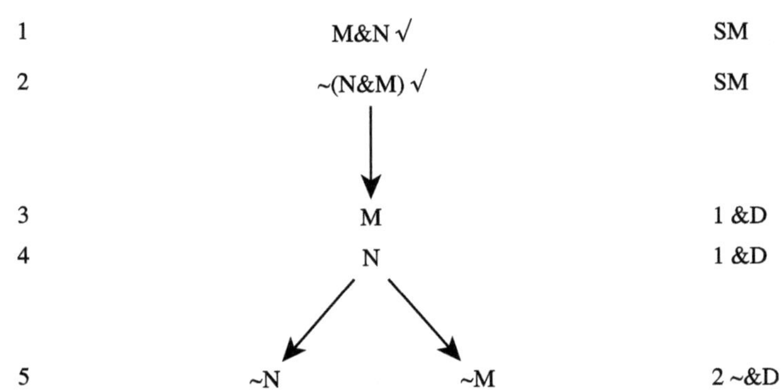

Fig. 18.3 Stage 3 of the construction of a tableau for M&N ∴ N&M

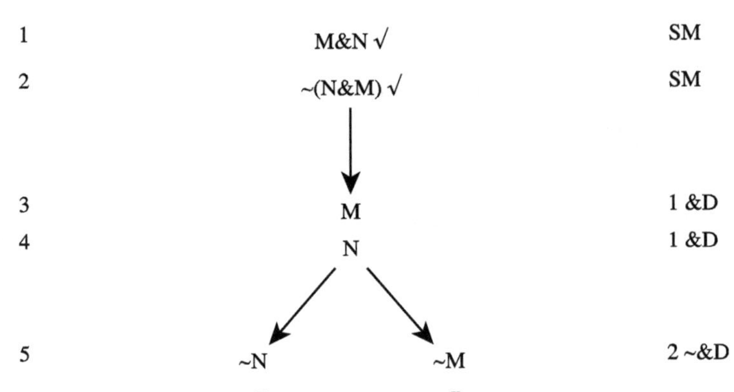

Fig. 18.4 Stage 4 of the construction of a tableau for M&N ∴ N&M

information with us during the extension steps so that we can close a branch merely by looking at the label on the relevant node (instead of having to search the branch). Then, later and independently, the label will be used to construct the linear proof. The new labeling system has labels, which are sets or lists of formulas; and the formulas themselves, in these labels, are signed.

There is an additional novelty in the transformed tableau. In it, only leaf nodes can be extended. (A leaf node is a node without children.) To clarify this, using the above example. In the transition from Stage 2 to Stage 3 the formula ~(N&M) of line 2 is extended to produce the two formulas of line 5. Line 2, or the node corresponding to it, is *not* a leaf node; it has line 3, or M, or that node, as its child. This type of extension of 'internal' non-leaf nodes, to all branches, containing them is standard in tableau techniques. But here only leaf nodes are extended. This makes two kinds of differences. The first is an algorithmic one. With standard tableau techniques, there is a bounded indeterminism in the choice of which formula to extend, and then, with that choice made, a determination has to be made of all (open) branches that contain the formula, as the extension goes to all of the branches containing it. But any leaf is in only one branch, a leaf determines its own branch uniquely, so any extension is easier. The indeterminism then becomes an indeterminism over which leaf to extend (if there are several available leaves). It turns out that this is irrelevant, so the leaves can be processed breadth or depth first. The indeterminism surfaces in what is done with a particular leaf, but more of that later. The second kind of differences arises with how this 'leaf-extension-only' promotes the desired transformation to the linear derivation.

The new labeling is as follows. The (single) root node is labeled

$P_1, P_2, \ldots P_n \therefore C$

the Ps are the premises (or signed true formulas) and the C is the conclusion (or a signed false formula).[4] So the label is the set of signed formulas

$\{True(P_1), True(P_2), \ldots True(P_n), False(C)\}$

but, for heuristic purposes, it is clearer to introduce the therefore symbol '∴' and to think of the label as

$P_1, P_2, \ldots P_n \therefore C$

The labeling for each of the lower nodes is obtained by transformation of the relevant parent labels. Most of the time the individual formula used for the tree extension step itself will be dropped from the set of formulas in the label for the extension, and the products for the extension will be incorporated (the exception is extension from the universal quantifier where both the formula and its product will be in the label for any children). That the original extended formula is dropped is just a book-keeping implementation issue. It is standard with ordinary tableau

[4] Although the premises are being depicted here as something of a list, conceptually they are a set. (There is no need to complicate this with 'structural' rules.)

techniques to mark a formula as 'dead' when it has been used. This is merely a non-essential counterpart to that.

Schematically, the label for a 'straight ahead' extension might look something like

P_1 & P_2, P_3, ... P_n ∴ C

Extended to

P_1, P_2, P_3, ... P_n ∴ C (*note here that the P_1&P_2 is dropped from the P_i*)[5]

And the labels for 'splitting' extension might look something like

P_1 v P_2, P_3, ... P_n ∴ C

Extended to

P_1, P_3, ... P_n ∴ C (*note here that the PjvPk is dropped from the Pi*)
P_2, P_3, ... P_n ∴ C (*note here that the PjvPk is dropped from the Pi*)

Some extensions steps can similarly transform C, the signed false formula. Here is a 'splitting' example

P_1, P_2, ... P_n ∴ (C_1 & C_2)

Extended to

P_1, P_2, ... P_n ∴ C_1
P_1, P_2, ... P_n ∴ C_2

In an ordinary tableau, a branch can, and should, be closed when it contains a formula that tries to be both true and false. With this new labeling scheme it is not so much branches that are closed, but instead individual leaf nodes are closed. That can occur in two ways: a formula can be signed true in one place in a label and signed false elsewhere, or a formula and its negation can both be signed true. So a label for a 'closure leaf' has form, either

P_1, P_2, P_i ... P_n ∴ P_i (*P_i signed both true and false.*)

or

P_1, P_2, P_i, ~P_i ... P_n ∴ < anything > (*P_i and ~P_i both signed true*)

In the example

M&N ∴ N&M

is the label for the root node

[5] Here the notation 'P_1&P_2' is being used as a pattern to show a formula with & as its main connective and P_1 and P_2 as subformulas. The premises are a set, but writing the formula that is being processed first helps draw attention to it.

Fig. 18.5 Newly constructed tableau for M&N ∴ N&M

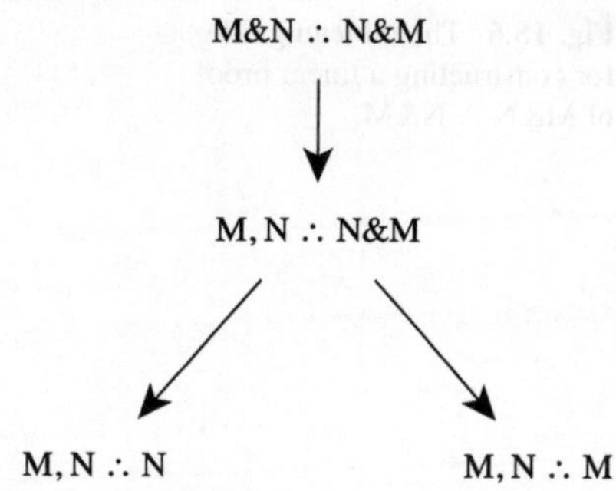

M, N ∴ N&M

is the label for the single node for lines 3 and 4.

M, N ∴ N   (* N is signed true as a premise and false as a conclusion *)

and

M, N ∴ M   (* M signed true and false*)

are the labels for the two nodes of line 5.

And here is the newly constructed tableau depicted in full (Fig. 18.5).

Individual labels can be large and complex. A particular label could easily have 50 'premise' formulas left of the therefore symbol. This scheme is not designed for humans to write out with a pencil on a sheet of paper. This is part of a computer program. And a computer can very easily copy or share a label, or parts of a label.

A remark is perhaps in order here on the relationship between this labeling and ordinary tableaux. An ordinary tableau is a set of branches and, in turn, a branch is a set of formulas. Here a tableau is a set of branches, each branch has a unique leaf, and, in turn, a leaf is a set of (signed) formulas.

Another observation should be made about the process of extension. In an ordinary tableau, each node usually has a unique extension (exceptions being identity, Universal Instantiation, etc.). And there is indeterminism as to which node in a branch to extend. With the new labeling system, usually a leaf node can be extended in several different ways. For example, the leaf node

Fa&Gb, (∀x)(Fx ⊃ Hx) ∴ Gb&Ha

can be extended via the main connective *and* or the main connective *the universal quantifier* from among the 'premises' or via the main connective *and* of the 'conclusion'. The indeterminism of extension re-asserts itself at the leaf extension level.

The new labels are intended also to play a role depicting some Gentzen sequents (Gentzen 1935). A Gentzen sequent has the form

A1, A2, . . . An ∴ S1, S2, . . . Sn

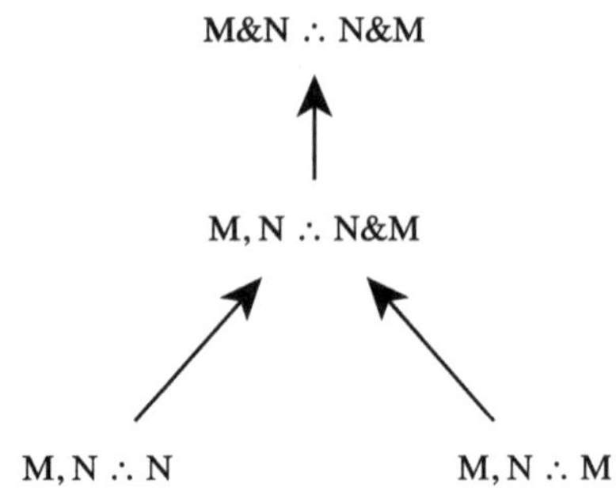

Fig. 18.6 The tableau guide for constructing a linear proof of M&N ∴ N&M

1	M	Ass
2	N	Ass
3	N	2 R

Fig. 18.7 Linear proof of the leaf M, N ∴ N

Where the A_i are 'antecedents' and Si 'succedents'. Without loss of generality for what we are going to do, we can assume that there is no more than one succedent. There can be zero or more antecedents.

The Gentzen sequents for the example are, going down the tree,

M&N ∴ N&M
M, N ∴ N&M
M, N ∴ N and M, N ∴ M

Then, as a result or provable theorem,

(a) there is an immediate (and trivial) linear derivation of the sequent corresponding to any closure leaf (either a premise is identical to the conclusion, or there is a Reductio proof from two premises which contradict), and
(b) if there are linear derivations of the sequent labels showing the results of the children of a single tableau decomposition step, then, from these, a linear derivation can be constructed of the sequent corresponding to the parent node used in that decomposition.

Thus, constructed derivations can be driven in the reverse direction up the tree, from the leaves to the root, and this provides a derivation of the root sequent (which is what was sought).

Conceptually, the linear derivations are assembled as follows (Figs. 18.6, 18.7, 18.8):

In the example, the two leaf nodes have linear proofs

Then, these two linear proofs can be merged together to form a linear proof of

M, N ∴ N&M

1	M	Ass
2	N	Ass
3	M	1 R

Fig. 18.8 Linear proof of the leaf M, N ∴ M

1	M	Ass
2	N	Ass
3	N	2 R
4	M	1 R
5	N&M	3,4 &I

Fig. 18.9 Merging the leaf linear proofs to obtain a linear proof of M, N ∴ N&M

1	M&N	Ass
2	N	1 &E
3	M	1 &E
4	N&M	2,3 &I

Fig. 18.10 Converting the linear derivation of M, N ∴ N&M to one for the root M&N ∴ N&M

the sequent for the node for lines 3 and 4 (Fig. 18.9).

Finally a linear derivation for the sequent M, N ∴ N&M can be converted into a linear derivation for M&N ∴ N&M, which is the sequent corresponding to the root node (Fig. 18.10).

This is a very simple example. But the technique has been proved to be appropriate and correct for the full rule set including quantifiers and sub-proofs, and it has been implemented and used over many years (Copeland and Murdoch 1991; Frické 1989a, b) with many different tableau rules and many different linear rules (you can see it at http://softoption.us in use on a variety of rule sets).

The technique is not tied to any particular Gentzen sequent calculus or its associated rule set. But the framework of thinking in terms of sequents is convenient for identifying assumptions and the nature of the proof steps.

There is some benefit to comparing this technique with ordinary sequent calculus proofs as devised in (Gentzen 1935) and illustrated, for example, in (Fitting 1996, p. 94). A sequent calculus proof or derivation is tree of sequents in which the leaves are axioms, the root is the theorem, and the derivation proceeds from the leaves to the root. It is also true that to an extent Gentzen used sequent trees as an intermediary in proof transformations from one system to another. However, (a) the present labeling scheme rests on semantics not syntax, i.e. whether the formulas can have truth values as signed, (b) sequent tree proofs can extend from internal nodes, the present labeling extends only from leaves, (c) the leaf closing via premise contradiction could not normally be done in a Gentzen sequent proof because that is not an axiom (although it could be added as an axiom), (d) the present construction is not, and is not intended to be, a proof in itself, instead it is a stepping stone to the generation of proofs (be they Hilbert-Frege proofs, Natural Deduction Proofs, or Sequent proofs). In sum, standard Gentzen sequent proofs are a historical antecedent of the present labeling transformation, but what is being undertaken here is not identical to producing sequent proofs. What is being offered here is an algorithm.

Quantifiers add a level of complexity. In the running example programs, Skolem functions are used for Existential Quantifiers, a standard technique (Fitting 1996; Gallier 1986; Hähnle 2001). And Unification, and/or caching of prior instantiations, are used for Universal Quantifiers (Fitting 1996; Gallier 1986; Hähnle 2001). So, a Universal Instantiation-like tableau step is the following

$(\forall x) < \text{scope } x>, P_2, \ldots P_n \therefore C$

Extended to

$(\forall x) < \text{scope } x> ,<\text{scope } t/x>, P_2, \ldots P_n \therefore C$ (*note here that the $(\forall x) < \text{scope } x >$ is retained in the P_i*)

The Universal formula is retained in its entirety, and a legitimate substitution instance of its scope is added. There is an implementation question that arises here. When Universal Instantiation is done in a standard tableau, when trying to close that tableau,

> there is a need to instantiate the Universal formula at least once
> there is never any need to instantiate it twice to the same closed term,
> and there is never any need to instantiate to anything other than the closed terms that occur in its branch.

A common difficulty is that, for a computer program, it is often unclear, at the time of the desired instantiation, which instance is required. A response is to do the instantiation 'lazily', that is, to instantiate initially to a free variable (Fitting 1996), and then to allow Unification to indicate later what that instantiation should be. So for example, a label/node like this

$(\forall x)Fx, {\sim}Fc \therefore C$

might be instantiated to

$(\forall x)Fx, Fx, {\sim}Fc \therefore C$

and then the leaf closing algorithm will detect, via Unification, that were the free variable x to be valued as the constant term c, the leaf can be closed. So, the earlier Universal instantiation needs to be to the constant c. There is computer work in the background here. At the end of the day, though, a Universal Instantiation extension is to one specific term, and the explanation offered in this paper will proceed on that basis.

Little has been said about either the tableau rules sets or the linear rules sets. The fact is: the technique is general; providing the tableau and linear sets are complete, a minimal adequacy requirement, the technique can be used. Actual running programs that have been written usually use 20 tableau extension rules: one each for {&, v, ⊃, ≡, ∀, ∃} and {~~, ~&, ~v, ~⊃, ~≡, ~∀, ~∃} signed true (i.e. from among the premises) and similarly, with some optimizations, for those types of formulas signed false (i.e. as the conclusion). Many of these rules are very similar, in terms of what they do, so there is common algorithmic code, but they are tagged individually to assist with the later generation of the linear proofs. Several different examples of linear rule sets have been used (often with slight tuning of the tableaux rules due to different requirements with, for example, Universal Generalization).

This construction will yield linear proofs, even though the underlying theorem prover is a tableau prover. And, actually, the same tableau data structure can generate different linear proofs for the different linear rule sets— essentially a filter can be applied to produce the desired output.

18.5 Improving the Linear Proofs

What has been done thus far will produce linear proofs, but not necessarily 'good' proofs.

Certain improvements can be made blind, so to speak, for example, redundant and unused lines can be removed. For example, a raw proof of

M&N ∴ N

might come out with the mildly inelegant derivation (Fig. 18.11).

What has happened here is that the tableau extension step takes by rote both conjuncts when extending a conjunction (it takes both M and N when extending M&N), and the conversion to linear proof retains these. Apparently there are two ways around this: to pay attention to goals when extending tableau (e.g. to extend only to N in this case) or to clean up afterwards. Implemented systems do the latter. They do a dependency analysis, after the case. Line 3, the conclusion line, depends on line 1. Nothing depends on line 2, so line 2 will be discarded to yield (Fig. 18.12).

In addition to dependency analysis, there are deeper, more structural issues, to consider.

There are some other fairly obvious optimization steps that can be made. For example, if the theorem prover is doing a Reductio proof and it has available a negated compound formula, for instance ~(M&N), it will be alert to the possibility that this formula is the Negative Horn of the contradiction and thus

1	M&N	Ass
2	M	1 &E
3	N	1 &E

Fig. 18.11 An inelegant derivation of M&N ∴ N

1	M&N	Ass
2	N	1 &E

Fig. 18.12 Discarding line 2 to improve the derivation of M&N ∴ N

consider whether (M&N) can be proved as the Positive Horn (as opposed to, say, breaking ~(M&N) apart eventually to get a contradiction in terms of literals).

So-called β-extensions, or splitting extensions (Smullyan 1968; Fitting 1996), also merit some special attention. For example, (A v B) among the premises can be split into two branches, one starting with A and the other with B. An output derivation corresponding to this step might well be Modus Ponens (if ~A can be proved on its own), Modus Tollens (if ~B can be proved on its own), or Or-Elimination (if neither ~A nor ~B is available as a standalone proof). Modus Ponens is common, easy, and quick; Modus Tollens is less common, but also easy and quick. Genuine Or-Elimination is somewhat involved. So, β-extension is more attractive if there is a known closing lemma to one of the branches. The tableau algorithm tests for this. The technique here is related to 'cut' or the use of lemmas (Smullyan 1968; Fitting 1996; Gallier 1986). The lemmas are not lemmas from outside the proof under consideration, using new predicates, constants or formulas. They are lemmas using sub-formulas of what is there. Small trees, produced on the side, are spliced in to close some of the branches. So one example heuristic here is: if you are considering extending from $A \supset B$, do it if you have A or if you have ~B, otherwise just wait on it – you might not need to do it, and other possibilities may emerge later. (Remember the parable. The King was going to execute a man. The man said 'If you do not execute me now, I will teach your horse to whistle within a year'. (Who knows what will happen in a year? The King may die. The horse may die. The horse may learn to whistle.))

Any proof can be done in different ways. Even with tableaux, there are usually different tableaux that can be built from the same root (and these different tableaux will usually produce different linear proofs). A certain amount of massaging can take place on the rule sets and good rewrites, using 'laws of thought', can help (for example, a rewrite, or replacement, de Morgan law can cut 12 lines out of many proofs).

Of course, for most roots, once it is known that a root node leads to a closed tableau, it is easy to generate most all the closed tableaux the root can have, and thus most all linear proofs, available within the constraints of the construction. Then the British Museum algorithm (Black 2005), exhaustive search, would produce the 'best' proof (however 'best' was construed). This approach would run into complexity issues—it overwhelms computing resources for large cases. (The qualifications. 'most roots', arise because some roots can have infinitely many closed tableau, for example, when there is a loop generator in there or when there is pointless instantiation of a Universal Quantifier.) A researcher always aspires to produce a short cut to a British Museum algorithm. Can this be done here?

There is an answer, perhaps a surprising one. The notion of best proof has an empirical component. All possible theorems, or all possible proofs, distribute evenly, in some sense or other, among the possible uses of inference steps. As examples, there are as many Modus Tollenses as Modus Ponenses, there are as many inferences from A, B to A&B as there are inferences from A, B to B&A, and there are as many inferences from A to AvB as there are inferences from A to BvA. However, with the proofs of interest to humans this uniformity of distribution does not hold. Humans favor Modus Ponens over Modus Tollens; humans do not like Reductio; and there are many other preferences. The author's research has taken many sample questions and proofs from logic texts, scholastic tests, and similar sources, and analyzed the best proofs for them ('best' here usually meaning shortest). This empirical research gives rise to some suggestions on heuristic, which will be given in Appendix A. Whether this particular empirical data is respected is secondary to the general point: humans are attuned to certain forms inference, empirical research can tell us what those are, and that empirical research can educate as to how tableau theorem provers, and their symbiotic liners counterparts, should run.

18.6 Conclusion

Tableau theorem provers, coupled with transformations to linear proofs and empirically sourced heuristic, can provide transparent and accessible theorem proving.

Appendix A: The Extension Heuristic

This is a heuristic (and algorithm) for the construction of tableaux that seems to do well, on average, on the linear proofs that it will generate, across a wide range of proofs. Details can differ, depending on what rules are available. But this is the core.

The overriding control principle looks through everything available, as extendable formulas in a leaf sequent label, in a series of 'sweeps'.[6]

[6] The extension steps correspond to one or more lines in a linear proof.

If any sweep produces a step or extension that can be made, that step is made and the whole series of sweeps is started again on the resulting formulas.

Sweep 1:

any formula signed true that has **implication** as its main connective, provided that its 'if clause' can be proved (this is tested for, recursively). This is Modus Ponens.

Sweep 2:

any formula signed true that is a **double negation**, an **and**, or an **equivalence**. These steps are Double Negation Elimination, And Elimination, and Equivalence Elimination.

Sweep 3:

any formula signed false that is an **implication**, a **negation**, or an **equivalence**. These steps are Conditional Proof, Reductio, and Equivalence Introduction

Sweep 4:

any formula signed true that is an **or**. This step is **Or Elimination**.

Sweep 5:

any formula signed true that has the **existential quantifier** as its main connective, and which can be instantiated **without change of bound variable**. This step is **Existential Instantiation (without change of variable).**

Sweep 6:

any formula signed true that has the **existential quantifier** as its main connective. This step is **Existential Instantiation** (with change of variable—sweep 5 picks up the ones that do not need it).

Sweep 7:

any formula signed false that has the **universal quantifier** as its main connective, and which can be generalized to **without change of bound variable**. This step is **Universal Generalization (without change of variable).**

Sweep 8:

any formula signed false that has the **universal quantifier** as its main connective. This step is **Universal Generalization** (with change of variable—sweep 7 picks up the ones that do not need it).

Sweep 9:

any formula signed false which is a **double negation**, an **and**, or an **or**. These steps are **Double Negation Introduction**, **And Introduction**, and **Or Introduction**.

Sweep 10:

any formula signed true that has the **universal quantifier** as its main connective, and which can be instantiated **without change of bound variable**. This step is **Universal Instantiation (without change of variable)**.

Sweep 11:

any formula signed false that has the **existential quantifier** as its main connective. This step is **Existential Generalization**.

Sweep 12:

any formula signed true that has the **universal quantifier** as its main connective. This step is **Universal Instantiation**.

Sweep 13:

any formula signed true that is the **negation of a universally quantified formula** or the **negation of an existentially quantified formula**. This step is **Negation of Quantifier** (usually a rewrite).

Sweep 14:

any formula signed true that is an **implication**, the **negation of an and, the negation of an or, the negation of an equivalence** or **the negation of a implication**. These steps are usually complex lemmas.

Sweep 15:

any formula signed true EXCEPT atomic and the negation of atomic formulas. (The default catch-all.)

Sweep 16:

any formula signed false EXCEPT atomic and the negation of atomic formulas. (The default catch-all.)

Appendix B: A Small Example

Here is one running implementation being asked to permute existential quantifiers (Fig. 18.13).

And this is the output when using Introduction-Elimination rules (Fig. 18.14).

And here is the same theorem/proof provided using Copi-style linear rules (Fig. 18.15).

These are quite different in that Copi brings in equivalence by definition, or replacement, from conjoined conjunctions, and it has its own version of Existential Instantiation (which it calls 'Elimination'). The Intro-Elim rule set makes much more extensive use of sub-proofs.

1	?	?<<
2	(∃x)(∃y)Fxy≡(∃y)(∃x)Fxy	?

Fig. 18.13 Starting to automatically derive (∃x)(∃y)Fxy ⊃ (∃y)(∃x)Fxy

1	(∃x)(∃y)Fxy	Ass Auto
2	(∃y)Fay	Ass Auto
3	Fab	Ass Auto
4	(∃x)Fxb	3 ∃I Auto
5	(∃y)(∃x)Fxy	4 ∃I Auto
6	(∃y)(∃x)Fxy	2,5 ∃E Auto
7	(∃y)(∃x)Fxy	1,6 ∃E Auto
8	(∃y)(∃x)Fxy	Ass Auto
9	(∃x)Fxa	Ass Auto
10	Fba	Ass Auto
11	(∃y)Fby	10 ∃I Auto
12	(∃x)(∃y)Fxy	11 ∃I Auto
13	(∃x)(∃y)Fxy	9,12 ∃E Auto
14	(∃y)(∃x)Fxy	8,13 ∃E Auto
15	(∃x)(∃y)Fxy≡(∃y)(∃x)Fxy	7,14 ≡I Auto

Fig. 18.14 An automatic derivation of (∃x)(∃y)Fxy ⊃ (∃y)(∃x)Fxy using Introduction-Elimination Rules

This is a small example. The point of it is not that one rule set is better than another. Rather it is that the User can choose whatever they would like, and suitable software can provide it for them.

The reader of this paper can try examples of their own derivations at
http://softoption.us/content/node/165 for Intro-Elim
and
http://softoption.us/content/node/286 for Copi

1	(∃x)(∃y)Fxy	AP Auto
2	(∃y)Fxy	1 ∃E Auto
3	Fxy	2 ∃E Auto
4	(∃x)Fxy	3 ∃I Auto
5	(∃y)(∃x)Fxy	4 ∃I Auto
6	(∃x)(∃y)Fxy⊃(∃y)(∃x)Fxy	5 CP Auto
7	(∃y)(∃x)Fxy	AP Auto
8	(∃x)Fxy	1 ∃E Auto
9	Fxy	2 ∃E Auto
10	(∃y)Fxy	3 ∃I Auto
11	(∃x)(∃y)Fxy	4 ∃I Auto
12	(∃y)(∃x)Fxy⊃(∃x)(∃y)Fxy	11 CP Auto
13	((∃x)(∃y)Fxy⊃(∃y)(∃x)Fxy).((∃y)(∃x)Fxy⊃(∃x)(∃y)Fxy)	6,12 Conj Auto
14	(∃x)(∃y)Fxy≡(∃y)(∃x)Fxy	13 Equiv Auto

Fig. 18.15 The same derivation using Copi style rules

References

Bergmann, M., J. Moor, and J. Nelson. 1998. *The logic book*, 4th ed. New York: McGraw-Hill.

Beth, E.W. 1969. Semantic entailment and formal derivability. In *The philosophy of mathematics*, ed. J. Hintikka, 9–41. London: Oxford University Press.

Black, P.E. 2005. *British Museum technique*. Dictionary of algorithms and data structures. Retrieved July 2009, from http://www.itl.nist.gov/div897/sqg/dads/HTML/britishMuseum.html.

Copeland, B.J., and D.R. Murdoch. 1991. The Arthur Prior memorial conference: Christchurch 1989. *The Journal of Symbolic Logic* 56(1): 372–382.

D'Agostino, M., et al. (eds.). 1999. *Handbook of tableau methods*. Dordrecht: Kluwer Academic Publishers.

Fitting, M. 1996. *First-order logic and automated theorem proving*, 2nd ed. Berlin: Springer.

Fitting, M. 1998. Introduction. In *Handbook of tableau methods*, ed. M. D'Agostino et al. Dordrecht: Kluwer Academic Publishers.

Frické, M. 1989a. *Derivation planner*. Dunedin: Unisoft.

Frické, M. 1989b. *Deriver plus*. Ventura: Kinko's Academic Courseware Exchange.

Gallier, J.H. 1986. *Logic for computer science: Foundations of automatic theorem proving*. New York: Harper Row.

Gentzen, G. 1935. Investigations into logical deduction. In *The collected papers of Gerhard Gentzen*, ed. M.E. Szabo. Amsterdam: North-Holland.

Hähnle, R. 2001. Tableaux and related methods. In *Handbook of automated reasoning*, ed. J.A. Robinson and A. Voronkov. Cambridge: MIT Press.
Howson, C. 1997. *Logic with trees: An introduction to symbolic logic*. London: Routledge.
Jeffrey, R.C. 1967. *Formal logic: Its scope and limits*. New York: McGraw Hill.
Letz, R. 1999. First-order tableau methods. In *Handbook of tableau methods*, ed. M. D'Agostino et al. Dordrecht: Kluwer Academic Publishers.
Lindstrom, P. 1969. On extensions of elementary logic. *Theoria* 35: 1–11.
Manzano, M. 1996. *Extensions of first-order logic*. Cambridge: Cambridge University Press.
Mill, J.S. 1869. II. Of the liberty of thought and discussion. In *On liberty*, ed. J.S. Mill. London: Longman, Roberts & Green.
Quine, W.V. 1970. *Philosophy of logic*, 2nd ed. Oxford: Oxford University Press.
Robinson, J.A., and A. Voronkov (eds.). 2001. *Handbook of automated reasoning*. Cambridge: MIT Press.
Sieg, W., and J. Byrnes. 1998. Normal natural deduction proofs (in classical logic). *Studia Logica* 60: 67–106.
Smullyan, R. 1968. *First order logic*. Berlin: Springer.
Wolfram, S. 2002. *A new kind of science*. Champaign: Wolfram Media, Inc.
Zalta, E.N. 2009. Achieving Leibniz's goal of a computational metaphysics. *The 2009 North American Conference on Computing and Philosophy*. Bloomington.
Zalta, E.N., B. Fitelson, and P. Oppenheimer. 2011. *Computational metaphysics*. Retrieved 18 November 2011, from http://mally.stanford.edu/cm/.

Chapter 19
Is Imperative Inference Impossible?

The Argument from Permissive Presuppositions

Hannah Clark-Younger

19.1 Introduction

Standard definitions of validity are designed to preserve truth, from premises to conclusion. This makes sense when we are interested in declarative sentences – the sort of sentences that aim to describe the world. Declarative sentences such as "the sky is blue" and "all cats are green" aim to describe the world, and are capable of being true or false. "The sky is blue" is true, "all cats are green" is false. But this is not the only sort of sentence there is. The sentence "open the door!" does not appear to describe the world at all, but rather tells somebody to do something. This sort of sentence is in the imperative mood, and expresses a command, a request, or something similar. Can imperative sentences be premises or conclusions of valid arguments? How is this possible? Here are some examples:

If it starts to rain, bring in the washing.
It has started to rain.
Therefore, Bring in the washing.
and:
Buy me either a sandwich or some sushi.
Don't buy me a sandwich.
Therefore, Buy me some sushi.

Both of these certainly seem like valid arguments; the first looks like an instance of modus ponens, the second an instance of disjunctive syllogism.

However, validity only applies to arguments with premises and conclusions that are capable of being true or false. This argument contains commands, which are not capable of being true or false. So, according to standard definitions of validity, these arguments are not valid. This is the problem of imperative inference.

H. Clark-Younger (✉)
Department of Philosophy, University of Otago, Dunedin, New Zealand
e-mail: hannah.cy@gmail.com

J. Maclaurin (ed.), *Rationis Defensor: Essays in Honour of Colin Cheyne*, Studies in History and Philosophy of Science 28, DOI 10.1007/978-94-007-3983-3_19,

19.2 Preliminaries

19.2.1 Imperatives

By 'imperatives', I mean sentences in which the main verb is in the imperative mood. These are typically used to issue commands, to make requests or even pleas, or to offer pieces of advice. "Shut the door!" is in the imperative mood, and it expresses the command, or request, that you shut the door. It is not the only way of getting that message across, though. For example, "I'm cold!" could in some contexts be taken as a command, or at least a request, to shut the door. So too could "would you mind shutting the door?" or for that matter "were you born in a tent?" or even a non-verbal action such as pointing at the door. In each instance of this type of case, the addressee understands that they have been commanded or requested to, for example, shut the door, and the speaker intended for them to understand it this way. In a sense the command or request "shut the door!" has been issued, but not uttered. So, why not include all instances of commands and requests as the subject matter of imperative inference? Compare non-literal indicative statements such as "my mouth is on fire", or "she tries her best at mathematics". The first of these could be metaphorical, the second could be a diplomatic way of saying "she is not very good at mathematics". Logic does not deal with these 'real' or intended meanings, but rather deals with the literal meaning of sentences; it is not valid to derive "this food is spicy" from "my mouth is on fire". Since it is these literal meanings that logic deals with, I will take as my subject matter sentences in the imperative mood, and not every instance of an intended command or request.

19.2.2 Inference

By inference, I mean some process of 'moving' from some set of sentences to another sentence, without adding any extra information. It is a process whereby what we 'get out' at the end (the conclusion) 'follows from' or 'is entailed by' what we 'put in' (the premises). To define inference precisely without reference to truth is difficult, but to include truth in our general definition of inference would (potentially) be to beg the question against the possibility of imperative logic. But as, perhaps, a *special case* of inference: in indicative logics, inference is a truth-preserving process. Truth is preserved in the move from the premises to the conclusion; so that it is impossible for the premises to be true and the conclusion false. Inference, in general, then, is taken to be a process that preserves some property of the premises, so that if the premises had it, so would the conclusion. That is, you (in some sense) *cannot be led off course* when you 'move' from the premises to the conclusion.

19.2.3 Implicature

I should also mention here what I do *not* mean by inference. Paul Grice (1989, originally published 1975) pointed out that we can often glean information from a person's utterance that goes beyond the purely literal meaning of the sentence that is uttered. This further information is called the *implicature* of a statement, which we can distinguish from *implication*. The important difference is that implication occurs when the conclusion follows from *the content of the statement*, whereas implicature occurs when something follows from *the fact that somebody has uttered the statement*. Grice distinguishes two kinds of implicature: conventional and conversational.

Conventional implicature is the sort that follows from the fact that someone has used one nuanced word over another with the same literal function. A clear example of this is the use of the word 'but' instead of 'and'. For instance, "Sam is intelligent, but is also good at sports" means the same thing as "Sam is intelligent, and is also good at sports" (in the sense that, necessarily, one is true if and only if the other is true – they have the same truth conditions). In both cases, two claims are being made about Sam. However, there is a sense in which the 'but' in the first sentence conveys extra (or at least different) information. It seems that if a person utters this sentence in favour of the 'and' version, he is in some sense saying that it is unusual for someone to be both intelligent and good at sports; that it is a surprising fact that Sam, being intelligent, is also good at sports. This extra meaning is an example of conventional implicature. It is conventionally implicated, but not implied, from "Sam is intelligent, but is also good at sports" that it is unusual for an intelligent person to also be good at sports.

Conversational implicature is the sort that follows from the presupposition that the utterer of a sentence is following *conversational maxims*. These maxims are based on a general 'Cooperative Principle': "Make your conversational contribution such as is required, at the stage at which it occurs, by the accepted purpose or direction of the talk exchange in which you are engaged" (Grice 1989, p. 26). This is broken down into specific maxims, which Grice lists as:

Maxim of Quantity – make your contribution as informative as is required (for the current purposes of the exchange), and do not make your contribution more informative than is required.

Maxim of Quality – be truthful; do not say what you believe to be false, and do not say that for which you lack adequate evidence.

Maxim of Relation – be relevant.

Maxim of Manner – avoid obscurity of expression, avoid ambiguity, be brief, and be orderly.

So, for example, if someone said, "the dairy is on either George Street or Great King Street", then he would be violating the maxim of Quantity if he knew it was on George Street (even though what he said was true), because he would not be providing the most informative contribution that he was in a position to provide. Thus, when we presume he is conforming to the maxims, it is *conversationally implicated* that he does not know which of the two streets the dairy is on. If, on the other hand, he were to specify further when he did not know, he would be violating the maxim of Quality, in that he would say something for which he lacked the appropriate evidence. It is not implied by the *content* of the disjunctive statement "the dairy is on either George Street or Great King Street" that the speaker does not know which street the dairy is on, but it is conversationally implicated that he does not.

So, there are a number of ways that we can get further information out of a sentence: conventional or conversational implicature, or implication. Inference, then, for the purposes of this discussion, is meant to include only implication – that is, what can be extracted from the content of the sentence/s alone.

19.2.4 Imperative Inference

Imperative inference, then, if it exists, is some process of moving from one or more premise/s to a conclusion – at least one (non-redundant) premise, or the conclusion, must be a sentence in the imperative mood. If the inference is a 'good' one, the conclusion must in some sense 'follow from' or be 'entailed by' the content of the premises. So, the problem of imperative inference is to figure out how, and if, imperatives can possibly be constituent parts of inferences.

There are still two different questions we could be interested in, however. First, whether we do in fact make imperative inferences, and second, whether we could in principle make imperative inferences. Even in standard declarative logic we sometimes express things that we would not normally say in English – because they sound clumsy, or worse, are not grammatical. For example: "~(A v (B ≡ (C&D))) v ((B&E) v (A ≡ F))". This would come out as "it is either not the case that either A, or B if and only if C and D, or it is the case that either B and E, or A if and only if F". We do not, in cases such as this, want to say we have no inference because it involves a statement we could not (or would not) say in English. So, I am interested in how, and if, we can *possibly* construct or formulate a formal logic of imperatives.

19.2.5 The Problem of Imperative Inference

19.2.5.1 Jorgensen's Dilemma

Jorgen Jorgensen put forward the most influential set-up of the problem of imperative inference in his paper *Imperatives and Logic* (Jorgensen 1938). In this paper, Jorgensen explains that the usually accepted notion of validity requires the premises and the conclusion of the inference to be capable of being true or false (that is, they must be 'truth-apt').

Jorgensen then explains that imperatives cannot be true or false – to ask of "be quiet" or "do your duty" whether it is true or false would be to ask a meaningless question. These, and other, imperative sentences can be obeyed or not, accepted or not, and considered justified or not, but to ask whether they are true or false seems without sense.

On the other hand, examples can be given of instances that *seem to be* clear examples of valid arguments. I have already offered two examples. Jorgensen's examples are:

Keep your promises.
This is a promise of yours.
Therefore, Keep this promise.
and:
Love your neighbour as yourself.
Love yourself.
Therefore, Love your neighbour.

In the first case, the first premise and the conclusion are in the imperative mood, and in the second case, both the premises and the conclusion are in the imperative mood. Yet, the conclusion seems just as *inescapable* as the conclusion in any valid argument containing only sentences in the indicative mood. That is, it seems correct that if the premises were 'accepted' (whatever that means), the conclusion could not be 'denied' (whatever that means). So, they *seem* to be valid, but they cannot be valid according to standard definitions of validity, because "keep your promises", "love your neighbour" and so forth are not capable of being true or false; they are not truth-apt.

So, Jorgensen arrives at this (paraphrased) dilemma:

(a) Validity requires the premises and conclusions of arguments to be truth-apt, so there are no valid arguments containing imperatives.
(b) There are valid arguments containing imperatives.

19.2.5.2 Trilemma

These two claims are only inconsistent if (that is, there is no dilemma unless) we assume a third claim, that imperatives are not truth-apt. This claim is just as important for producing the inconsistency. Without it, the first part of (a) does not get us to the second part. Without it, there is no contradiction with (b), because the first part of (a) does not contradict (b).

I propose, therefore, that we treat the problem not as a dilemma, but as a *trilemma* of three plausible claims (an inconsistent triad):

T1. There are valid arguments containing imperatives.
T2. Imperatives are not truth-apt.
T3. Validity requires the premises and conclusions of arguments to be truth-apt.

T1 is Jorgensen's (b). Jorgensen's (a) is separated into T2 and T3. These three claims cannot all be true. Any two of the three are consistent with each other and entail the negation of the third. That is, T1 & T2 entail the negation of T3; T2 & T3 entail the negation of T1; and T1 & T3 entail the negation of T2. This trilemma, I claim, is the problem of imperative inference. There are three claims, all of which seem to be correct, which together are inconsistent. Solving the problem, then, requires rejecting one of these three claims. We could reject T1, that there are valid arguments containing imperatives, and instead claim either that all imperative arguments are invalid, or that there are no arguments containing imperatives (that imperatives cannot be constituent parts of arguments). Or, we could reject T2, that imperatives are not truth-apt, and instead claim that they are (somehow) capable of being true or false. Or, finally, we could reject T3, that validity requires the premises and conclusions of arguments to be truth-apt, and instead formulate a new account of validity.

19.3 Williams' Argument

I will discuss an argument, offered by Bernard Williams (1963), against the possibility of imperative arguments. If successful, it would lead us to favour the rejection of T1. I claim that not only does Williams' argument fail, but also that imperative inference is possible by Williams' own lights. Williams claims that there is only one logical relation between imperatives: that of contradiction. Two commands contradict when it is logically impossible for both commands to be obeyed. However, he claims, this is the only relation that can hold between commands, and there is nothing that can be generally called 'imperative inference'.

Williams claims that strings of commands cannot form anything analogously like an argument because they are *essentially discontinuous*. That is, a second command uttered after a first command represents a change of mind on the part of the commander. As arguments require all the premises to be accumulated and accepted at once, strings of commands cannot form the premises of an argument.

To illustrate this accumulation point, suppose a person, Alice, said "I want an icecream" then changed her mind and said "actually, I want some orange juice". The two statements "Alice wants icecream" and "Alice wants orange juice" cannot be used together as premises of an argument, for example to the conclusion "Alice wants icecream and orange juice" because she changed her mind between uttering them, and thus they are not meant to both be considered true at the same time. Williams claims that commands uttered in succession always exhibit the same kind of discontinuity.

He asks us to consider the following example:

Argument A

A1. Do x or do y
A2. Do not do x

Therefore

A3. Do y

Compare (where p and q are declarative sentences):

Argument B

B1. p or q
B2. not-p

Therefore

B3. q

Argument A seems to be a parallel of Argument B, which is an instance of disjunctive syllogism. It seems intuitively to be valid; that is, it seems that A3 genuinely follows from A1 and A2 in much the same way that B3 follows from B1 and B2. But, Williams argues, we have to imagine it in a context – as commands are not commands unless they are commanded by somebody to somebody. He claims that the commander makes no inference, and the agent makes an inference but it is not an imperative one. In this paper I will only address Williams' argument against the possibility that a *commander* ever makes an imperative inference. I also will not discuss his claim that commands cannot be abstracted from their context.

Suppose a commander utters A1, then some time later utters A2. This commander, according to Williams, must change his mind between the two utterances, as he gives conflicting permissions in commanding each; namely he gives permission to do x and then denies permission to do x.

It is reasonably straightforward that command A2 denies permission to do x. "Do not do x" and "you are not permitted to do x" seem to at least confer identical information to the hearer, if they are not synonymous (whether a command can be synonymous with any indicative is a live issue, but unfortunately not one I have time to discuss in this paper. For a discussion of the possibility of a non-cognitivist theory of imperatives, see Parsons (2012)). The more interesting claim here is that the command "do x or do y" implicitly gives permission to do x, and permission to do y. Williams demonstrates this by insisting that when a commander commands "do x or do y", he permits the agent not to do x, if he does y, and he permits him not to do y, if he does x. This, in turn, permits the agent to do x, and permits the agent to do y, because " "I permit you not to do x, if you do y, but I do not permit you to do x" does not confer or admit any permission at all" (Williams 1963, p. 31). When the commander commands A1, what he effectively does is give the agent a choice. That is, he says "you can choose between x and y, but make sure you do one of them". So, Williams argues, this shows that the permissive presuppositions of the disjunctive command "do x or do y" include permission to do x and y themselves.

However, to command A2, "do not do x" is to explicitly express that x is not permitted. Thus, the same commander utters A1, which gives permission to do x, then utters A2, which disallows x. So he in fact reneges on the permission given when he uttered A1.

In order to give permission to do x and then take it away, the commander must have changed his mind about what he wants the agent to do. So it follows that the speaker is not making anything that can be called an imperative inference, as he does not endorse the premises all together at any time.

To demonstrate this point, Williams asks us to suppose that the commander utters the second command a few moments too late – the agent has already trotted off to obey the first command, and fails to hear the second command. Then the agent comes back, having done x but not y. He has obeyed the command which he heard, but the commander would be disappointed (or even angry). This indicates a change of mind or desire.

To further illustrate Williams' claim that moving from "do x or do y" to "do not do x" represents a change of mind, suppose that A1 and A2 were commanded in the reverse order; that is, the commander first says "do not do x" then after some time says "do x or do y". For example, suppose the commander first says "don't go into the lounge" then a few moments later says "go into the lounge or the kitchen". This seems to more clearly demonstrate a change of mind. It would even seem rather deceitful of someone to do this, and to expect their first command to remain binding. I think that this demonstrates Williams' description of the permissive presuppositions of the respective commands. That is, if "do not do x" is commanded, and then "do x or do y" is subsequently commanded, and we take this to nullify the first command, then the permissive presuppositions Williams outlined must be correct; namely that "do x or do y" gives permission to do x (and to do y), otherwise there would be no conflict.

I offer four objections to Williams:

1. Permissive presuppositions are analogous to conversational implicature, and so should be ignored when formulating imperative arguments.
2. There are other valid argument forms, which Williams does not consider, that do not suffer from different permissive presuppositions.
3. There are explanations for the change in permissive presuppositions in Williams' example other than a change of mind, so the different permissive presuppositions do not have to prevent accumulation.
4. Williams accepts enough logical relations between imperatives (contradiction and a form of negation) for a definition of a valid imperative inference to follow naturally.

19.4 Permissive Presuppositions and Conversational Implicature

Consider again the parallel, indicative case (Argument B) of disjunctive syllogism:

B1. p or q
B2. not-p

Therefore

B3. q

In this case, the move from B1 to B2 seems like it has an analogous change of mental state. For the utterer (or hearer) of these two premises to sincerely leave open “p or q”, he must not know that not-p. But the second utterance (of not-p) obviously demonstrates that he does know that not-p. So is this not relevantly like the imperative case?

Williams claims that it is not, because it represents a movement from ignorance to knowledge. That is, the first claim displays less definite knowledge than the second. “p or q” can only sincerely be asserted if the speaker genuinely does not know which. Stating “p” some time later simply means that the speaker has gained more knowledge. This does not upset the consistency of a set of premises. This, he claims, is due to the fact that the truth of a statement of fact depends on whether it is in accordance with the way the external world is, not on whether or not the utterer knows it. That is, statements of fact (assertions) attain their fulfilment value when they are correctly modelled on the world. In contrast, commands are obeyed/complied with when the world is in accordance with them. That is, commands attain their fulfilment value by the world being modelled on them. What is commanded depends entirely on the mental state of the commander. Which commands are appropriate depends on what the commander wills; it is up to him, at each stage in time, to decide what changes he wants initiated by means of his commands. If he is content with either x or y, he can command “do x or do y”, but if he wants “y and not x” then he is free to command this. They represent different desires or wills, and so if he moves from the first to the second, then he has to have changed his mind about what he wants. So, although the indicative case may represent an analogous change of mental state to the imperative case, it makes a difference only to the imperative case. More specific knowledge can be gained over time without the consistency of the premises being compromised, but a more specific command cannot be uttered without a change of mind and thus disjointed premises.

Consider the preparatory statement “I am going to command you to do x or to do y”. This prepares the agent for a later command, but is itself not a command. The indicative disjunction, “p or q”, is in one relevant way more like the preparatory statement than the disjunctive command: it allows the speaker to add “and I may be able to tell you later which of the two”. This is not true of the disjunctive command, “do x or do y”, as this in effect tells the agent to choose; to decide for himself whether to do x or to do y.

Williams’ claim that in the indicative case the move from B1 to B2 represents a move from ignorance to knowledge relies on B1 coming first and B2 coming second. So we apply the same reversal technique: now we have someone uttering “not-p” and then “p or q”. This cannot be thought of as a move from ignorance to more specific knowledge, as by Williams’ own admittance “p or q” cannot seriously be left open if it is known that not-p. Consider, as an illustration, someone uttering “Mark is not a doctor” and then a few moments later uttering “Mark is either a doctor or a nurse". In this sort of case, it seems clear (at least in some contexts) that we would take the second utterance to cancel the first utterance.

In the indicative case, at least, we surely agree that the order of utterance makes no difference to their being premises of an inference. My reversal example, however, shows that the order makes a difference to the overall intended (and understood) meaning. If logic does not preference order, then it must be making a difference to the *non-logical* meaning.

In particular, the order seems to make a difference because of Gricean conversational implicature. Uttering "p or q" *conversationally implicates* that the speaker does not know which of p and q is true, because if they *did* know which one was true then they would be violating the maxim of Quantity by not being as specific as possible. It is because of this conversational implicature that reversing the order of "p or q" and "not-p" makes a difference to what we 'conclude' from these two utterances. When "p or q" is uttered first, followed by "not-p", it is plausible that the speaker could have gained some more specific knowledge. When "not-p" is uttered first, followed by "p or q", it seems that the speaker must have made some sort of mistake.

We ignore this conversational implicature when we accumulate utterances into a set of premises. If we can strip indicative statements of their presuppositions, context, and so forth in order to make them into premises of an argument, then can we not do the same to commands?

19.5 Other Valid Argument Forms

There is another serious problem with Williams' argument against the possibility of imperative arguments. Even if Williams' analysis of the apparent imperative disjunctive syllogism is correct, he does not consider other cases of apparently valid imperative arguments. His analysis of it as essentially discontinuous due to a change of mind on the part of the commander seems like it could be specific to this form of argument, and will not hold in cases where there is no change of mind. The simplest example of this is the inference to a command from itself. For example:

Argument C

C1. Do x!

Therefore

C2. Do x!

In this case, the permissive presuppositions are the same in both C1 and C2, so there is no change of mind required to command C2 after commanding C1. Williams makes no mention of this kind of case, so perhaps on his behalf we can point out that this is a very strange inference, even for its indicative counterpart. There are not really two commands here, but one – uttered twice. No inference needs to occur to get from C1 to C2, because they are one and the same command.

Perhaps 'inference' is a process that must occur between at least two different statements. That is, "X entails Y" requires that X is distinct from Y.

So, Williams could say that there is no inference going on in Argument C because there is no movement from one command to another. There is just the same command twice. He could say the same thing, presumably, about the indicative case; there is no inference from "p" to "p".

There are other forms of inference, however, that have two or more different commands and are not considered by Williams. For example, consider the following cases:

Argument D

D1. Do x and do y!

Therefore

D2. Do x!

And, similarly:

Argument E

E1. Do x!
E2. Do y!

Therefore

E3. Do x and y!

These represent cases of conjunction elimination and conjunction introduction, respectively. There does not seem to be any way of construing these cases as changes of mind. In Argument D, furthermore, there is no accumulation of premises required as it only has one premise. In Argument E, there are two premises, but it does not seem impossible that both could be endorsed by the commander at the same time and thus form a set of two premises. Suppose that somebody said to you "open the window!" and then said "open the door!" It seems like in some cases, he may have changed his mind about what he wanted you to do, but not in every case. It seems perfectly reasonable that he simply wants you to do both.

So, a commander who uttered E1 and then E2 some time later *may* have changed his mind about what he wanted the agent to do, but he *need* not. For Williams' account to be correct it must *always* be the case that the commander changes his mind between commanding E1 and E2.

The only way I can see for Williams to plausibly deny that these are arguments is to deny that the conjunction of two commands is itself a command. He could instead claim that a conjunction of commands is still two commands. So, for example, "do x and do y" is the same as "do x" and "do y". If this is the case, then the inference from "do x and do y" to "do x" is really an inference from "do x" to "do x", which is the same as Argument C and thus, according to my proposed analysis, not a real inference. Similarly with Argument E; the conclusion is just restating the premises: it moves from "do x" and "do y" to "do x" and "do y", which is no movement at all. So again, there is no real inference.

Note that if this line was taken here, Williams would not be committed to denying conjunction elimination and conjunction introduction in ordinary indicative inferences. The claim here is that the conjunction of two commands is not itself a (different) command. This does not require that the conjunction of two indicatives is not a separate statement. Williams could still maintain standard indicative logical inferences and deny that conjunctive commands exist. As for mixed conjunctions, like "p & do x!", Williams will have to group them in with conjunctions of two commands. That is, "p & do x!" is not a distinct sentence from "p" and "do x!". Otherwise, he would not be able to explain why it was not an inference to go from "p & do x!" to "do x!". Consequently, though, he will have to also accept that there is no inference from "p & do x!" to "p".

What about equivalent commands, that nonetheless seem like they are different? For example, consider two De Morgan-like imperative equivalences:

Argument F

F1. Don't do either x or y!/Do neither x nor y!

Therefore,

F2. Don't do x and don't do y!

Argument G

G1. Don't do both x and y!

Therefore,

G2. Don't do x, or don't do y!

Williams will have to say that F1 and F2 are not really different commands, but the same command issued using different words. They have the same permissive presuppositions as each other, namely "you are not permitted to do x and you are not permitted to do y". Similarly, he will have to say that G1 and G2 are the same command. Again, they have the same permissive presuppositions, namely "you are permitted to do x, as long as you don't do y, and you are permitted to do y, as long as you don't do x".

Perhaps this is enough for them to be one thought of as one and the same command. The plausibility of this becomes somewhat strained when we modify F2 to only include one of the conjuncts (eg F2* "don't do x!"). It seems that F1 and F2* are different enough to be considered distinct commands. However, when combined with the claim that a conjunction is not a different command to its two conjuncts separately, this response does not seem implausible (if somewhat ad hoc).

Next, consider arguments with conditionals:

Argument H

H1. If do x then do y!
H2. Do x!

Therefore

H3. Do y!

H1 is not grammatically acceptable; examples simply do not occur in English. It would be something along the lines of "if I command you to go to bed, then I command you to brush your teeth first", where "I command" must be taken as performative, not descriptive. However, because subordinating conjunctions (such as "if") cannot precede imperative sentences (that is, imperatives cannot occur as subordinate clauses), this is impossible to express (at least in English) (for an argument from this fact to the impossibility of imperative inferences, see Wedeking (1970)).

So we can turn our attention to *conditional commands* such as "if there is a stop sign, then stop!" or "if the weather is fine, then attack at dawn!" or even "attack at dawn only if the weather is fine!" (these examples are due to Josh Parsons). There are two main forms of inferences involving conditional commands:

Argument I

I1. Attack at dawn only if the weather is fine!
I2. Attack at dawn!

Therefore

I3. The weather is fine.

This can be generalised to:

Argument J

J1. Do x only if p
J2. Do x!

Therefore

J3. p

And also:

Argument K

K1. Attack at dawn if the weather is fine.
K2. The weather is fine.

Therefore

K3. Attack at dawn.

Which can be generalised to:

Argument L

L1. Do x if p
L2. p

Therefore

L3. Do x!

The first case, Argument I, can be analysed in Williams' terms. There does seem to be a change of mind – that is, after commanding "attack at dawn only if the weather is fine", in commanding I2 the commander seems to be saying something like "scrap that, attack at dawn regardless of the weather". This can be demonstrated by means of permissive presuppositions: in the case of I1, the commander is saying "you are not permitted to attack at dawn except if the weather is fine", or "you are not permitted to attack at dawn if the weather is anything but fine". On the other hand, in the case of I2, there is no condition placed upon the command, it simply says "attack at dawn (regardless of the weather – or any other considerations)". In permissive presupposition terms, "you are not permitted to not attack at dawn" – there is no escaping the duty.

The reversal technique also demonstrates that these permissive presuppositions are at work. Suppose we reverse the order of these premises in the same way as we reversed A1 and A2 previously. Suppose someone were to command I2 and then I1: that is, "attack at dawn!" and some time later, "attack at dawn only if the weather is fine!" It seems clear, here, that we are not expected to infer "the weather is fine" from this (nor does it seem that the commander ought to or could do so). Rather, we (under normal circumstances) have been given a new, overriding command.

The second case, Argument K, is however not so straightforward for Williams' theory. There is certainly no change of mind required to explain the move from "attack at dawn if the weather is fine" and "the weather is fine".

Actually, K1 "Attack at dawn if the weather is fine" is potentially ambiguous between:

(a) "At dawn: if the weather is fine, attack!" (or "if the weather is fine at dawn then attack!"); and
(b) "If the weather is fine (now) then attack at dawn".

This is important because if it is the first (as it seems reasonable to suppose is intended in this case), then it is an instruction to consider a certain command; "attack at dawn" binding at a future time if the weather is fine at that time, and dismiss the command, that is, consider no command to be binding, if the weather is not fine.

There are other instances of the same form which seem to clearly be cases of the second. For example: "cover your eyes if you are afraid" or "don't eat it if you don't like it". These seem to be of the second kind, that is, they are not instructions for a future time. The antecedent is about the mental state of the agent; "if you are afraid (now), then cover your eyes".

In response to these examples, generally, perhaps Williams could respond that the commander never holds both L1 and L2 (to stand for any of the examples given) in his mind at the same time. If he knew p to be true, and he wanted the agent to do x if p was true, then he should just command "do x". On the other hand, if he does not know (or even believe) that p is true, he would not assert it. In the examples above, the commander does not know that the antecedent is true either because it is a statement about some future time, or because it is about the mental state of the agent. In either of these scenarios, the commander is in no position to make the inference as he does not know p.

This seems like a plausible account of what might be going on, at least some of the time. Suppose, however, that the commander says "if you are afraid, come here" and the agent says "I am afraid". Now, surely, the commander knows p to be true, so if he was to command "well then, come here!" then it seems that he has made a genuine inference.

19.6 Alternative Explanations for the Different Permissive Presuppositions

Another objection to Williams' analysis of (any instance of) Argument A is that he only considers one explanation for the different permissive presuppositions of A1 and A2. He assumes that the commander must have to change his mind in order to command A1 and then A2. This is important because the change of mind is crucial to his position that the commander does not endorse the two at the same time, so they cannot be accumulated into a set of premises. If there is another adequate explanation for the move from A1 to A2, and if that explanation does not involve a change of desire or will on the part of the commander, then Williams' argument that the commander never makes an imperative inference fails. So, do any such explanations exist? I shall offer and discuss four candidates.

1. Perhaps the commander is logically incompetent, and does not realise that there is anything wrong with his commands because he cannot see the inconsistency.

This would not require a change of mind. It also, however, is no help with attributing an imperative inference to him, because it is problematic to ascribe any kind of inference to a logically inept person.

2. Perhaps he always wanted the agent to do y (A3), and he uttered the premises A1 and A2 for rhetorical flourish or emphasis. For example, consider the following instance of Argument A:

   Argument M

   M1. Bow to me, or leave my court!
   M2. Stay in my court!

   Therefore

   M3. Bow to me!

   If M1 and M2 were commanded for rhetorical flourish, then the commander is not intending the permissive presuppositions that Williams outlines, and neither is the agent meant to interpret them that way. If a disjunctive command is taken as Williams suggests to mean a command to make a choice, then the commander does not even utter a disjunctive command, as such, but something that grammatically resembles one.

This is no more helpful, as the commander still cannot be said to be making a genuine inference. He is, quite possibly, asking the agent to make an inference, but he has not arrived at the conclusion through any inference of his own. Rather he starts by desiring the conclusion, and merely commands it in an unusual way for effect.

3. Perhaps, instead of being logically incompetent or not intending the disjunctive command to be taken seriously, the commander simply does not realise that his two commands carry the permissive presuppositions that it is claimed they carry. So the commander commands A1 and A2 of Argument A, but does not realise that "do x or do y" presupposes that x and y are both permitted by this command. In this case, he would not need to change his mind to subsequently command A2, as he would not realise that it would renege the permission he had just given.

Again, this is not helpful, because if Williams is correct about the presuppositions inherent in a disjunctive command, then what this means is that a disjunctive command by definition includes its permissive presuppositions, and thus the commander does not realise he is commanding a disjunctive command. This means either that he is not really commanding a disjunctive command (if intention determines or is a necessary factor in determining what is uttered), or (if intention is *not* necessary in determining what is uttered) that he has such a lack of understanding of what he is commanding that he cannot be said to be making an inference.

4. Perhaps the commander actually desires two conflicting things. That is, he fully understands his that his commands carry the permissive presuppositions that they do, he intends them to do so, but he really does desire both things. He wants the agent to choose between x and y, but he also wants the agent not to do x. So he both gives permission and denies permission to do x. It certainly seems possible that someone could desire conflicting things, even when he understood that they conflicted.

The problem with this scenario, however, is that the conclusion, "do y" does not seem to satisfactorily follow from his desires. If the commander really does want the agent to choose between x and y, then the commander does not want to end up effectively commanding that he must do y. Perhaps it could be said that the command "do y" is an unpleasant consequence of his two somewhat conflicting desires, but a consequence nonetheless.

Consider the following hypothetical instance of Argument M; a king (the commander of Argument M) may know that it is a very strict rule that his subjects must bow to him or be cast out of his court, and furthermore he must command this, but he also might rather like this particular subject and wish him to remain. So he says "Bow to me or leave my court!" in his kingly, authoritative way, and then softly, perhaps pleadingly, "stay in my court, your grace!" This instance certainly does not seem to demonstrate a change of mind, nor does it seem to be given for emphasis or flourish.

In fact, what appears to be going on here is that the second imperative, "stay in my court, your grace", is not a command, but a request or even a plea. However, as they are both are in the imperative mood, this is still an instance of imperative inference. In particular, it is an instance of Argument A in which the commander utters the premises (with their contradictory permissive presuppositions) but does not change his mind between uttering the first and uttering the second. Because there is no change of mind involved, Williams' argument that the premises cannot be accumulated into a set of premises fails, even for his own paradigm case.

A similar example of this is given by Geach (1963) in his reply to Williams: suppose you are seeking counsel from somebody, and they give you the advice "do x or do y". You then give him a list of reasons why you should not do x, and he replies "very well, do not do x, so do y". It is not required that the commander in this case change his mind. He may still endorse the disjunctive command; that is, his advice to choose between doing either x or y may still stand. However, in light of some new information, he also (perhaps for quite a different reason) endorses the advice not to do x.

19.7 Inference in Terms of Contradiction and Negation

A more fundamental objection to Williams' argument is that he contradicts himself: he claims that there is no such thing as imperative inference, but he accepts enough about the logical relations between imperatives for an account of imperative inference to follow naturally.

Williams admits that "there are certain logical relations between imperatives". They can contradict each other: "two imperatives may be said to be inconsistent, if and only if it is logically impossible that they should both be obeyed" (Williams 1963, p. 30). He also accepts a negation-like operator. In his example (Argument A), the second premise was "do not do x", which he takes as denying permission to do x. This is a kind of negation of "do x", which he takes as denying permission *not* to do x. It may be objected that negation is defined by its truth-table, and so, by definition, is something that operates only on truth-apt sentences. In that case, call this one 'imperative-negation' or 'i-negation' instead.

An account of imperative inference can be formulated simply in terms of contradiction and i-negation: "an imperative conclusion follows from a set of imperative premises if and only if the set of sentences consisting of the premises and the i-negation of the conclusion contains a contradiction". In other words (using Williams' definition of imperative contradiction): "an imperative conclusion follows from a set of imperative premises if and only if it is logically impossible for the premises and the i-negation of the conclusion to all be obeyed".

I do not wish to claim that this is the best possible account of imperative validity, nor even that it is adequate (for instance, it will not account for inferences with some imperative and some indicative sentences, like instances of Arguments J and L). I simply claim that it is an account of imperative inference that uses only the tools that Williams accepts. Without demonstrating that this proposed account is impossible, and because he accepts that imperatives can be contradictory and that a form of imperative negation is possible, Williams cannot consistently hold that there is no such thing as imperative inference.

References

Geach, P.T. 1963. Imperative inference II. *Analysis Supplement* 23: 37–42.
Grice, P. 1989. *Studies in the way of words*. Cambridge, MA: Harvard University Press.
Jorgensen, J. 1938. Imperatives and logic. *Erkenntnis* 7: 288–296.
Parsons, J. (2012) Cognitivism about imperatives. *Analysis* 72(1): 49–54.
Wedeking, G.A. 1970. Are there command arguments? *Analysis* 30(5): 161–166.
Williams, B.A.O. 1963. Imperative inference I. *Analysis Supplement* 23: 30–36.

About the Editor

James Maclaurin is the Head of the Department of Philosophy at the University of Otago. He holds a doctorate in philosophy from the Australian National University. James has taught at the Australian National University, the University of Melbourne and Victoria University of Wellington. He has also been a Marsden Postdoctoral Fellow at Victoria University.

He is primarily a philosopher of biology, but his interests extend to philosophy of science more broadly construed, the scientific study of religion, and metaphilosophy. He has co-authored, with Kim Sterelny, *What is Biodiversity?* (University of Chicago Press 2008). James lives in Dunedin with his wife Kristen and son George.

J. Maclaurin (ed.), *Rationis Defensor: Essays in Honour of Colin Cheyne*, Studies in History and Philosophy of Science 28, DOI 10.1007/978-94-007-3983-3,

Index

J. Maclaurin (ed.), *Rationis Defensor: Essays in Honour of Colin Cheyne*, Studies in History and Philosophy of Science 28, DOI 10.1007/978-94-007-3983-3,

MIX
Papier aus verantwortungsvollen Quellen
Paper from responsible sources
FSC® C105338

If you have any concerns about our products,
you can contact us on
ProductSafety@springernature.com

In case Publisher is established outside the EU,
the EU authorized representative is:
Springer Nature Customer Service Center GmbH
Europaplatz 3, 69115 Heidelberg, Germany

Printed by Libri Plureos GmbH
in Hamburg, Germany